Analytical Similarity Assessment in Biosimilar Product Development

Analytical Similarity Assessment in Biosimilar Product Development

Shein-Chung Chow
Duke University School of Medicine
Durham, North Carolina

CRC Press
Taylor & Francis Group
Boca Raton London New York

CRC Press is an imprint of the
Taylor & Francis Group, an **informa** business

A CHAPMAN & HALL BOOK

CRC Press
Taylor & Francis Group
6000 Broken Sound Parkway NW, Suite 300
Boca Raton, FL 33487-2742

First issued in paperback 2020

© 2019 by Taylor & Francis Group, LLC
CRC Press is an imprint of Taylor & Francis Group, an Informa business

No claim to original U.S. Government works

ISBN 13: 978-0-367-73383-4 (pbk)
ISBN 13: 978-1-138-30733-9 (hbk)

Visit the Taylor & Francis Web site at
http://www.taylorandfrancis.com

and the CRC Press Web site at
http://www.crcpress.com

Contents

Preface

Biologic drug products are therapeutic moieties that are manufactured using a living system or organism. These are important life-saving drug products for patients with unmet medical needs. They also comprise a growing segment in the pharmaceutical industry. In 2007, worldwide sales of biological products reached $94 billion US dollars, accounting for about 15.56% of the pharmaceutical industry's gross revenue. Meanwhile, many biological products face losing their patents in the next decade. Attempts have been made therefore to establish an abbreviated regulatory pathway for approval of biosimilar drug products, i.e., follow-on (or subsequently entered) biologics of the innovator's biologic products, in order to reduce cost. However, due to the complexity of the structures of biosimilar products and the nature of the manufacturing process, biological products differ from the traditional small-molecule (chemical) drug products. Although the concepts and principles for bioequivalence and interchangeability could be the same for both chemical generics and biosimilar products, scientific challenges remain for establishing an abbreviated regulatory pathway for approval of biosimilar products due to their unique characteristics.

This is intended to be the first book entirely devoted to the design and analysis of analytical similarity assessment, including tests for similarity in critical quality attributes at various stages of the biosimilar product development manufacturing process. It covers all of the statistical issues that may occur at various stages of biosimilar product research and development. The goal of this book to provide a useful desk reference and a state-of-the art examination of this subject area to (i) scientists and researchers engaged in pharmaceutical/clinical research and development of biologic products, (ii) those in government regulatory agencies who have to make decisions on the review and approval process of biological regulatory submissions, and (iii) biostatisticians who provide the statistical support to the assessment of analytical similarity of biosimilar products. It is my hope that this book can serve as a bridge between the pharmaceutical/biotechnology industry, government regulatory agencies, and academia.

The scope of this book is restricted to scientific factors and practical issues that are commonly encountered in biosimilars research and development. This book contains 12 chapters. Chapter 1 provides background for biosimilar product development. Also included in this chapter is a brief introduction to some commonly seen scientific factors and practical issues in biosimilar product development. Chapter 2 reviews the regulatory approval pathway of biosimilar products in the United States with special focus on the regulatory requirements for analytical similarity assessment. Chemistry, manufacturing, and control (CMC) requirements, including manufacturing process validation, quality control and assurance, and stability analysis for biosimilar product development are discussed in Chapter 3. Chapter 4 covers assay development and process and validation, including analytical method validation and the evaluation of reliability, repeatability, and reproducibility. Based on mechanisms of action or pharmacokinetics and pharmacodynamic behaviors, Chapter 5 introduces the identification and classification of critical quality attributes for functional/structural characterization under the appropriate statistical model. Chapter 6 introduces FDA's recommended tiered approach (i.e., equivalence tests for Tier 1 attributes, a quality range approach for Tier 2 attributes, and raw data and graphical presentations for Tier 3 attributes) for testing similarity of the critical quality attributes identified in Chapter 5. Also included are some practical issues and recent developments regarding analytical similarity assessment. Chapter 7 discusses sample size requirements for analytical similarity assessment under various considerations of controlling variability, adjusting for equivalence acceptance margin, the optimal allocation ratio, and achieving the desired degree of reproducibility. The issue of testing the analytical similarity for multiple references is discussed in Chapter 8. The derivation of a proposed simultaneous confidence interval is also included in this chapter. Chapter 9 studies the issue of extrapolation across indications (the conditions of use) and proposes statistical methods for assessment of extrapolation in terms of the sensitivity index for generalizability between patient populations (related to conditions of use). Chapter 10 discusses a couple of recent regulatory submissions (Avastin biosimilar sponsored by Amgen and Herceptin biosimilar sponsored by Mylan) reviewed at the Oncologic Drugs Advisory Committee (ODAC) held on July 13th, 2017 within the FDA in Silver Spring, Maryland. Practical and challenging issues that are commonly encountered during analytical similarity assessment in biosimilar product development are outlined in Chapter 11. Chapter 12 summarizes recent developments in biosimilar product development. A brief review of a recent FDA draft guidance on analytical similarity assessment is also provided in this chapter.

From Taylor & Francis, I would like to thank Mr. David Grubbs for providing me with the opportunity to work on this book. I wish to express my gratitude to my wife Dr. Annpey Pong for her understanding, encouragement, and support during the preparation of this book. I also wish to thank my colleagues from Duke University School of Medicine (as this book was done while I was at Duke) and many friends from academia, the pharmaceutical industry, and regulatory agencies for their support and constructive discussions during the preparation of this book.

Finally, the views expressed are those of the author and not necessarily those of Duke University School of Medicine. I am solely responsible for the content and any errors in this book. Any comments and suggestions for future editions of this book are very much appreciated.

Shein-Chung Chow, PhD
Duke University School of Medicine

Author

Shein-Chung Chow, Ph.D, is currently an Associate Director at Office of Biostatistics, Center for Drug Evaluation and Research, United States Food and Drug Administration (FDA). Prior to joining FDA, Dr. Chow was a Professor at Duke University School of Medicine, Durham, NC. He was also a special government employee (SGE) appointed by the FDA as an Advisory Committee member and statistical advisor to the FDA. Prior to that, Dr. Chow also held various positions in the pharmaceutical industry such as Vice President at Millennium, Cambridge, MA, Executive Director at Covance, Princeton, NJ, and Director and Department Head at Bristol-Myers Squibb, Plainsboro, NJ. Dr. Chow is the Editor-in-Chief of the *Journal of Biopharmaceutical Statistics* and the Editor-in-Chief of the *Biostatistics Book Series* at Chapman and Hall/CRC Press, Taylor & Francis Group. He was elected Fellow of the American Statistical Association and an elected member of the ISI (International Statistical Institute). Dr. Chow is the author or co-author of over 300 methodology papers and 29 books including *Designs and Analysis of Bioavailability and Bioequivalence Studies, Sample Size Calculations in Clinical Research, Adaptive Design Methods in Clinical Trials, Translational Medicine, Design and Analysis of Clinical Trials,* and *Quantitative Methods for Traditional Chinese Medicine Development.*

1

Introduction

1.1 Background

When a brand-name chemical (small molecular) drug product is going off patent protection, pharmaceutical/generic companies may file an abbreviated new drug application (ANDA) for generic approval. In 1984, the United States (US) Food and Drug Administration (FDA) was authorized to approve generic drug products under the *Drug Price Competition and Patent Term Restoration* Act. For approval of generic drug products, the FDA requires that evidence in average of bioavailability in terms of rate and extent of drug absorption be provided through the conduct of bioavailability (BA) and bioequivalence (BE) studies. The assessment of bioequivalence as a surrogate for evaluation of drug safety and efficacy is based on the *Fundamental Bioequivalence Assumption* that if two drug products are shown to be bioequivalent in average bioavailability in terms of drug absorption, it is assumed that they will reach the same therapeutic effect, or that they are therapeutically equivalent. Under the Fundamental Bioequivalence Assumption, regulatory requirements, study design (e.g., a two-sequence, two period or replicated crossover design), criteria (e.g., 80/125 rule based on log-transformed data), and statistical methods (e.g., Shuirmann's two one-sided tests) for assessment of bioequivalence are well established and accepted by most regulatory agencies worldwide (see, e.g., Shuirmann, 1987; EMEA, 2001; FDA 2001, 2003a, 2003b; WHO, 2005; Chow and Liu, 2013).

Unlike small molecular (chemical) drug products, the concept for development of *generic versions* of biologic products is different. The generic version of biologic products is usually referred to as follow-on biologics by the US FDA, biosimilars by the European Union (EU) European Medicines Agency (EMA), and subsequent entered biologics (SEB) by the Public Health Agency (PHA) of Canada. The generic versions of biologic products are viewed as similar biological drug products (SBDP). The SBDP are *not* generic drug products, which are drug products that contain *identical* active ingredient(s) as the innovative drug product. Webber (2007) defines follow-on (protein) biologics as products that are intended to be sufficiently similar to an approved product to permit the applicant to rely on certain existing scientific knowledge about safety and efficacy of an approved reference product. Under this definition, follow-on products are not only intended to be similar to the reference product, but also intended to be interchangeable with the reference product. As defined in 351(k) of Public Health Service (PHS) Act, the term *biological product* means a virus, therapeutic serum, toxin, antitoxin, vaccine, blood, blood component or derivative, allergenic product, protein (except any chemically synthesized polypeptide), or analogous product, or arsphenamine or derivative of arsphenamine (or any other trivalent organic arsenic compound), applicable to the prevention, treatment, or cure of a disease or condition of human beings. The term *biosimilar* is in reference to a biological product that is the subject of an application under subsection 351(k), That is, the biological product is highly similar to the reference product, notwithstanding minor differences in clinically inactive components; and there are no clinically meaningful differences between the biologic product and the reference product in terms of the safety, purity, and potency of the product.

Biosimilars are fundamentally different from generic chemical drugs (Table 1.1). Important differences include the size and complexity of the active substance and the nature of the manufacturing process. Unlike classical generics, biosimilars are not identical to their originator products and therefore should not be brought to market using the same procedure applied to generics. This is partly a reflection of the complexities of manufacturing and safety and efficacy controls of biosimilars when compared to their small-molecule generic counterparts (see, e.g., Schellekens, 2004; Chirino and Mire-Sluis, 2004; Crommelin et al., 2005; Roger and Mikhail, 2007). Biologic products are usually recombinant protein molecules manufactured in living cells (Kuhlmann and Covic, 2006). Thus, manufacturing processes for biologic products are highly complex and require hundreds of specific isolation and purification steps. In practice, it is impossible to produce an identical copy of a biologic product, as changes to the structure of the molecule can occur with changes in the production process. Since a protein can be modified (e.g., side chain may be added, structure may have changed due to

TABLE 1.1

Fundamental Differences between Chemical Drugs and Biologic Drugs

Chemical Drugs	Biologic Drugs
Made by chemical synthesis	Made by living cells
Defined structure	Heterogeneous structure
	Mixtures of related molecules
Easy to characterize	Difficult to characterize
Relatively stable	Variable
	Sensitive to environmental conditions such as light and temperature
No issue of immunogenicity	Issue of immunogenicity
Usually taken orally	Usually injected
Often prescribed by a general practitioner	Usually prescribed by specialists

protein misfolding, and so on) during the process, different manufacturing processes may invariably lead to structural differences in the final product, which may result in differences in efficacy and may have a negative impact on patient immune responses.

In 2009, the *Biologics Price Competition and Innovation* (BPCI) *Act* (as part of the Affordable Care Act) was passed by the US Congress, giving the FDA the authority to approve biosimilar products. Following the passage of the BPCI Act, in order to obtain input on specific issues and challenges associated with the implementation of the BPCI Act, the US FDA conducted a two-day public hearing on *Approval Pathway for Biosimilar and Interchangeability Biological Products* held on November 2–3, 2010 at the FDA in Silver Spring, Maryland, USA. As a number of biologic products are due to expire in the next few years, the subsequent production of biosimilar products has aroused interest within the pharmaceutical industry as biosimilar manufacturers strive to obtain part of an already large and rapidly-growing market. The potential opportunity for price reductions versus the originator biologic products remains to be determined, as the advantage of a slightly cheaper price may be outweighed by the hypothetical increased risk of side-effects from biosimilar molecules that are not exact copies of their originators.

In practice, it is recognized that standard methods established for bioquivalence assessment (both *in vivo* and *in vitro* bioequivalence testing) cannot be directly and appropriately applied to biosimilarity assessment for biosimilar products due to fundamental differences between generic drug products and biosimilar products (Table 1.1). To provide a better understanding, a comparison between bioequivalence testing and biosimilarity testing is given in Table 1.2.

TABLE 1.2

Comparison of Various Types of Equivalence Testings

Characteristics	Bioequivalence (Generic Drug Products)		Biosimilarity (Biosimilar Drug Products)	
	In vitro BE Testing	*In vivo* BE Testing	Analytical	PK/Clinical
Fundamental Assumption	Yes[a]	Yes[b]	No	No
Log-data	No	Yes	No	No
Primary focus	Mean	Mean	Mean	Mean
Variability	<10%	20–30%	Vary	40–50%
Criterion	(90%, 111%)	(80%, 125%)	$EAC = \pm 1.5 * \sigma_R$	SABE?[c]
Analysis	Profile/non-profile	Hypothesis/CI	Hypothesis/CI	Hypothesis/CI

[a] Drug release/delivery is predictive of drug absorption.
[b] Drug absorption is predictive of clinical outcomes.
[c] SABE is proposed criterion by the FDA for highly variable drug products (i.e., intra-subject CV is greater than 30%).

As a result, three draft guidances were circulated for public comments on February 9, 2012. These guidances were finalized in 2015. As indicated in the guidance *Scientific Considerations in Demonstrating Biosimilarity to a Reference Product*, the FDA recommends a *stepwise approach* for obtaining totality-of-the-evidence for demonstrating biosimilarity between a proposed biosimilar product and a reference product. The stepwise approach starts with analytical similarity assessment for structural and functional characterizations of critical quality attributes (CQAs) at various stages of manufacturing process that are relevant to clinical outcomes. Although the stepwise approach also includes animal studies for toxicity, pharmacokinetics and pharmacodynamics (PK/PD) for pharmacological activities, and clinical studies for assessment of immunogenicity, safety/tolerability, and efficacy, the primary focus of this book will be placed on analytical similarity assessment for CQAs that are relevant to clinical outcomes.

In analytical studies for structural and functional characterizations of CQAs, the results are obtained through the conduct of *in vitro* assays. Thus, the concept of analytical similarity assessment is similar to that of *in vitro* bioequivalence testing. In this case, it is an interesting question to the principal investigator whether standard methods established for *in vitro* bioequivalence testing can be directly applied for analytical similarity assessment. In the next section, past experience for *in vitro* bioequivalence testing, including regulatory requirements, criteria, statistical methods, and sample size requirement are briefly described. A brief description of the tiered approach recommended by the FDA for analytical similarity assessment is given in Section 1.3. Section 1.4 provides discussion of scientific factors and practical issues that are commonly encountered when performing analytical similarity assessment for identified CQAs. The aim and scope of the book is given in Section 1.5.

1.2 Past Experience for *In Vitro* Bioequivalence Testing

For evaluation and approval of small-molecule chemical generic drug products, bioequivalence testing is considered as a surrogate for clinical evaluation of the therapeutic equivalence of drug products based on the *Fundamental Bioequivalence Assumption* that when two drug products (e.g., a brand-name drug and its generic copy) are equivalent in bioavailability (in terms of drug absorption), they are assumed to reach the same therapeutic effect, or they are therapeutically equivalent. For approval of most small-molecule generic drug products, FDA requires evidence in bioavailability be provided through the conduct of *in vivo* bioequivalence studies. For local acting and/or delivery drug products, however, the FDA indicates that bioequivalence may be assessed, with suitable justification, by *in vitro* bioequivalence studies alone (see, e.g., Part 21 Codes of Federal Regulations Section 320.24). In 2003, FDA published guidance on Bioavailability and Bioequivalence Studies for Nasal Aerosols and Nasal Sprays for Local Action (FDA, 2003). The 2003 FDA guidance indicates that *in vitro* bioequivalence can be established through seven *in vitro* tests. These *in vitro* tests include tests for (i) single actuation content through container life, (ii) droplet size distribution by laser diffraction, (iii) drug in small particles/droplets, or particle/droplet size distribution by cascade impactor, (iv) drug particle size distribution by microscopy, (v) spray pattern, (vi) plume geometry, and (vii) priming and re-priming. The FDA classifies statistical methods for these seven tests as either the non-profile analysis or the profile analysis.

1.2.1 Study Design and Data Collection

According to the FDA, three lots (or sub-lots) from each product are required to be tested for *in vitro* emitted dose uniformity, droplet size distribution, spray pattern, plume geometry, priming/re-priming, and tail-off profile. For each *in vitro* test, ten samples are randomly drawn from each lot. Samples are randomized for *in vitro* tests. The analysts will not have access to the randomization codes. An automated actuation station with a fixed setting (actuation force, dose time, return time, and hold time) is usually used for the *in vitro* tests. In this section, brief descriptions of the recommended study design and data collection for each of the seven *in vitro* tests are given below.

Emitted dose uniformity, priming, priming/re-priming, and tail-off profile

Following the FDA's recommendations, the priming, emitted dose uniformity, priming/re-priming, and tail-off tests may be tested in the following setting. Three individual lots of test product and reference product are evaluated. For each lot, ten samples are then tested for pump priming, unit spray content through life, and tail-off studies. Then, additional samples for each lot are evaluated for the prime hold study (re-prime study).

For each sample unit, spray samples are collected for sprays 1-8 and analyzed in order to determine the minimum number of actuations required before the pump delivers the labeled dose of drug (sprays 1-8). To characterize emitted dose uniformity at the beginning of unit life, spray 9 is collected. Sprays 10-15 are wasted by the automatic actuation station. Spray 16 is collected in the middle of unit life. Sprays 17-20 are wasted. Sprays 21-23 are collected at the end of the unit life. Additional sprays after spray 23 are collected and analyzed to determine the tail-off profile.

Ten additional samples are drawn randomly from each lot of drug product for the pump prime hold study. For each unit, the first 12 sprays (sprays 1-12) are wasted. Sprays 13 and 14 are collected as fully primed sprays. The unit is then stored undisturbed for 24 hours. Within each lot, five samples are placed in the upright position, while the other five samples are placed in a side position. After that, sprays 15-17 are collected. The unit is then stored undisturbed in its former position for another 24 hours. After that, the doses emitted by sprays 18-20 are collected. All spray samples are weighted in order to obtain re-priming characteristics.

Spray pattern

A spray pattern produced by a nasal spray pump evaluates in part the integrity and the performance of the orifice and pump mechanism in delivering a dose to its intended site of deposition. Measurements can be made on the diameter of the horizontal intersection of the spray plume at different distances from the actuator tip. Spray patterns are usually measured at three distances (e.g., 1, 2, and 4 cm) at both the beginning (sprays 8-10) and the end (sprays 17-19) of unit life. As a result, a total of six spray patterns is collected for each sample unit. For each spray pattern image, the diameters (the longest and shortest diameters) and the ovality (which is defined by the ratio of the longest to the shortest diameters) are measured.

Droplet size distribution

For a test of droplet size distribution, methods of laser diffraction and cascade impaction are commonly used. These methods are briefly described below.

For the method of laser diffraction, each sample unit is first primed by actuating the pump eight times using an automatic actuation station. Droplet size distribution is then determined at three distances (e.g., 3, 5, and 7 cm) from the laser beam and at the beginning, the middle, and the end of unit life. At each distance, three measurements of delay times (plume, formation, start of dissipation, and intermediate measurements) and overall evaluation are used to characterize the droplet size. As a result, a total of 36 measurements are recorded for each sample unit.

For the method of cascade impaction, when the spray pump is actuated in the nasal cavity, a fine mist of droplets is generated. Droplets that are greater than 9 in diameter are considered non-respirable and are therefore useful for nasal deposition. As recommended in the 1999 FDA draft guidance, the data should be reported as follows

Group 1: Adaptor (expansion chamber, i.e., 5-L flask), rubber gasket, throat, and Stage 0.

Group 2: Stage 1

Group 3: Stage 2 to filter.

Each sample unit is first primed by actuating the pump seven times using an automatic actuation station. Droplet size distribution is then determined at the beginning and the end of the life of the sample. Thus, a total of six groups of results are reported for each spray unit.

Plume geometry

Plume geometry is performed on the nasal spray plume, which is allowed to develop into an unconstrained space that far exceeds the volume of nasal cavity. It represents a frozen moment in spray plume development that is viewed from two axes perpendicular to the axis of plume development. The samples should be actuated vertically. Prime the pump with 10 actuations until a steady fine mist is produced from the pump. A fast-speed video camera is placed in front of the sample bottle and starts recording. Repeat the test by rotating the actuator 90 degree to the previous actuator placement so that two side views are at 90 degrees to each other (two perpendicular planes) and, relative to the axis of the plume of the spray, are captured when actuated into space. Spray plumes are characterized at three stages: early upon formation, as the plume starts to dissipate, and at some intermediate time. Longest vertical distance (LVD), widest horizontal distance (WHD), and plume angle (ANG) are recorded and analyzed.

1.2.2 Bioequivalence Limit

Similar to the assessment of individual bioequivalence and population bioequivalence, the 1999 FDA draft guidance recommends the following criterion for bioequivalence limit be used:

$$\frac{(\text{average BE limit in natural log scale})^2 + \text{variance terms offset}}{\text{scaling variance}}$$

As it can be seen, in order to obtain the BE limit, there are three quantities that need to be specified. They are (i) average BE limit, (ii) variance terms offset, and (iii) scaling variance, respectively. The FDA guidance indicates that the final specification of those parameters should be based on the results of the on-going simulation study. However, the following values are recommended in the FDA's draft guidance.

Due to the small variability of *in vitro* measurements, at the present time, the FDA recommends that the ratio of geometric means should fall within 0.90 and 1.11. As a result, a value of 0.90 is recommended as the average BE limit for *in vitro* data. The objective of variance terms offset is to allow some difference among the total variances that may be inconsequential. As a result of the low variability of *in vitro* measurements, the FDA recommends that a value of 0 should be taken based on the guidance of population and individual bioequivalence. In practice, however, a value of 0.01 may be accepted by the FDA for variance terms offset depending upon the nature of the drug products under investigation. The purpose of scaling variance is to adjust the BE criterion depending upon the reference product variance. When the reference variance is greater than the scaling variances, the limit is widened.

On the other hand, the limit is narrowed when reference variance is less than scaling variance. The FDA indicates that the choice of the scaling variance should be at least 0.1. As a result, the specification of 0.90 for the average BE limit, 0.0 for the variance offset, and 0.10 for scaling standard deviation gives the following BE limit:

$$\theta_{BE} = \frac{\ln(0.9)^2 + 0}{(0.1)^2} = 1.11 \tag{1.1}$$

More specifically, let y_T, y_R, and y'_R be independent *in vitro* bioavailabilities, where y_T is from the test product and y_R y'_R are from the reference product. The two products are said to be *in vitro* bioequivalent if $\theta < \theta_{BE}$, where

$$\theta = \frac{E(y_R - y_T)^2 - E\left(y_R - y'_R\right)^2}{\max\left\{\sigma_0^2, \dfrac{E\left(y_R - y'_R\right)^2}{2}\right\}}, \tag{1.2}$$

θ_{BE} is a pre-specified equivalence limit, and σ_0^2 is a pre-specified constant. Values of σ_0^2 and θ_{BE} can be found in the 1999 FDA draft guidance. According to the FDA draft guidance, *in vitro* bioequivalence can be claimed if the hypothesis that $\theta \geq \theta_{BE}$ is rejected at the 5% level of significance, provided that the ratio of geometric means between the two drug products is within 0.90 and 1.11.

1.2.3 Statistical Methods

For assessment of bioequivalence for the six in vitro tests, in addition to so-called non-comparative analysis, the FDA classifies statistical methods as either non-profile analysis or profile analysis (see also Wang et al., 2000; Chow, Shao, and Wang, 2003), which are briefly described below.

Non-comparative analysis

For each *in vitro* test, the FDA requires that a non-comparative analysis be performed. Non-comparative analysis refers to the statistical summarization of the bioavailability data by descriptive statistics. As a result, means, standard deviations, and coefficients of variation (CVs) in percentage of the six *in vitro* tests should be documented. More specifically, the overall sample means for a given formulation should be averaged over all samples (e.g., bottle/canisters), life stages (except for priming and re-priming evaluations), and lots or batches. In addition to the overall means, means at each life stage for each batch averaged over all bottles/canisters and for each life stage averaged over all lots (or batches) should be presented. For profile data, means, standard deviations, and percent CVs should be reported for each stage. The between-lot (or batch), within-lot (or batch) between-sample (e.g., bottle or canister), and within-sample (e.g., bottle or canister) between-life stage variability should be evaluated through appropriate statistical models.

Non-profile Analysis

The FDA classifies statistical methods for assessment of the six *in vitro* bioequivalence tests for nasal aerosols and sprays as either the non-profile analysis or the profile analysis. In this paper we focus on the non-profile analysis, which applies to tests for dose or spray content uniformity through container life, droplet size distribution, spray pattern, and priming and re-priming. Non-profile analysis applied to emitted dose or spray content uniformity, through container life, droplet size distribution, spray pattern, and priming/re-priming. For non-profile analysis, commonly used criterion for assessment of *in vitro* bioequivalence is given in (1.1).

Suppose that m_T and m_R canisters from respectively the test and the reference products are randomly selected for *in vitro* bioequivalence testing and one observation from each canister is obtained. The data can be described by the following model:

$$y_{jk} = \mu_k + \varepsilon_{jk}, \quad j = 1, ..., m_k, \tag{1.3}$$

where $k = T$ for the test product, $k = R$ for the reference product, μ_T and μ_R are fixed product effects, ε_{jk} are independent random measurement errors distributed as $N(0, \sigma_k^2)$, $k = T, R$. Under model (1.3), the parameter θ in (1.2) becomes

$$\theta = \frac{(\mu_T - \mu_R)^2 + \sigma_T^2 - \sigma_R^2}{\max\{\sigma_0^2, \sigma_R^2\}}, \tag{1.4}$$

and $\theta < \theta_{BE}$ if and only if $\xi < 0$, where

$$\xi = (\mu_T - \mu_R)^2 + \sigma_T^2 - \sigma_R^2 - \theta_{BE} \max\{\sigma_0^2, \sigma_R^2\}. \tag{1.5}$$

To test bioequivalence at level 5%, it suffices to construct a 95% upper confidence bound for ξ. Under model (1.3), the best unbiased estimator of $\delta = \mu_T - \mu_R$ is given by

$$\hat{\delta} = \bar{y}_T - \bar{y}_R \sim N\left(0, \frac{\sigma_T^2}{m_T} + \frac{\sigma_R^2}{m_R}\right),$$

where $\bar{y}_k$ is the average of y_{jk} over j for a fixed k. The best unbiased estimator of σ_k^2 is

$$s_k^2 = \frac{1}{m_k - 1} \sum_{j=}^{m_k} \left(y_{jk} - \bar{y}_k \right)^2 \sim \frac{\sigma_k^2 \lambda_{m_k-1}^2}{m_k - 1},$$

where $k = T, R$ and χ_t^2 denotes the central chi-square distribution with t degrees of freedom. Using the method in Hyslop, Hsuan, and Holder (2000) for individual bioequivalence testing, an approximate 95% upper confidence bound for ξ in (1.5) is

$$\tilde{\xi}_U = \hat{\delta}^2 + s_T^2 - s_R^2 - \theta_{BE} \max\left\{ \sigma_0^2, s_R^2 \right\} + \sqrt{U},$$

where U is the sum of the following three quantities:

$$\left[\left(|\hat{\delta}| + z_{0.95} \sqrt{\frac{s_T^2}{m_T} + \frac{s_R^2}{m_R}} \right)^2 - \hat{\delta}^2 \right]^2,$$

$$s_T^4 \left(\frac{m_T - 1}{x_{0.05;m_T-1}^2} - 1 \right)^2,$$

and

$$(1 + c\theta_{BE})^2 s_R^4 \left(\frac{m_R - 1}{x_{0.95;m_R-1}^2} - 1 \right)^2,$$

$c = 1$ if $s_R^2 \geq \sigma_0^2$, $c = 0$ if $s_R^2 < \sigma_0^2$, z_a is the ath quantile of the standard normal distribution, and $x_{t;a}^2$ is the ath quantile of the central chi-square distribution with t degrees of freedom. *In vitro* bioequivalence can be claimed if $\tilde{\xi}_U < 0$. This procedure is recommended by the FDA guidance.

As indicated in the FDA draft guidance, the FDA requires that m_k be at least 30. However, $m_k = 30$ may not be enough to achieve a desired power of the bioequivalence test in some situations. Increasing m_k can certainly increase the power, but in some situations, obtaining replicates from each canister may be more practical, and/or cost-effective. With replicates from each canister, however, the previously described test procedure is necessarily modified in order to address the between- and within-canister variabilities.

Profile Analysis

As indicated in the FDA draft guidance, profile analysis using a confidence interval approach should be applied to cascade impactor or multistage liquid impringer (MSLI) for particle size distribution. Equivalence may be assessed based on chi-square differences. The idea is to compare the profile difference between test product and reference product samples to the profile variation between reference product samples. More specifically, let y_{ijk} denote the observation from the jth subject's ith stage of the kth treatment. Given a sample (j_0) from test product and two samples (j_0, j_1) from reference products and assuming that there is a total of S stages, the profile distance between test and reference is given by

$$d_{TR} = \sum_{i=1}^{S} \frac{\left(y_{ij_0 T} - 0.5(y_{ij_1 R} + y_{ij_2 R})\right)^2}{\left(y_{ij_0 T} + 0.5(y_{ij_1 R} + y_{ij_2 R})\right)}.$$

Similarly, the profile variability within reference is defined as

$$d_{RR} = \sum_{i=1}^{S} \frac{(y_{ij_1 R} - y_{ij_2 R})^2}{0.5(y_{ij_1 R} + y_{ij_2 R})}.$$

For a given triplet sample of (Test, Reference 1, Reference 2), the ratio of d_{TR} and d_{RR}, i.e.,

$$rd = \frac{d_{TR}}{d_{RR}}$$

can then be used as a bioequivalence measure for the triplet samples between the two drug products. For a selected sample, the 95% upper confidence bound of $E(rd) = E(d_{TR}/d_{RR})$ is then used as a bioequivalence measure for the determination of bioequivalence. In other words, if the 95% upper confidence bound is less than the bioequivalence limit, then we claim that the two products are bioequivalent. The 1999 FDA draft guidance recommends a bootstrap procedure to construct the 95% upper bound for $E(rd)$. The procedure is described below.

Assume that the samples are obtained in a two-stage sampling manner. In other words, for each treatment (test or reference), three lots are randomly sampled. Within each lot, ten samples (e.g., bottles or canisters) are sampled. The following is quoted from the 1999 FDA draft guidance regarding the bootstrap procedure to establish profile bioequivalence.

For an experiment consisting of three lots each of test and reference products, and with 10 canisters per lot, the lots can be matched into six different combinations of triplets with two different reference lots in each triplet. The 10 canisters of a test lot can be paired with the 10 canisters of each of the two reference lots in (10 factorial)^2 = 3,628,800)^2 combinations in each of the lot triplets. Hence a random sample of the N canister pairing of the six Test-Reference 1-Reference 2 lot triplets is needed. The value of *rd* is estimated by the sample mean of the *rd*s calculated for the triplets in 10 selected samples of N. Note that the FDA recommends that N = 500 be considered.

1.2.4 Sample Size Requirement

In view of the fact that the FDA requires $m_k \geq 30$ and that $m_k = 30$ and $n_k = 1$ may not produce a test with sufficient power, Chow, Shao, and Wang (2003) proposed a procedure for determining sample sizes as follows.

As a typical approach, Chow, Shao, and Wang choose $m = m_T = m_R$ and $n = n_T = n_R$ so that the power of the bioequivalence test reaches a given level β (say 80%) when the unknown parameter vector $\psi = \left(\delta, \sigma^2_{BT}, \sigma^2_{BR}, \sigma^2_{WT}, \sigma^2_{WR}\right)$ is set at some initial guessing value $\tilde{\psi}$ for which the value of ξ is negative. Let U be given in the definition of $\hat{\xi}_U$ and U_β be the same as U but with 5% and 95% replaced by $1 - \beta$ and β, respectively. Since

$$P\left(\hat{\xi}_U < \xi + \sqrt{U} + \sqrt{U_\beta}\right) \approx \beta,$$

the power of the bioequivalence test, $P\left(\hat{\xi}_U < 0\right)$, is approximately larger than β if $\xi + \sqrt{U} + \sqrt{U_\beta} \leq 0$. Let $\tilde{U}$ and $\tilde{U}_\beta$ be U and U_β, respectively, with $\left(\hat{\delta}, s^2_{BT}, s^2_{BR}, s^2_{WT}, s^2_{WR}\right)$ replaced by $\tilde{\Psi}$. Then, the sample sizes $m = m_T = m_R$ and $n = n_T = n_R$ that produce a test with power approximately β should satisfy

$$\tilde{\xi} + \sqrt{\tilde{U}} + \sqrt{\tilde{U}_\beta} \leq 0. \tag{1.6}$$

Note that having a large m and a small n is an advantage when mn, the total number of observations for one treatment, is fixed. Thus, the following procedure is useful.

Step 1. Set $m = 30$ and $n = 1$. If (1.6) holds, stop, and the required sample sizes are $m = 30$ and $n = 1$; otherwise, go to step 2.

Step 2. Let $n = 1$ and find a smallest integer m_* such that (1.6) holds. If $m_* \leq m_+$ (the largest possible number of canisters in a given problem), stop, and the required sample sizes are $m = m_*$ and $n = 1$; otherwise, go to step 3.

Step 3. Let $m = m_+$ and find a smallest integer n_* such that (1.6) holds. The required sample sizes are $m = m_+$ and $n = n_*$.

If in practice it is much easier and inexpensive to obtain more replicates than to sample more canisters, then Steps 2–3 in the previous procedure can be replaced by

Step 2′. Let $m = 30$ and find a smallest integer n_* such that (1.6) holds. The required sample sizes are $m = 30$ and $n = n_*$.

1.3 Analytical Similarity Assessment

As can be seen from the previous section, unlike *in vivo* bioequivalence testing (mainly pharmacokinetic studies) based on the assessment of average bioequivalence (ABE), *in vitro* bioequivalence testing adopted the concept of population bioequivalence (PBE) which takes variabilities associated with both the test product and the reference product into consideration. Along this line, under the same framework, the FDA recommends a tiered approach for analytical similarity assessment. For analytical similarity assessment, the FDA suggests the sponsors should first identify critical quality attributes (CQAs) that are relevant to clinical outcomes via the study of mechanism of action (MOA) and pharmacokinetics/pharmacodynamics. Then, the identified CQAs will be assigned to three different tiers (say Tier 1, Tier 2, and Tier 3) according to their criticality or risking ranking which are assessed under certain appropriate statistical models. The FDA further recommends equivalence test for CQAs in Tier1, a quality range approach for CQAs in Tier 2, and raw data and graphical presentation for CQAs in Tier 3. Criteria and statistical methods for analytical similarity assessment for CQAs from different tiers are briefly described below.

1.3.1 Tier 1 Equivalence Test

For CQAs in Tier 1, the FDA recommends that an equivalency test be performed to assess analytical similarity. As indicated by the FDA, for a given CQA, we may test for equivalence by the following interval (null) hypothesis:

$$H_0 : \mu_T - \mu_R \leq -\delta \ \text{ or } \ \mu_T - \mu_R \geq \delta,$$

where $\delta > 0$ is the equivalence limit (or similarity margin), and μ_T and μ_R are the mean responses of the test (the proposed biosimilar) product and the reference product lots, respectively. Analytical equivalence (similarity) is concluded if the null hypothesis of non-equivalence (*dis*-similarity) is rejected. Note that Yu (2004) defined inequivalence as when the confidence interval falls entirely outside the equivalence limits. Similarly, to the confidence interval approach for bioequivalence testing under the raw data model, analytical similarity would be accepted for a quality attribute if the (1-2α)100% two-sided confidence interval of the mean difference is within (−δ, δ). The FDA further recommended that the equivalence acceptance criterion (EAC), δ = EAC = 1.5 ∗ σ_R, where σ_R is the variability of the reference product, be used, based on extensive simulation studies and internal scientific input. Chow (2015) provided statistical justification for the selection of $c = 1.5$ in EAC following the idea of scaled average bioequivalence (SABE) criterion for highly variable drug products proposed by the FDA.

For the establishment of EAC, the FDA made the following assumptions. First, the FDA assumes that the true difference in means is proportional to σ_R, i.e., $\mu_T - \mu_R$ is proportional to σ_R. Second, FDA adopts the similarity limit as EAC = 1.5 ∗ σ_R and recommended that σ_R be estimated by the sample standard deviation of test values from reference lots (one test value from each lot). Third, in the interest of achieving a desired power of the similarity test, FDA further recommends that an appropriate sample size be selected by evaluating the power under the alternative hypothesis at $\mu_T - \mu_R = \frac{1}{8}\sigma_R$. The assumption that $\mu_T - \mu_R$ is proportional to σ_R, the selection of $c = 1.5$, and the allowed mean shift of $\mu_T - \mu_R = \frac{1}{8}\sigma_R$ has generated tremendous discussion among the FDA, biosimilar sponsors, and academia, and they are debatable.

To provide a better understanding of the debatable issues and the FDA's proposal, we would like to point out the following which may be helpful to resolve the debatable issues: (i) unlike the traditional bioequivalence test, the FDA's intention is to take variability into consideration by considering the effect size adjusted for variability, i.e.,

$$\text{effect size} = eff = \frac{\mu_T - \mu_R}{\sigma_R} = \frac{\frac{1}{8}\sigma_R}{\sigma_R} = \frac{1}{8} = 0.125,$$

which is half-way between 1 and 1.25 (unity to the upper equivalence limit of 125%), (ii) the EAC for effect size adjusted for variability becomes fixed, i.e., $EAC = c = 1.5$, and (iii) if the true difference falls on the half-day between 1 and 1.25, the worst possible observed difference could fall on the 1.25 (this may happen if the worst possible reference lot is selected for comparison). In this case, the original EAC (bioequivalence testing for generic drug products with $\mu_T = \mu_R$) is necessarily shifted by 0.25. Thus, the upper limit is shifted from 1.25 to $c = 1.25 + 0.25 = 1.5$.

1.3.2 Tier 2 Quality Range Approach

For CQAs in Tier 2, the FDA suggests that analytical similarity be performed based on the concept of quality ranges, i.e., $\pm x \ast \sigma_R$, where σ_R is the standard deviation of the reference product and x a constant which should be appropriately justified. Thus, the quality range of the reference product for a specific quality attribute is defined as $(\hat{\mu}_R - x\hat{\sigma}_R, \hat{\mu}_R + x\hat{\sigma}_R)$. Analytical similarity would be accepted for the quality attribute if a sufficiently large percentage of test lot values fall within the quality range. Under normality assumption, if $x = 1.645$, we would expect 90% of the test results from reference lots to lie within the quality range. If x is chosen to be 1.96, we would expect that about 95% of test results of reference lots will fall within the quality range. Thus, the selection of x could have an impact on the width of the quality range and consequently the percentage of test lot values that will fall within the quality range.

At the 2015 Duke-Industry Statistics Symposium (DISS) held on the Duke Campus on October 22–23, one of FDA speakers indicated that x should be selected between 2 and 3 to guarantee that majority of test values of the test lots will fall within the quality range established, based on test values of the reference lots. Under the normality assumption, in practice, we would expect that 95% of data would fall below and above 2 (i.e., $x = 2$) standard deviations (SD) of the mean and about 99.7% of data would fall within ±3 SDs (i.e., $x = 3$) of the mean. Under the normality assumption, the FDA-recommended quality range approach is considered a reasonable approach only under the assumption that $\mu_T \approx \mu_R$ and $\sigma_T \approx \sigma_R$. In this case, it can be expected that the majority of test values obtained from the test lots will fall within ±x SD of the range established based on the test values of the reference lots.

In practice, however, the assumptions that $\mu_T \approx \mu_R$ and $\sigma_T \approx \sigma_R$ are usually not true due to the nature of biosimilar products. Thus, one of the major criticisms of the quality range approach is that it ignores the fact that there are differences in population mean and population standard deviation between the proposed biosimilar product and the reference product, i.e., $\mu_T \neq \mu_R$ and $\sigma_T \neq \sigma_R$. In practice, it is recognized that biosimilarity between a proposed biosimilar product and a reference product could be established even under the assumption that $\mu_T \neq \mu_R$ and $\sigma_T \neq \sigma_R$. Thus, under the assumption that $\mu_T \approx \mu_R$ and $\sigma_T \approx \sigma_R$, the quality range approach for analytical similarity assessment for CQAs from Tier 2 is considered more stringent as compared to equivalence testing for CQAs from Tier 1 (most relevant to clinical outcomes); regardless, they are mild-to-moderately relevant to clinical outcomes. This is because equivalence testing allows a possible mean shift of $\frac{\sigma_R}{8}$, while the quality range approach does not. In practice, there are several possible scenarios that include the cases where (i) $\mu_T \approx \mu_R$ or there is a significant mean shift (either a shift to the right or a shift to the left), and (ii) $\sigma_T \approx \sigma_R$, $\sigma_T > \sigma_R$, or $\sigma_T < \sigma_R$.

Thus, one of the most controversial issues for the quality range approach for CQAs in Tier 2 is that the approach does not reflect the real practice that $\mu_T \neq \mu_R$ and $\sigma_T \neq \sigma_R$. As a result, the test results are somewhat misleading and not reliable.

1.3.3 Tier 3 Raw Data and Graphical Comparison

For CQAs in Tier 3 with lowest risk ranking, the FDA recommends an approach that uses raw data/graphical comparisons. The examination of similarity for CQAs in Tier 3 is by no means less stringent, which is acceptable because they have least impact on clinical outcomes in the sense that a notable dis-similarity will not affect clinical outcomes.

An evaluation based on raw data and graphical presentation is not only somewhat subjective, but also biased. The Tier 1 equivalence test and the Tier 2 quality range similarity test is supposed to be the more rigorous than Tier 3 raw data and graphical comparison. That is, passing the Tier 1 equivalence test and the Tier 2 quality range similarity test will pass the Tier 3 raw data graphical comparison test. In practice, however, there is no guarantee that a given CQA which passes the Tier 1 equivalence test or the Tier 2 quality range similarity test will pass Tier 3 raw data graphical comparison test and vice versa. Since CQAs in Tier 3 are considered least relevant to clinical outcomes, it is necessary that all Tier 3 CQAs pass the test. If not, it is of interest to know about what percentage of CQAs needs to pass in order to pass Tier 3 test. Figures 1-3 exhibit graphical comparison for the cases where (i) $\mu_T = \mu_R$ and $\sigma_T \neq \sigma_R$, (ii) $\mu_T \neq \mu_R$ and $\sigma_T = \sigma_R$, and (iii) $\mu_T \neq \mu_R$ and $\sigma_T \neq \sigma_R$, respectively.

1.4 Scientific Factors and Practical Issues

1.4.1 Fundamental Similarity Assumption

For small molecule drug products, as indicated by Chow and Liu (2008), bioequivalence studies are necessarily conducted for regulatory review and approval of small molecule generic drug products. This is because it constitutes legal basis (from the *Hatch-Waxman* Act) under the *Fundamental Bioequivalence Assumption*, which states that

> If two drug products are shown to be bioequivalent, it is assumed that they willreach the same therapeutic effect or they are therapeutically equivalent.

Under the Fundamental Bioequivalence Assumption, bioavailability (defined as the rate and extent of drug absorbed into the blood stream and become available) serves as surrogate endpoint for clinical outcomes (safety and efficacy). Thus, under the Fundamental Bioequivalence Assumption, an *approved* generic drug product can serve as a substitute to the innovative (brand-name) drug product. Although this Fundamental Bioequivalence

Assumption constitutes legal basis, it has been challenged by many research-ers. In practice, there are four possible scenarios:

1. Drug absorption profiles are similar, and they are therapeutic equivalent;
2. Drug absorption profiles are not similar, but they are therapeutic equivalent;
3. Drug absorption profiles are similar, but they are not therapeutic equivalent;
4. Drug absorption profiles are not similar, and they are not therapeutic equivalent.

The Fundamental Bioequivalence Assumption is considered scenario (1). Scenario (1) works if the drug absorption (in terms of the rate and extent of absorption) is predictive of clinical outcome. In this case, PK responses such as AUC (area under the blood or plasma concentration-time curve for measurement of the extent of drug absorption) and Cmax (maximum concentration for measurement of the rate of drug absorption) serve as surrogate endpoints for clinical endpoints for assessment of efficacy and safety of the test product under investigation. Scenario (2) is the case that generic companies use to argue for generic approval of their drug products especially when their products fail to meet regulatory requirement for bioequivalence. In this case, it is doubtful that there is a relationship between PK responses and clinical endpoints. The innovator companies usually argue with the regulatory agency against generic approval with scenario (3). However, more studies are necessarily conducted in order to verify scenario (3). There are no arguments with respect to scenario (4).

It should be noted that a generic drug contains *identical* active ingredient(s) as the brand-name drug. Thus, it is reasonable to assume that generic (test) drug products and the brand-name (reference) drug have identical means, i.e., $\mu_T = \mu_R$. In addition, bioequivalence testing focuses on mean difference (i.e., $\mu_T - \mu_R$) or ratio of means (i.e., $\frac{\mu_T}{\mu_R}$) and ignores heterogeneity in variability of the test and reference product (i.e., $\sigma_T \neq \sigma_R$). As a result, a generic drug product may fail the bioequivalence testing when σ_R is relatively large (say > 30 %) even when $\mu_T = \mu_R$.

Follow similar idea, SSAB (2010) proposed the following Fundamental Similarity Assumption:

> When a follow-on biologic product is claimed to be biosimilar to an innovator product in some well-defined study endpoints, it is assumed that they will reach similar therapeutic effects, or they are therapeutically equivalent.

Although the above proposed Fundamental Similarity Assumption does not constitute legal basis, the FDA seems to adopt the assumption for analytical similarity assumption without verifying the validity of the assumption. In other words, the FDA assumes that analytical similarity in terms of CQAs identified at various stages such as functional and structural characterization of the manufacturing process is predictive of clinical outcomes.

Unlike small-molecule drug products, biosimilar products are large molecule drug products which are made of living cells or living organisms. As a result, it is expected that $\mu_T \neq \mu_R$, i.e., $\mu_T = \mu_R + \Delta$, where Δ is the true mean difference. Chow et al (2011) indicated that there are fundamental differences between small-molecule drug products and biosimilar products. For example, biosimilar products are often very sensitive to environmental factors during the manufacturing process. A small change and variation may translate to a huge change in clinical outcomes. Consequently, biosimilar products are expected to have much larger variability as compared to that of generic drug products. In this case, statistical methods for similarity assessment following the concept of bioequivalence testing (i.e., focusing on mean difference or ratio of means but ignore variability) for assessing biosimilarity of biosimilar products may not be appropriate. Table 1.2 provides a comparison between bioequivalence tests for generic drug products and biosimilarity tests for biosimilar products.

1.4.2 Primary Assumptions for Tiered Approach

Suppose there are n_R and n_T lots for analytical similarity assessment. For a given reference (test) lot, assume that the test value follows a distribution with mean $\mu_{Ri}(\mu_{Ti})$ and variance σ^2_{Ri} (σ^2_{Ti}). The FDA's recommended approach assumes that $\mu_{Ri} = \mu_{Rj}$ and $\sigma^2_{Ri} = \sigma^2_{Rj}$ for $i \neq j$, $i, j = 1, \ldots, n_R$ and $\mu_{Ti} = \mu_{Tj}$ and $\sigma^2_{Ti} = \sigma^2_{Tj}$ for $i \neq j$, $i, j = 1, \ldots, n_T$ for equivalence test in Tier 1 and quality range approach in Tier 2. Now let σ^2_R and σ^2_T be the variabilities associated with the reference product and the test product, respectively. Thus, we have

$$\sigma^2_R = \sigma^2_{WR} + \sigma^2_{BR} \text{ and } \sigma^2_T = \sigma^2_{WT} + \sigma^2_{BT},$$

where $\sigma^2_{WR}, \sigma^2_{BR}$, and $\sigma^2_{WT}, \sigma^2_{BT}$ are the within-lot variability and between-lot (lot-to-lot) variability for the reference product and the test product, respectively. In practice, it is very likely that $\sigma^2_R \neq \sigma^2_T$ and often $\sigma^2_{WR} \neq \sigma^2_{WT}$ and $\sigma^2_{BR} \neq \sigma^2_{BT}$ even when $\sigma^2_R \approx \sigma^2_T$. This has posed a major challenge to the FDA's proposed approaches for the assessment of analytical similarity for CQAs from both Tier 1 and Tier 2, especially when there is only one test sample from each lot from the reference product and the test product. The FDA's proposal ignores within-lot variability, i.e., when $\sigma^2_{WR} = 0$ or $\sigma^2_R = \sigma^2_{BR}$. In other words, sample variance based on x_i, $i = 1, \ldots, n_R$ from the reference product may underestimate the true σ_R^2, and consequently may not provide a fair and reliable assessment of analytical similarity for a given quality attribute.

In practice, it is well recognized that $\mu_{Ri} \neq \mu_{Rj}$ and $\sigma_{Ri}^2 \neq \sigma_{Rj}^2$ for $i \neq j$, where μ_{Ri} and σ_{Ri}^2 are the mean and variance of the ith lot of the reference product. A similar argument is applied to the proposed biosimilar (test) product. As a result, the selection of reference lots for the estimation of σ_R is critical for the proposed approach. The selection of reference lots has an impact on the estimation of σ_R and consequently on the EAC. Assuming that $n_R > n_T$, the FDA suggested using the remaining $n_R - n_T$ lots to establish EAC to avoid selection bias. It sounds a reasonable approach if $n_R \gg n_T$. In practice, however, there might be a few lots available. Alternatively, it is suggested that all of the n_R lots be used to establish EAC.

The assumption that $\mu_{Ri} = \mu_{Rj}$ and $\sigma_{Ri}^2 = \sigma_{Rj}^2$ for $i \neq j$; $i,j = 1,\ldots,n_R$ and $\mu_{Ti} = \mu_{Tj}$ and $\sigma_{Ti}^2 = \sigma_{Tj}^2$ for $i \neq j$; $i, j = 1,\ldots, n_T$ is a strong assumption which does not reflect real practice. Since biosimilar products are made of living cell and/or living organisms, it is expected that that

$$\mu_{Ri} \neq \mu_{Rj},\ \sigma_{Ri}^2 \neq \sigma_{Rj}^2 \text{ for } i \neq j;\ i, j = 1,\ldots, n_R;$$
$$\mu_{Ti} \neq \mu_{Tj},\ \sigma_{Ti}^2 \neq \sigma_{Tj}^2 \text{ for } i \neq j;\ i, j = 1,\ldots, n_T.$$

Heterogeneity within lots and across lots between the test product and the reference product has posed the following controversial issues in analytical similarity assessment. First, suppose two extreme reference lots, one lot with smallest within-lot variability and one with the largest within-lot variability, are randomly selected for analytical similarity assessment. In this case, chances are that the reference product (the two selected extreme lots) may not even pass the equivalence test itself. Thus, analytical similarity between a test product and the reference product is not comprehensive. The other controversial issue is that if reference lots selected for establishment of EAC are extreme lots with smallest variability, the established EAC could be too narrow to penalize good test products.

1.4.3 Fixed Approach for Margin Selection

In the tiered approach, the FDA seems to prefer a *fixed margin approach* for EAC by treating $s = \sigma_R$ as the true σ_R (i.e., $1.5 * \sigma_R$) for Tier 1 CQAs. The fixed margin approach is also referred to as a *fixed standard deviation (SD) approach* for similarity quality range by treating the estimate of SD (which is obtained based on test values of reference lots) as the true σ_R (i.e., $x * SD$) for Tier 2 CQAs. The fixed margin or SD is in fact a statistic, which is a random variable' rather than a fixed constant. In other words, it may vary depending upon the selected reference lots for tiered testing. The fixed SD approach is a conditional approach rather than an unconditional approach. In practice, if reference lots with *less* variability are selected for tiered testing, the proposed biosimilar product is most likely to fail the test. As a result, the scientific validity of the fixed SD approach is questionable.

One of the major criticisms of the fixed approach for margin/range selection in tiered analysis is that the fixed approach does not take into consideration the variability of the estimate of the standard deviation. Thus, it is considered *bad luck* to the biosimilar sponsors if reference lots with less variability are selected for a Tier 1 equivalence test. Another criticism is that the reference product cannot pass the Tier 1 equivalence test itself if we divide all of the reference lots into two groups: one group with less variability and the other group with large variability. In this case, the group with large variability may not pass a Tier 1 equivalence test with EAC established based on test values from the reference lots with less variability. It would be a concern that a reference product cannot pass a Tier 1 equivalence test when compared to itself.

1.4.4 Inconsistent Test Results between Tiered Approaches

As indicated in Tsong (2015), the Tier 1 equivalence test is considered more rigorous than the Tier 2 quality range approach, which is in turn more rigorous than the Tier 3 raw data and graphical comparison. The primary assumptions for these tired approaches, however, are different. Thus, under different assumptions, there is no guaranteed that passing a Tier 1 equivalence test will pass a Tier 2 quality range approach and vice versa, although these tests are conducted based on the same data set collected from the test and reference lots under study. In practice, it is of interest to evaluate inconsistencies regarding the passages between the Tier 1 equivalence test and Tier 2 quality range approach.

For a given CQA, the inconsistencies between the Tier 1 equivalence test and Tier 2 quality range approach can be assessed by means of clinical trial simulation as follows. Let p_{ij} be the probability of passing the ith tier test given that the CQA has passed the jth tier test. Thus, we have the following $2x2$ contingency table (Table 1.3) for comparison between the Tier 1 equivalence test and Tier 2 quality range approach.

Let $\mu_T = \mu_R + \Delta$ and $\sigma_T = C\sigma_R$. It is then suggested that the inconsistencies between the Tier 1 equivalence test and Tier 2 quality range approach be evaluated at various combinations of (i) $\Delta = 0$ (no mean shift), $\frac{1}{8}\sigma_R$ (FDA recommended mean shift allowed), and $\frac{1}{4}\sigma_R$ (the worst possible scenario),

TABLE 1.3

Probabilities of Inconsistencies

Tier 1 Equivalence Test	Tier 2 Quality Range Approach	
	Pass	Fail
Pass	p_{11}	p_{12}
Fail	p_{21}	p_{22}

and (ii) $C = 0.8$ (deflation), 1.0, and 1.2 (inflation) to provide a complete picture of the relative performance of the Tier 1 equivalence test and Tier 2 quality range approach.

1.4.5 Sample Size Requirement

One of the most commonly asked questions for analytical similarity assessment is probably how many reference lots are required for establishing an acceptable EAC for achieving a desired power. For a given EAC, formulas for sample size calculation under different study designs are available in Chow, Shao, and Wang (2008). In general, sample size (the number of reference lots, k) required is a function of (i) overall type I error rate (α), (ii) type II error rate (β) or power ($1 - \beta$), (iii) clinically or scientifically meaningful difference (i.e., $\mu_T - \mu_R$), and (iv) the variability associated with the reference product (i.e., σ_R) assuming that $\sigma_T = \sigma_R$. Thus, we have

$$k = f(\alpha, \beta, \mu_T - \mu_R, \sigma).$$

In practice, we select an appropriate k for achieving a desired power of $1 - \beta$ for detecting a clinically meaningful difference of $\mu_T - \mu_R$ at a pre-specified level of significance α assuming that the true variability is σ. If α, $\mu_T - \mu_R$, and σ are fixed, the above equation becomes $k = f(\beta)$ We can then select an appropriate k for achieving the desired power. The FDA's recommendation attempts to control all parameters at the desired levels (e.g., $\alpha = 0.05$ and $1 - \beta = 0.8$) by *knowing* that $\mu_T - \mu_R$ and σ are varying. In practice, it is often difficult, if not impossible, to control (or find a balance point among) α (type I error rate), $1 - \beta$ (power), $\mu_T - \mu_R = \Delta$ (clinically meaningful difference), and σ (variability in observing the response) *at the same time*. For example, controlling α at a pre-specified level of significance may be at the risk of decreasing power with a selected sample size.

1.4.6 Relationship between Similarity Limit and Variability

As it can be seen from Table 1.2, similarity (equivalence) limit for assessment of similarity (equivalence) depends upon the variability associated with the drug product. For example, if the variability is less than 10%, (90%, 111%) similarity limit is recommended, while (80%, 125%) similarity limit is used for drug product with variability between 20% and 30%. For drug products that exhibit high variability such as highly variable small-molecule drug products or large-molecule biological products including biosimilars, it is suggested that a scaled similarity limit adjusted for variability be considered. As an alternative to the scaled similarity limit and in the interest of a one-size-fits-all criterion, some researchers suggest (70%, 143%) be considered. The selection of similarity limits based on the associated variability is somewhat arbitrary without scientific/statistical justification. In what follows, we attempt to describe the relationship between similarity limit and variability for achieving a desired probability of claiming similarity.

Let (δ_L, δ_U) be the similarity limits for evaluation of similarity between a test product (T) and a reference product (R). Current regulation indicates that we can claim similarity if the 90% confidence interval of difference in mean (i.e., $\mu_T - \mu_R$) falls entirely within the lower and upper similarity limits. Let (L, U) be the 90% confidence interval for $\mu_T - \mu_R$. Define

$$p = P\left\{ (L, U) \subset (\delta_L, \delta_U) \big| \sigma_R^2, \sigma_T^2 \right\},$$

where σ_R^2 and σ_T^2 are the variabilities associated with the reference product and the test product, respectively, and p is a desired probability of claiming similarity. Thus, for a given set of p, σ_R^2 and σ_T^2, appropriate (δ_L, δ_U) can be determined.

1.4.7 Regulator's Current Thinking on Scientific Input

In his recent presentation, Tsong (2015) indicated that the FDA's current thinking for establishment of EAC is to consider $1.5 * \sigma_R + \Delta$, where Δ is a regulatory allowance depending upon *scientific input*. From statistical perspective, we may interpret the scientific input as scientific justification for accounting for the worst possible reference lot (i.e., a reference lot with extremely large variability) in establishment of EAC. Thus, $1.5 * \sigma_R + \Delta$, can be rewritten as $1.5 * \sigma_R'$, where $\sigma_R' = \sigma_R + \varepsilon$. Thus, we have

$$\text{EAC} = \pm 1.5 * (\sigma_R + \varepsilon).$$

The FDA's original proposal is to estimate σ_R' using sample standard deviation (s) of the test values obtained from the reference lots, assuming that there is only one single test value per lot. Although Wang and Chow (2015) showed that s is an unbiased estimate of σ_R, it underestimates σ_R' because it does not take the variability associated with s into consideration. Thus, Chow (2015) suggested using the 95% upper confidence bound to estimate σ_R', i.e.,

$$\hat{\sigma}_R' = \sqrt{\frac{n_R - 1}{\chi_{\alpha/2, n_R - 1}^2}} \, s.$$

This leads to

$$\varepsilon = \left(\sqrt{\frac{n_R - 1}{\chi^2_{\frac{\alpha}{2}, n_R - 1}}} - 1 \right) s.$$

One of the controversial issues for establishment of EAC is whether the margin should be fixed. FDA seems to recommend using the estimate of σ_R as the true σ_R without taking into consideration of the variability associated with the observed sample variance. The variability associated with the observed sample variance depends upon the sample size (i.e., the number of reference lots) used for analytical similarity assessment. As a result, the result of equivalence test may not be *reproducible*.

1.4.8 A Proposed Unified Tiered Approach

For biosimilar products, their population means and population variances are expected to be different. The relationship between a proposed biosimilar (test) product and an innovative (reference) product can be described as $\mu_T = \mu_R + \Delta$ and $\sigma_T = C\sigma_R$, where Δ is a measure of a possible shift in population mean and C is an inflation factor. When there is a significant shift in mean (e.g., $\Delta \gg 0$) or notable heterogeneity between the proposed biosimilar product and the reference product (e.g., either $C >> 1$ or $C << 1$), the validity of the equivalence test for Tier 1 CQAs and the quality range approach for Tier 2 CQAs is questionable because the equivalence test and/or quality range approach are unable to handle a major shift in population mean and a significant change in population variance.

To overcome the problems of major mean shift and significant change in variability, alternatively, we may consider the equivalence test and quality range approach be applied to standardized test values (or *effect size adjusted for standard deviation*) rather than apply to untransformed raw data. For simplicity and without loss of generality, consider the quality range approach for Tier 2 CQAs.

Define $eff_R = \dfrac{\mu_R}{\sigma_R}$ and $eff_T = \dfrac{\mu_T}{\sigma_T}$ and let y_i, $i = 1,\ldots,n_R$ be the test values of the reference lots, where n_R is the number of reference lots considered for the test. Also, let z_i, $i = 1,\ldots,n_T$ be the test values of the reference lots, where n_T is the number of test lots considered for the test. FDA's quality range approach is applied on $\{y_i, i = 1,\ldots,n_R\}$ to establish the *quality range* with appropriate selection of x. Instead, we suggested the quality range approach be applied to $\left\{ \dfrac{y_i}{\hat{s}_R}, i = 1,\ldots,n_R \right\}$, where $\hat{s}_R$ is the sample standard deviation of the test values of the reference lots. The quality range can then be established with the following adjustment on the selected x for achieving $eff_R \approx eff_T$. For simplicity and illustration purpose, consider that case that $\sigma_R \approx \sigma_T$ (i.e., $C \approx 1$). In this case, under the assumption that $eff_R \approx eff_T$, we have

$$\frac{\mu_R}{\sigma_R} \approx \frac{\mu_T}{\sigma_T} = \frac{\mu_R + \Delta}{\sigma_T}.$$

This leads to

$$\sigma_T = \sigma_R \left(1 + \frac{\Delta}{\mu_R} \right) \approx \sigma_R,$$

where $1 + \dfrac{\Delta}{\mu_R}$ is referred to as the adjustment factor for selection of x.

Note that the left-hand side of the above equation is for test lots and the right-hand side is referred to the reference lots. Various selection of x and adjusted x when there is a shift in mean are given in Table 1.4.

As can be seen from the above table, under the assumption that $eff_R \approx eff_T$, if there is a 20% shift in mean, i.e., $\dfrac{\Delta}{\mu_R} = 0.2$, the selection of $x = 2.5$ is equivalent to the selection of $x = 3$ without a shift in population mean.

Similar idea can be applied to the case where there is a shift in scale parameter (i.e., $C \neq 1$). It should also be noted that the above proposal is similar to the justification (based on % of coefficient of variation). It should be noted that for biosimilar products, the assumption that $\mu_T \approx \mu_R$ and $\sigma_T \approx \sigma_R$ is usually not true. In practice, it is reasonable to assume that $eff_R \approx eff_T$. The above proposal accounts for a possible shift in population mean and heterogeneity in variability.

TABLE 1.4

Adjustment on Selection of x When There Is a Shift in Mean

$\dfrac{\Delta}{\mu_R}$	x	$x(adj)$
0.1	2	2.2
	2.5	2.75
	3	3.3
0.2	2	2.4
	2.5	3
	3	3.6

1.4.9 Practical Issues

Since there are many critical quality attributes to a potential patient's response in biosimilar products, for a given critical attribute, valid statistical methods are necessarily developed under a valid study design and a given set of criteria for similarity, as described in the previous section. Several practical issues can be identified for developing appropriate statistical methodologies for the assessment of biosimilarity between a proposed biosimilar product and an innovative biological product. These practical issues include, but are not limited to:

How Similar is Similar?

Current criteria for assessment of bioequivalence/biosimilarity are useful for determining whether a biosimilar product is similar to a reference product on average. However, it does not provide additional information regarding the *degree* of similarity. As indicated in the BPCI Act, a biosimilar product is defined as a product that is *highly similar* to the reference product. However, little or no discussion regarding the degree of similarity for highly similar was provided. Besides, the following is also of concern to the sponsor: "What if a biosimilar product turns out to be superior to the reference product (in other words, the proposed biosimilar product is bio*better*)?" A simple answer to the concern is that superiority (i.e. bio*better*) is not biosimilar.

Criteria for biosimilarity

Current criteria for assessment of *in vitro* bioequivalence or biosimilarity (e.g., from 90% to 111%) and *in vivo* bioequivalence or biosimilarity (e.g., from 80% to 125%) focus on average bioequivalence or biosimilarity (ABE). One of the major criticisms is that current criteria do not take variabilities associated with both the test product and the reference product into consideration. Thus, it is a concern that good product (e.g., the proposed biosimilar product with relatively small variability) may be penalized by not passing the bioequivalence or biosimilarity testing, even though both products have identical means. To overcome this drawback, it is suggested a criterion that is able to account for both average and variability be considered. As a result, Haidar et al (2008) proposed the use of scaled average criterion (SABE) which adjusts for the variability associated with the reference product. Chow et al. (2015) further suggested the use of a scaled criterion for drug interchangeability (SCDI) which adjusts for both variabilities associated with the test product and the reference product and the variability due to subject-by-drug interaction. In addition, Chow and Liu (2013) suggested that comparison in distribution of the response between the test product and the reference product.

Criteria for interchangeability

In practice, it is recognized that drug interchangeability is related to the variability due to subject-by-drug interaction. However, it is not clear whether criterion for interchangeability should be based on the variability due to subject-by-drug interaction or the variability due to subject-by-drug interaction adjusted for intra-subject variability of the reference drug. It is also not clear whether criterion for interchangeability should be based on aggregated criterion or disaggregated criterion.

Bridging studies for assessing biosimilarity

As most biosimilars studies are conducted using a parallel design rather than a replicated crossover design, independent estimates of variance components such as the intra-subject and the variability due to subject-by-drug interaction are not possible. In this case, bridging studies may be considered.

Other practical issues include (i) the use of a percentile method for the assessment of variability, (ii) comparability in biologic activities, (iii) assessment of immunogenicity, (iv) consistency in manufacturing processes (see, e.g., ICH, 1996, 1999, 2005), (v) stability testing for multiple lots and/or multiple labs (see, e.g., ICH, 1996), (vi) the potential use of sequential testing procedures and multiple testing procedures, (vii) assessing biosimilarity using a surrogate endpoint or biomarker such as genomic data (see, e.g., Chow, Shao, and Li, 2004).

Further research is needed in order to address the above mentioned scientific factors and practical issues recognized at the FDA Public Hearing.

1.4.10 Remarks

The concept of stepwise approach recommended by the FDA is well taken. The purpose is to obtain the totality-of-the-evidence in order for demonstration of biosimilarity between a proposed biosimilar product and an innovative biological product. The totality-of-the-evidence consists of evidence from analytical studies for characterization of the molecule, animal studies for toxicity, pharmacokinetics and pharmacodynamics for pharmacological activities, clinical studies for safety/tolerability, immunogenicity, and efficacy. FDA, however, does not indicate whether these evidences should be obtained sequentially or simultaneously. It is a concern that the sequential approach may kill good products early purely by chance alone.

The stepwise approach starts with structural and functional characterization of critical quality attributes that may be relevant to clinical outcomes. FDA's recommended tiered approach is to serve the purpose. The recommended tired approach depends upon the classification of identified CQAs based on their criticality (or risk ranking) relevant to clinical outcomes. The assessment of criticality, however, is somewhat subjective and often lack of scientific/statistical justification.

The FDA recommended tiered approach have raised a number of scientific and/or controversial issues. These controversial issues are related to difference in population means and heterogeneity within and across lots within and between the test product and the reference product. In practice, the primary assumption that that $\mu_T \approx \mu_R$ and $\sigma_T \approx \sigma_R$ is usually not true. In this case, it is reasonable to assume that $\mathit{eff}_R = \dfrac{\mu_R}{\sigma_R} \approx \dfrac{\mu_T}{\sigma_T} = \mathit{eff}_T$ so that assessment of similarity between the proposed biosimilar product and the reference product is possible.

As indicated by Tsong (2015), the Tier 1 equivalence test is supposed to be more rigorous than the Tier 2 quality range approach. Thus, we would expect passing the Tier 1 test will pass the Tier 2 test. In practice, however, there is no guarantee that a given CQA which passes the Tier 1 test will pass the Tier 2 test and vice versa. This may be due to difference in primary assumptions made for the Tier 1 equivalence test and Tier 2 quality range approach. Since there may be a large number of CQAs in both Tier 1 and Tier 2, "Does FDA require all CQAs at either Tier pass the corresponding test in order to claim totality-of-the-evidence?" is probably the most commonly asked question. If not, are there any rules to follow?

Liao and Darken (2013) indicated that a good study design than can include different reference lots manufactured at different times with different shelf-lives should be used in order to accurately and reliably quantitate different sources of variability for estimation of σ_R. Under a valid study design, appropriate statistical model depending upon the nature of the CQAs (e.g., paired or non-paired) should be employed. A proposed biosimilar product with relatively smaller variability as compared to its innovative biological product should be rewarded (Liao and Darken, 2013).

1.5 Aim and Scope of the Book

This is intended to be the first book entirely devoted to analytical similarity assessment for CQAs that are relevant to clinical outcomes in biosimilar product development. It covers challenging and/or practical/statistical issues that may occur at various stages of the manufacturing process of biosimilar products. It is our goal to provide a useful desk reference and state-of-the-art examination of analytical similarity assessment for scientists engaged in biosimilar product research and development, those in government regulatory agencies who have to make decisions on biosimilarity between a proposed biosimilar product and an innovative (reference) drug product, and to biostatisticians who provide the statistical support for biosimilar studies and related clinical projects. More importantly we would like to provide graduate students in pharmacokinetics, clinical pharmacology, biopharmaceutics, and biostatistics an advanced textbook in biosimilar studies. We hope that this book can serve as a bridge among the pharmaceutical/biotechnology industry, government regulatory agencies, and academia.

This book is organized as follows. In this chapter, some background, past equivalence of *in vitro* bioequivalence testing, definition, and scientific factors and/or challenging issues have been discussed. In the next chapter, current regulatory requirement and/or guidance are reviewed. Chapter 3 deals with regulatory requirement regarding chemistry, manufacturing, and control (CMC) for biosimilar product development. Also included in this chapter are manufacturing processes for biosimilar products, United States Pharmacopedia (USP) tests, technology transfer, and the concept of quality by design. In Chapter 4, statistical methods for assay development and validation and manufacturing process validation are discussed. Chapter 5 provides statistical methods for identification and classification of critical quality attributes (CQAs) that are relevant to clinical outcome at various stages of the manufacturing process. Chapter 6 introduces the FDA's recommended tiered approach (i.e., equivalence test for CQAs in Tier 1, quality range approach for CQAs in Tier 2, and raw data and graphical presentation for CQAs in Tier 3) for obtaining totality-of-the evidence for demonstrating biosimilarity of a proposed biosimilar product to an innovative biologic product. Sample size requirements based on different criteria such as maintaining treatment effect, controlling variability, achieving desired conditional power, and reaching a certain degree of reproducibility are discussed in Chapter 7. Chapter 8 focuses on analytical similarity assessment when there are multiple references (e.g., EU-approved reference and US-licensed product). Chapter 9 covers some controversial issues regarding extrapolation across different indications and/or populations. Case studies regarding analytical similarity assessment based on recent regulatory submissions to the FDA are discussed in Chapter 10. These cases include proposed biosimilar products from Amgen and Mylan. Chapter 11 discusses some practical and challenging issues that are commonly encountered in analytical similarity assessment. Some recent developments are given in the last chapter of this book.

2

Regulatory Approval Pathway
of Biosimilar Products

2.1 Introduction

As indicated in the previous chapter, a biosimilar product is a similar biological product such as protein product, vaccine, or blood product whose active drug substance is made of a living organism or derived from a living organism. The term *similar* is in the sense that it is similar to an innovator drug product in terms of safety, purity, and potency. In some cases, the term biosimilar has been used in an inappropriate way, and therefore it is important to review differences in definitions of biosimilar products in different regions (see Table 2.1).

The WHO defines SBP as a biotherapeutic product, which is similar in terms of quality, safety, and efficacy to an already licensed reference biotherapeutic product (WHO, 2009). Health Canada defines biosimilar to be a biologic drug that enters the market subsequent to a version previously authorized in Canada, and with demonstrated similarity to a reference biologic drug (HC, 2010). As indicated in the Biologics Price Competition and Innovation (BPCI) Act passed by the US Congress and enacted on March 23, 2010, a biosimilar product is defined as a product that is *highly similar* to the reference product, notwithstanding minor differences in clinically inactive components, and for which there are no clinically meaningful differences in terms of safety, purity, and potency from the reference product. The EMA, on the other hand, did not provide the definition of biosimilars in their original guidelines; however, a recently published concept paper on the revision of the guidelines on similar biological medicinal product indicated that it might be prudent to discuss if a definition of biosimilar, in extension of what is in the legislation and relevant CHMP (Committee for Medicinal Products for Human Use) guidance, is necessary (EMA, 2011).

Based on these different definitions, there are three determinants in the definition of the biosimilar product: (i) it should be a biologic product; (ii) the reference product should be an already licensed biologic product; (iii) the demonstration of high similarity in safety, quality, and potency (efficacy) is necessary (Wang and Chow, 2012). Besides, it is well recognized that the

TABLE 2.1

Definitions of Biosimilar Products

Term	By	Definition
SBP	WHO	A biotherapeutic product similar to an already licensed reference biotherapeutic product in terms of quality, safety, and efficacy
FOB	US FDA	A product highly similar to the reference product without clinically meaningful differences in safety, purity, and potency
SEB	Canada	A biologic drug that enters the market subsequent to a version previously authorized in Canada with demonstrated similarity to a reference biologic drug
Biosimilar	KFDA	Biological products which have demonstrated their equivalence to an already approved reference product with regard to quality, safety, and efficacy

similarity should be demonstrated using a set of comprehensive comparability exercise at the quality, non-clinical, and clinical level. Products not authorized by this comparability regulatory pathway cannot be called biosimilars.

In practice, generic drug products are expected to have same mean and standard deviation as the brand-name drug product because they contain the same active ingredient(s) as the brand-name drug product. Biosimilar drug products, on the other hand, are expected to have different mean and standard deviation as compared to the innovative biologic drug product because they are made of or derived from living cells or organisms. Thus, the criteria and standard methods developed for the bioequivalence assessment of generic drug products may not be appropriately directly applied to the assessment of biosimilarity of biosimilar products, due to fundamental differences (see also Chow et al., 2011 and Figure 2.1).

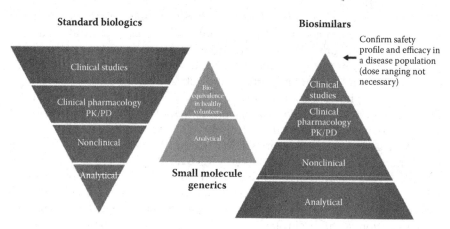

FIGURE 2.1
Standard biologics and biosimilars development.

In its guidance on scientific consideration (FDA, 2015), the US FDA recommends a stepwise approach for obtaining totality-of-the-evidence for demonstrating biosimilarity of proposed biosimilar product to an innovative biologic drug product (FDA, 2015). The stepwise approach starts with analytical studies for functional and structural characterization of critical quality attributes (CQAs) at various stages of manufacturing process followed by animal studies for the assessment of toxicity, clinical pharmacology for pharmacokinetics (PK), or pharmacodynamics (PD) studies, and clinical studies for the assessment of immunogenicity, safety/tolerability, and efficacy. To assist the sponsor in performing tests for analytical similarity, recently the FDA circulated a draft guidance on analytical similarity assessment for public comments in September, 2017.

The purpose of this chapter is to review regulatory requirements by focusing analytical similarity assessment for the approval pathway of biosimilar products worldwide including WHO and various regions, such as European Union (EU), United States (US), Canada, and Asian Pacific Region (e.g., Japan, Korea, and China). Comparison of these regulatory requirements, and some recommendations regarding global harmonization will be made. In the next section, regulatory requirements from different regions are briefly summarized. Also included in this section are some debatable issues in analytical similarity assessment requirements. In Section 2.3, specific requirements for analytical similarity assessment are provided. Recommendations on global harmonization of regulatory approval pathways are given in Section 2.4. Section 2.5 provides some concluding remarks.

2.2 Regulatory Requirements

As indicated in the previous section, the standard methods for the assessment of bioequivalence for generic drug products with identical active ingredient(s) may not be appropriately applied for biosimilarity assessment for biosimilar drug products due to the fundamental differences between generic (small-molecule) drug products and biologic (large-molecule) products (Chow et al., 2011). For the assessment of biosimilars, regulatory requirements from different regions such as the European Union (EU), United States, and Asian Pacific Region are similar and yet slightly different (Wang and Chow, 2012). In what follows, these regulatory requirements (especially on analytical similarity assessment) are briefly described.

2.2.1 World Health Organization (WHO)

As an increasingly wide range of SBPs are under development or are already licensed in many countries, the WHO formally recognized the need for guidance for their evaluation and overall regulation in 2007. *Guidelines on Evaluation of Similar Biotherapeutic Products (SBPs)* was developed and adopted by the 60th meeting of the WHO Expert Committee on Biological Standardization in 2009 (CPMP, 2009). The intention of the guidelines is to provide globally acceptable principles for licensing biotherapeutic products that are claimed to be similar to the reference products that have been licensed based on a full licensing dossier (WHO, 2009). The scope of the guidelines includes well-established and well-characterized biotherapeutic products that have been marketed for a suitable period of time with a proven quality, efficacy, and safety, such as recombinant DNA-derived therapeutic proteins.

Key Principles and Basic Concept – Key principles and basic concept for licensing SBPs have been explained in the WHO's guidelines. One of the most important principles of developing SBP is the stepwise approach starting with characterization of quality attributes of the product and followed by non-clinical and clinical evaluations. Manufactures should submit a full quality dossier that includes a complete characterization of the product, the demonstration of consistent and robust manufacture of their product, and the comparability exercise between the SBP and reference biotherapeutic product (RBP) in the quality part, which together serve as the basis for the possible reduction in data requirements in the non-clinical and clinical development. This principle indicates that the data reduction is only possible for the non-clinical and clinical parts of the development program, and significant differences between the SBP and the chosen RBP detected during the comparability exercise would result in requirement for more extensive non-clinical and clinical data. In addition, the amount of non-clinical and clinical data considered necessary is also depend on the class of products, which calls for a case by case approach for different classes of products.

Reference Biotherapeutic Product (RBP) – The choice of the RBP is another important issue covered in the WHO's guidelines. Traditionally, the National Regulatory Authorities (NRA) has required the use of a nationally-licensed reference product for licensing of generic medicines, but this may not be feasible for countries lacking nationally-licensed RBPs. Thus additional criteria to guide the acceptability of using a RBP licensed in another jurisdiction may be needed. Considering the choice of the RBP, the WHO requires that it should have been marketed for a suitable duration and have a volume of marketed use, and should be licensed based on a full quality, safety, and efficacy data. Besides, same RBP should be used throughout the development of SBP, and the drug substance, dosage form, and route of administration of SBP should be the same as that of RBP.

Quality – As mentioned in the former section, the comprehensive comparison showing quality similarity between SBP and RBP is a prerequisite for applying the clinical safety and efficacy profile of RBP to SBP, thus a full quality dossier for both drug substance and drug product is always required. To evaluate comparability, the WHO recommends the manufacturer conduct a comprehensive physicochemical and biological characterization of the SBP in head-to-head comparisons with RBP. The following aspects of product quality and heterogeneity should be assessed.

1. *Manufacturing Process*

The manufacture process should meet the same standards as required by the NRA for originator products, and implement Good Manufacturing Practices, modern quality control and assurance procedures, in-process controls, and process validation. The SBP manufacturer should assemble all available knowledge of the RBP with regard to the type of host cell, formulation and container closure system, and submit a complete description and data package delineating the whole manufacturing process, including obtaining and expression of target genes, the optimization and fermentation of gene engineering cells, the clarification and purification of the products, formulation and testing, and aseptic filling and packaging.

2. *Characterization*

Thorough characterization and comparability exercise are required, and details should be provided on primary and higher-order structure, post-translational modifications, biological activity, process- and product-related impurities, relevant immunochemical properties, and results from accelerated degradation studies, and studies under various stress conditions.

Non-Clinical and Clinical Studies – After demonstrating the similarity of SBP and RBP in quality, the proving of safety and efficacy of a SBP usually requires further non-clinical and clinical data. Non-clinical evaluations should be undertaken both *in vitro* (e.g., receptor-binding studies, cell-proliferation, cytotoxicity assays) and *in vivo* (e.g., biological/pharmacodynamic activity, repeat dose toxicity study, toxicokinetic measurements, anti-product antibody titers, cross reactivity with homologous endogenous proteins, product neutralizing capacity).

In terms of the clinical evaluation, the comparability exercise should begin with pharmacokinetic (PK) and pharmacodynamics (PD) studies followed by the pivotal clinical trials. PK studies should be designed to enable detection of potential differences between SBP and RBP. Single-dose, cross-over PK studies in homogenous population are recommended by the WHO. The manufacturer should justify the choice of single-dose studies, steady-state studies, or repeated determination of PK parameters, and the study population. Due to the lack of established acceptance criteria for the demonstration of similar PK between SBP and RBP, the traditional 80-125% equivalence range is often used. Besides, PD studies and confirmatory PK/PD studies may be appropriate if there are clinically-relevant PD markers. In addition, similar efficacy of SBP and RBP has to be demonstrated in randomized and well-controlled clinical trials, which should preferably be double-blind or at least observer-blind. In principle, equivalence designs (requiring lower and upper comparability margins) are clearly preferred for the comparison of efficacy and safety of SBP with RBP. Non-inferiority designs (requiring only one margin) may be considered if appropriately justified. WHO also suggest the pre-licensing safety data and the immunogenicity data should be obtained from the comparative efficacy trials.

In addition to the non-clinical and clinical data, applicants also need to present an ongoing risk management and pharmacovigilance plan, since data from pre-authorized clinical studies are usually too limited to identify all potential side effects of the SBP. The safety specification should describe important identified or potential safety issues for the RBP and any that are specific for the SBP.

In summary, the WHO guidelines on evaluating similar biotherapeutic products represent an important step forward in the global harmonization of biosimilar product evaluation and regulation, and provide clear guidance for both regulatory bodies and the pharmaceutical industry.

2.2.2 European Union (EU)

The European Union (EU) has pioneered a regulatory system for biosimilar products. The European Medicines Agency (EMA) began formal consideration of scientific issues presented by biosimilar products at least as early as January 2001, when an *ad hoc* working group discussed the comparability of medicinal products containing biotechnology-derived proteins as active substances (CPMP, 2001). In 2003, the European Commission amended the provisions of EU secondary legislation governing requirements for marketing authorization applications for medicinal products to establish a new category of applications for "similar biological medicinal products" (CP, 2003). In 2005, the EMA issued a general guideline on similar biological medicinal products, in order to introduce the concept of similar biological medicinal products, to outline the basic principles to be applied, and to provide applicants with a user guide, showing where to find relevant scientific information (EMA, 2005a). Since then, several biosimilar products have been approved by the EMA under the pathway. Among these biosimilar products, two of them are somatropins, five are epoetins, and six are filgrastims.

Key Principles and Basic Concept – Unlike the WHO's guideline, which seems to focus on recombinant, DNA-derived therapeutic proteins, the EMA's guidelines clearly indicate that the concept of a 'similar biological medicinal product' is applicable to a broad spectrum of products, ranging from biotechnology-derived therapeutic proteins to vaccines, blood-derived products, monoclonal antibodies, and gene and cell-therapy. However, comparability exercises to demonstrate similarity are more likely to be applied to highly purified products, which can be thoroughly characterized, such as biotechnology-derived medicinal products. Considering the amount of data submitted, the EMA also requires a full quality dossier, while the comparability exercise at the quality level may allow a reduction of the non-clinical and clinical data requirement compared to a full dossier. In 2011, a concept paper on the revision of the guideline on similar biological medicinal product was published by the EMA (EMA, 2011a), which emphasizes another main concept: that clinical benefit has already been established by the reference medicinal product, and that the aim of a biosimilar development program is to establish similarity to the reference product, not clinical benefit.

Reference Biotherapeutic Product – Similar to the WHO, the EMA requires that the active substance, the pharmaceutical form, strength, and route of administration of the biosimilar should be the same as that of the reference product. The same chosen reference medicinal product should be used throughout the comparability program for quality, safety, and efficacy studies during the development of the biosimilar product. One of the major differences between the WHO and the EMA in terms of the choice of reference product is that the EMA requires the chosen reference medicinal product to be a medicinal product authorized in the Community. Data generated from comparability studies with medicinal products authorized outside the Community may only provide supportive information.

Quality – In 2006, the guideline entitled *Guideline on Similar Biological Medicinal Products Containing Biotechnology-derived Proteins as Active Substance: Quality Issues* was adopted by the Committee for Medicinal Products for Human Use (CHMP) (EMA, 2005b), which addresses the requirements regarding manufacturing processes, the comparability exercises for quality, analytical methods, physicochemical characterization, biological activity, purity and specifications of the similar biological medicinal product. In 2011, EMA issued a concept paper on the revision of this guideline (EMA, 2011b). This concept paper proposes that the guideline published in 2006 needs refinements that take into account the evolution of quality profile during the product lifecycle, since in the context of a biotherapeutic product claiming or claimed to be similar to another one already marketed, the conclusion of a comparability exercise performed with a reference product at a given time may not hold true from the initial development of the biosimilar, through marketing authorization, until the product's discontinuation.

Non-clinical and Clinical Evaluation – The guideline entitled *Guideline on Similar Biological Medicinal Products Containing Biotechnology-derived Proteins as Active Substance: Non-clinical and Clinical Issues* was published in 2006, which lays down the non-clinical and clinical requirements for a biological medicinal product claiming to be similar to another one already marketed (EMA, 2006).

The non-clinical section of the guideline addresses the pharmaco-toxicological assessment, and the clinical section addresses the requirements for pharmacokinetic, pharmacodynamics, and efficacy studies. Clinical safety studies as well as the risk management plan (with special emphasis on studying the immunogenicity of the biosimilar products) are also required. In 2011, EMA published a concept paper on the revision of this guideline (EMA, 2011c), which indicates several issues that need discussion for a potential revision. First, the EMA emphasizes the need to follow the principles of replacement, reduction, and refinement with regard to the use of animal experiments. Second, a revised version of the guideline will consider a risk-based approach for the design of an appropriate non-clinical study program. Third, the guideline should be clearer considering the need and acceptance of pharmacodynamics markers, and what measures should be taken in case relevant markers are not available.

Product Class-Specific Guidelines – The principles of biosimilar drug development discussed in the former sections apply in general to all biological drug products. However, there are no standard data sets that can be applied to the approval of all classes of biosimilars. Each class of biologic varies in its benefit/risk profile, the nature and frequency of adverse events, the breadth of clinical indications, and whether surrogate markers for efficacy are available and validated. Accordingly, the EMA has developed product class-specific guidelines that define the nature of comparative studies. So far, guidance for the development of biosimilar products has been developed for six different product classes, including erythropoietins, insulins, growth hormones, alfa interferons, granulocyte-colony stimulating factors, and low-molecular weight heparins (LMWH), with three more (beta interferons, follicle stimulation hormone, monoclonal antibodies) currently being drafted (EMA, 2005c-f; EMA, 2009a-b; EMA, 2010a-b; EMA, 2011d).

In summary, the EU has taken a thoughtful and evidence-based approach and has established a well-documented legal and regulatory pathway for the approval of biosimilar products distinct from the generic pathway. In order to grant a biosimilar product, the EMA requires comprehensive and justified comparability studies between the biosimilar and the reference in the quality, non-clinical, and clinical level, which are explained in detail in the EMA guidelines. The approval pathway of biosimilar products in the EU is based on case-by-case reviews, owing to the complexity and diversity of the biologic products. Therefore, besides the three general guidelines, the EMA also developed additional product class-specific guidelines on non-clinical and clinical studies. This approval pathway is now held up as one of the gold standards for authorizing biosimilar products.

2.2.3 North America (United States of America and Canada)

United States (US) – For the approval of follow-on biologics in the United States (US), current regulations depends on whether the biologic product is approved under the United States Food, Drug, and Cosmetic Act (US FD&C) or licensed under the United States Public Health Service Act (US PHS). For those biologic drugs marketed under the PHS Act, the BPCI Act amends the PHS Act to establish an abbreviated approval pathway for biological products that are highly similar or interchangeable with an FDA-authorized biologic drug and gives the FDA the authority to approve follow-on biologics under new section 351(k) of the PHS Act. Some early biologic drugs, such as somatropin and insulin were approved under the FD&C Act. In this case, biosimilar versions can receive approval for New Drug Applications (NDAs) under section 505 (b)(2) of the FD&C Act.

Following the passage of the BPCI, in order to obtain input on specific issues and challenges associated with the implementation of the BPCI Act from a broad group of stakeholders, the US FDA conducted a two-day public hearing on *Approval Pathway for Biosimilar and Interchangeability Biological Product* held on November 2–3, 2010 at the FDA in Silver Spring, Maryland. The scientific issues covered in this public hearing included, but were not limited to, criteria and design for biosimilarity and interchangeability, comparability between manufacturing processes, patient safety and pharmacovigilance, exclusivity, and user fees.

In practice, there is a strong industrial interest and desire for the regulatory agencies to develop review standards and an approval process for biosimilars rather than an *ad hoc* case-by-case review of individual biosimilar applications. For this purpose, the FDA has established three committees to ensure consistency in the FDA's regulatory approach of follow-on biologics. The three committees are the CDER/CBER Biosimilar Implementation Committee (BIC), the CDER Biosimilar Review Committee (BRC), and the CBER Biosimilar Review Committee. The CDER/CBER BRC will focus on the cross-center policy issues related to the implementation of the BPCI Act. The CDER BRC and CBER BRC are responsible for considering requests of applicants for advice about proposed development programs for biosimilar products, reviewing Biologic License Applications (BLAs) that are submitted under section 351(k) of the PHS Act, and managing related issues. Thus, the review process steps of CDER BRC and CBER BRC include: (i) applicant submits request for advice, (ii) internal review team meeting, (iii) internal CDER BRC (CBER BRC) meeting, (iv) internal post-BRC meeting, and (v) applicant meeting with CDER (CBER).

Another important issue aroused by the BPCI Act is the interchangeability of biosimilars. Once approved, standard generic drugs can be automatically substituted for the reference product without the intervention of the healthcare provider in many states. However, the automatic interchangeability cannot be applied to all biosimilars. In order to meet the higher standard of interchangeability, a sponsor must demonstrate that the biosimilar products can be expected to produce the same clinical result as the reference product in any given patient.

On February 9, 2012, the FDA announced the publication of three draft guidance documents to assist industry in developing biosimilar products, including (i) *Scientific Considerations in Demonstrating Biosimilarity to a Reference Product*, (ii) *Quality Considerations in Demonstrating Biosimilarity to a Reference Protein Product*, and (iii) *Biosimilars: Questions and Answers Regarding Implementation of the Biologics Price Competition and Innovation Act of 2009*, which were finalized in 2015 (FDA, 2015a-c). Most recently, the FDA circulated the following two guidances on interchangeability and analytical similarity assessment (FDA, 2017a-b): (i) Guidance for Industry – *Considerations in Demonstrating Interchangeability with a Reference Product* and (ii) Guidance for Industry – *Statistical Approaches to Evaluate Analytical Similarity*. Similar with requirements of the WHO and EMA, a number of factors are considered important by the FDA when assessing applications for biosimilars, including the robustness of the manufacturing process, the demonstrated structural similarity, the extent to which mechanism of action was understood, the existence of valid, mechanistically related pharmacodynamics assays, comparative pharmacokinetics and immunogenicity, and the amount of clinical data and experience available with the original products. The FDA is now seeking public comment on the guidance within 60 days of the notice of publication in the Federal Register. Even though the guidance does not provide clear standards for assessing biosimilar products, they are the first step toward removing the uncertainties surrounding the biosimilar approval pathway in the United States.

Canada (Health Canada) – Health Canada, the federal regulatory authority that evaluates the safety, efficacy, and quality of drugs available in Canada also recognizes that with the expiration of patents for biologic drugs, manufacturers may be interested in pursuing subsequent entry versions of these biologic drugs, which are called Subsequent Entry Biologics (SEB) in Canada. In 2010, Health Canada issued a guidance entitled *Guidance for Sponsors: Information and Submission Requirements for Subsequent Entry Biologics (SEBs)*, whose objective is to provide guidance on how to satisfy the data and regulatory requirements under the Food and Drugs Act and Regulations for the authorization of subsequent entry biologics (SEBs) in Canada (HC, 2010).

The concept of an SEB applies to all biologic drug products; however, there are additional criteria to determine whether the product will be eligible to be authorized as SEBs: (i) a suitable reference biologic drug exists that was originally authorized based on a complete data package, and has significant safety and efficacy data accumulated; (ii) the product can be well characterized by state-of-the-art analytical methods; (iii) the SEB can be judged similar to the reference biologic drug by meeting an appropriate set of pre-determined criteria. With regard to the similarity of products, Health Canada requires the manufacturer to evaluate the following factors: (i) relevant physicochemical and biological characterization data; (ii) analysis of the relevant samples from the appropriate stages of the manufacturing process; (iii) stability data and impurities data; (iv) data obtained from multiple batches of the SEB and reference to understand the ranges in variability; (v) non-clinical and clinical data and safety studies. In addition, Health Canada also has stringent post-market requirements, including the adverse drug reaction report, periodic safety update reports, and suspension or revocation of notice of compliance (NOC).

The guidance of Canada shares similar concepts and principles as indicated in the WHO's guidelines, since it is clearly mentioned in the guidance that Health Canada has the intention to harmonize as much as possible with other competent regulators and international organizations.

2.2.4 Asian Pacific Region (Japan, South Korea, and China)

Japan (MHLW) – The Japanese Ministry of Health, Labor and Welfare (MHLW) has also been confronted with the new challenge of regulating biosimilar/follow-on biologic products. Based on the similarity concept outlined by the EMA, Japan published a guideline for quality, safety, and efficacy of biosimilar products in 2009 (MHLW, 2009). The scope of the guideline includes recombinant plasma proteins, recombinant vaccines, PEGylated recombinant proteins, and non-recombinant proteins that are highly purified and characterized. Unlike the EU, polyglycans such as low-molecular weight heparin have been excluded from the guideline. Another class of product excluded is synthetic peptides, since the desired synthetic peptides can be easily defined by structural analyses and can be defined as generic drugs. Same as the requirements by the EU, the original biologic should be already approved in Japan. However, there are some differences in the requirements of stability test and toxicology studies for impurities in biosimilar between EU and Japan. A comparison of the stability of a biosimilar with the reference innovator products as a strategy for development of biosimilar is not always necessary in Japan. In addition, it is not required to evaluate the safety of impurities in the biosimilar product through non-clinical studies without comparison to the original product. According to this guideline, two follow-on biologics, *Somatropin* and *Epoetin alfa BS* have been recently approved in Japan.

Korea Ministry of Food and Drug Safety (MFDS) – In Korea, *Pharmaceutical Affairs Act* is the high-level regulation to license all medicines including biologic products. The Ministry of Food and Drug Safety (MFDS) of South Korea (a similar Food and Drug Administration organization in Korea) notifications serve as a lower-level regulation. Biological products and biosimilars are subject to the *Notification of the Regulation on Review and Authorization of Biological Products*. The MFDS takes an active participation in promoting a public dialog on the biosimilar issues. In 2008 and 2009, the MFDS held two public meetings and co-sponsored a workshop to gather input on scientific and technical issues. The regulatory framework of biosimilar products in Korea is a three-tiered system: (i) Pharmaceutical Affairs Act; (ii) Notification of the regulation on review and authorization of biological products; (iii) Guideline on evaluation of biosimilar products (Suh and

Park, 2011). As Korean guidelines for biosimilar products were developed along with that of the WHO's (HC, 2010), most of the requirements are similar, except for that of the clinical evaluation to demonstrate similarity. The MFDS requires that equivalent rather than non-inferior efficacy should be shown in order to open the possibility of extrapolation of efficacy data to other indications of the reference product. Equivalence margins need to be pre-defined and justified, and should be established within the range which is judged not to be clinically different from reference products in clinical regards.

China National Drug Administration (CNDA) – On October 29, 2014, the Center for Drug Evaluation (CDE) of the China National Drug Administration (CNDA) circulated a draft guidance on the evaluation of biosimilar products for public input and comment. Although the draft guidance is very similar to the draft guidance published by the US FDA (FDA 2015a), it is considered a bit more stringent. For example, the CDE/CNDA guidance requests that the intra-batch and inter-batch similarity be taken into consideration when demonstrating biosimilarity between a proposed biosimilar product and an innovative (reference) product.

2.2.5 Debatable Issues in Regulatory Requirements

In this section, we will focus on several critical and debatable issues in regulatory requirements worldwide. These debatable issues include, but are not limited to, (i) heterogeneity of reference product, (ii) the assessment of similarity via non-inferiority trials, (iii) the potential use of a scaled average bioequivalence criterion for biosimilarity assessment, and (iv) intra-batch versus inter-batch biosimilarity assessment, which are outlined below.

Heterogeneity of reference products – While both CDE/CNDA and US FDA recognize that there may be large variability associated with the reference product which may have an impact on the assessment of biosimilarity based on the current criterion for assessing biosimilarity for pharmacokinetic (PK) and pharmacodynamics (PD) parameters, the possible implications on the current criterion adjusted for variability were not discussed. As indicated in Chow (2013), variability associated with the response of the reference product could be high. For most small-molecule drug (e.g., generic) products, the variability is somewhere between 20% and 30%. For large-molecule (e.g., biosimilars) products, the variability can be expected to be as high as somewhere between 40% and 50%. In the United States, the FDA considers a drug product a highly variable drug product if its intra-subject CV is higher than 30%. As a result, many biosimilar drug products are considered highly variable drug products.

When the variability is high, the reference product may not be able to pass equivalence (similarity) testing when compared to itself (after log-transformation) with the current bioequivalence limits of 80% to 125%. In practice, reference products from different batches of the same manufacturing process (site) may differ due to the large variability associated with the response. With large variability, it is very likely that reference products randomly taken from a given batch could be in the lower end while another reference product randomly taken from the same batch could be in the higher end. The comparison between these two reference products may fail a similarity test due to large variability of the reference product (which could be due to intra- or inter-batch variability of the products during the manufacturing process). In this case, the following questions are raised: First, which reference product should be used as the reference for comparison? Second, what is the purpose of showing similarity between a biosimilar (test) product with the reference product if the reference product cannot pass the similarity test itself?

In practice, much important information regarding the reference product is generally not available. Thus, it is important to conduct a R-R study (i.e., a study comparing the reference product to itself). Note that for a R-R study, the reference product could come from two different manufacturing processes (locations) or different batches from the same manufacturing process in order to obtain important information regarding the reference product. The R-R study will not only provide the information regarding the variability associated with the reference product, but also establish baseline (i.e., similarity between R and R) for comparison (i.e., comparison between T vs R and R vs R).

For this purpose, prospectively, one may consider a three-arm parallel design (i.e., one arm with test product and two arms with the reference product) proposed by Kang and Chow (2013). For a post-study approach for a two-arm study, one could randomly split the reference arm into two groups, say R1 and R2, of equal sample size a large number of times. One can then establish a standard (or specification) for the reference product. Based on the established standards, the determination of the degree of similarity for highly similar is possible.

Assessment of similarity by a non-inferiority trial – For the assessment of biosimilarity between a biosimilar and an innovative biological drug product, a commonly asked question is whether biosimilarity can be demonstrated by a non-inferiority trial? This question has become critical to a sponsor who is seeking the approval of a proposed biosimilar product mainly because the establishment of non-inferiority requires a smaller sample size for achieving the same desired power assuming that the non-inferiority margin is the same as the biosimilarity limit. As can be seen from Table 1, however, the answer to the question is negative because the concept of non-inferiority includes the concepts of both equivalence and superiority. Statistically, one can start with a non-inferiority trial and test for non-inferiority first. Once the non-inferiority has been established, one can test

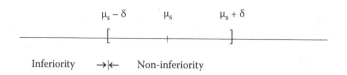

FIGURE 2.2
Statistical test for non-inferiority.

for the null hypothesis of non-superiority. If we fail to reject the null hypothesis, we can then conclude equivalence (similarity).

As indicated in the 2015 FDA guidance on scientific consideration, a non-symmetric alpha may be chosen for the assessment of biosimilarity via non-inferiority trial with proper clinical and/or statistical justification (Figure 2.2).

Potential use of scaled average bioequivalence in demonstrating biosimilarity – For the assessment of bioequivalence of generic drug products, FDA has recommended different bioequivalence limits. For example, for in vitro bioequivalence testing, FDA recommends that bioequivalence limits of 90.00% and 111.11% be used, while for in vivo bioequivalence testing, the FDA suggests that bioequivalence limits of 80.00% and 125.00% be considered. The recommended bioequivalence limits depend upon the variability associated with the reference product.

As noted earlier, many biological drugs have large variabilities when repeatedly administered to subjects. It is difficult to satisfy the regulatory requirements for comparing drug products with large within-subject variation (say, higher than 30%) unless a very large number of subjects is included in an investigation.

Therefore, the FDA has recommended an alternative procedure for the determination of bioequivalence between highly variable drug products. This procedure is stringent but requires fewer subjects. The approach of scaled average bioequivalence (SABE) standardizes, in the logarithmic scale, the difference between the two means (those of the test and reference products) by the standard deviation of the within-subject variation (Haidar et al., 2008).

Implementation of the SABE approach requires, at least, that the reference product be determined twice in each subject. Therefore, crossover studies with 3 or 4 periods have been recommended.

Other regulatory authorities, such as the European Medicines Agency, have recommended either SABE or a procedure closely related to it for the evaluation of bioequivalence of highly variable drugs. These approaches should be entertained also for the determination of biosimilar products.

Intra-batch versus inter-batch biosimilarity assessment – As indicated in the recent CFDA draft guidance on biosimilars, the following issues regarding within-batch and between-batch similarity were raised:

1. Clinical efficacy should be demonstrated via equivalence trials. If a non-inferiority design is used, the non-inferiority margin should be properly selected.

   The guidance seems to indicate that clinical efficacy should be demonstrated via equivalence trials. If a non-inferiority design is used, the non-inferiority margin should be properly selected. However, it should be noted that non-inferiority does not imply equivalence. If a non-inferiority test is to be performed, a test for non-superiority should be performed after the non-inferiority margin has been established.

2. The assessment of biosimilarity between a proposed biosimilar product and a reference product should take the within-batch and between-batch difference into consideration.

   This issue is related to the within-batch and between-batch heterogeneity of the reference product. It could be interpreted as that (i) reference products from different batches may be different (this difference may be due to batch-to-batch variation), and (ii) if reference products are from the same batch, the within-batch variability is confounded with the variability of the reference product (see Table 2.2). Suppose there are two reference products from the same batch. As indicated in Table 2.2, we assume that $\mu_{R1} = \mu_{R2} = \mu_R$ and $\sigma_{R1} = \sigma_{R2} = \sigma_R$. In this case, there is no problem to use μ_R and σ_R as the reference product for comparison. On the other hand, if reference products are from different batches, it is likely that $\mu_{R1} \neq \mu_{R2}$ and $\sigma_{R1} \neq \sigma_{R2}$. Consequently, it is not clear which batch, i.e., (μ_{R1} or μ_{R2}) or (σ_{R1} or σ_{R2}) should be used as the true reference for comparison. The Chinese guidance attempts to account for the within-batch and between-batch heterogeneity for a fair and reliable comparison between a proposed biosimilar product and a reference product. This concept can be applied to the care where there are two references (e.g., one is an US-licensed reference product and the other is an EU-approved reference product).

TABLE 2.2

Heterogeneity of Reference Product

Sources of Variability	Characteristics	QA/QC	Reference
Within-Batch[a]	$\mu_{R1} = \mu_{R2} = \mu_R$ $\sigma_{R1} = \sigma_{R2} = \sigma_R$	Assay validation Process Validation	μ_R and σ_R
Between-Batch[b]	$\mu_{R1} \neq \mu_{R2}$ $\sigma_{R1} \neq \sigma_{R2}$	Assay Validation Process Validation	(μ_{R1} or μ_{R2}) and (σ_{R1} or σ_{R2})

[a] Assume that the reference products are from the same batch.
[b] Assume that the reference products are from different batches.

To take into consideration the within-batch and between-batch heterogeneity for a fair and reliable assessment of biosimilarity, the study design and statistical method proposed by Kang and Chow (2013) are useful.

Analytical similarity assessment – In 2015, the FDA finalized the draft guidance on *Scientific Considerations in Demonstrating Biosimilarity to a Reference Product*. The FDA recommends that a stepwise approach be considered for providing the totality-of-the-evidence in order to demonstrate the biosimilarity of a proposed biosimilar product as compared to a reference product. The stepwise approach suggests beginning with the assessment of similarity in critical quality attributes at various critical stages of the manufacturing process, especially at the stages of structural and functional characterization. The stepwise approach continues with PK/PD studies, animal studies, clinical immunogenicity, and clinical studies for safety/tolerability and efficacy. Similarly, the Chinese guidance also recommends a similar stepwise approach for the demonstration of biosimilarity.

The stepwise approach starts with the assessment of similarity in critical quality attributes (CQAs) at various stages of the manufacturing process of the biosimilar product as compared to those of the reference product. The assessment of CQAs at various stages of the manufacturing process is also referred to as the assessment of analytical similarity because the comparisons are made mainly based on analytical test values from several lots (or batches) of the manufacturing process. This is to make sure that the information provided is sufficient to fulfill the FDA's requirement for providing totality-of-the-evidence for the demonstration of biosimilarity. However, little information regarding the assessment of analytical similarity in the stepwise approach was provided in either the 2012 FDA draft guidance or the Chinese guidance.

For the assessment of analytical similarity, most recently, the FDA has encouraged the sponsors to assess all relevant critical quality attributes (CQAs) by ranking them in terms of their critical relevance to clinical outcomes (Figure 2.3). Based on these critical rankings, the FDA suggests

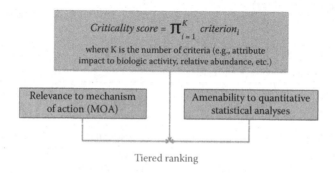

FIGURE 2.3
Score function for assessment of criticality or risking ranking.

classifying all relevant CQAs into three tiers. Tier 1 includes CQAs with the highest risk relevant to clinical outcomes. CQAs in Tier 1 would generally include assay(s) that evaluate clinically relevant mechanism(s) of action of the product for each indication for which approval is sought. Tier 2 includes CQAs with lower (say mild to moderate) risk associated with clinical outcomes, while Tier 3 contains CQAs with the lowest risk or least impact on clinical outcomes (see e.g., Chow, 2014).

Remarks – In principle, the Chinese guidance on *Development and Evaluation of Biosimilars* (both Chinese versions: one version was for comments and the other version is for current trial) is similar to the 2015 FDA guidance on *Scientific Considerations in Demonstrating Biosimilarity to a Reference Product*. Both guidances recommend the use of a stepwise approach for providing totality-of-the-evidence to demonstrate biosimilarity between a proposed biosimilar product with a reference product. Both guidances, however, suffer from some unsolved scientific considerations or practical issues such as (i) how similar is considered highly similar? (ii) criteria for biosimilarity at different structural or functional (non-clinical), toxicological and pharmacological (preclinical), and clinical (e.g., safety/tolerability, immunogenicity, and efficacy) areas, (iii) mean shift and heterogeneity among different lots within test and reference products and between test and reference products, and (iv) the issue of interchangeability in terms of the concepts of switching and/or alternating.

Note that the CNDA finalized the draft guidance and published a current trial version on February 28, 2015. In this version, one significant change is to relax the regulation for selection of reference products for comparison. In the version for comments, the CNDA requires the reference product to be the originator product authorized by CNDA. In the current trial version, the CNDA has removed this restriction. In other words, EU-approved and/or US-licensed reference products can be used as the reference products. In addition, for PK studies, the traditional 80–125% equivalence range is not mandatory. In the current trial version, CNDA indicates that an alternative equivalence range (e.g., SABE for highly variable drugs) can be used if scientifically justifiable.

2.3 Analytical Studies for Structural/Functional Characteristics

As indicated in the BPCI Act, a biosimilar product is a product that is *highly similar* to the reference product, notwithstanding *minor* differences in clinically inactive components, and there are no clinically meaningful differences in terms of safety, purity, and potency. In its guidance on scientific consideration (FDA 2015), the US FDA recommends a stepwise approach for obtaining totality-of-the-evidence for demonstrating biosimilarity of proposed biosimilar product to an innovative biologic drug product (FDA, 2015; Figure 2.4).

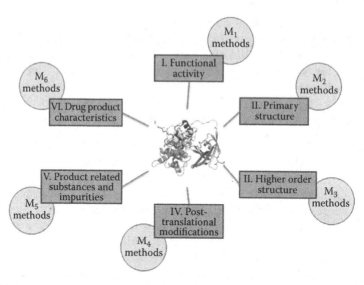

FIGURE 2.4
Functional and structural characterization.

Totality-of-the-evidence – Totality-of-the-evidence is referred to substantial evidence regarding safety, purity and potency of the proposed biosimilar product as compared to the innovative biologic product. The concept of totality-of-the-evidence includes (i) local biosimilarity (e.g., in a specific domain such as analytical studies, animal studies, PK/PD studies, immunogenicity studies, and clinical studies) versus global biosimilarity, (ii) degree of biosimilarity may vary from domain to domain. Each domain may carry different weights. The FDA seems to suggest a scoring system for measuring totality-of-the-evidence. More details regarding the interpretation of totality-of-the-evidence are given in Chapter 11.

Analytical Similarity Assessment – The stepwise approach starts with analytical studies for functional and structural characterization of critical quality attributes (CQAs) at various stages of the manufacturing process. Analytical similarity assessment involves three critical steps: (i) identification of CQAs relevant to clinical outcomes based on mechanism of action (MOA) or pharmacokinetics (PK) and pharmacodynamics (PD), (ii) the study of the relationship between the identified CQAs with clinical outcomes under appropriate statistical models, (iii) assignment of the identified CQAs to appropriate tiers according to their criticality and/or risk-ranking relevant to clinical outcomes. The goal is to demonstrate that there are no clinically meaningful differences for any given CQAs that are relevant to clinical outcomes for achieving totality of evidence of safety, purity, and efficacy. More details regarding the identification/classification of CQAs can be found in Chapter 5. The discussion regarding the FDA recommended tiered approach is given in Chapter 6.

Fundamental Biosimilarity Assumption – As indicated in the previous chapter, bioequivalence assessment for generic approval is possible only under the *Fundamental Bioequivalence Assumption* that "When a generic drug is claimed to be bioequivalent (in terms of drug absorption) to a brand-name drug, it is assumed that they are therapeutically equivalent." Under this Fundamental Bioequivalence Assumption, pharmacokinetic (PK) responses (for drug absorption) serve as surrogate endpoints for clinical outcomes (i.e., safety and efficacy). For analytical similarity assessment, however, we do not have such fundamental assumption. Along this line, SSAB (2009) proposed so-called *Fundamental Biosimilarity Assumption* that "When a follow-on biologic product is claimed to be biosimilar to an innovator product in some well-defined study endpoints, it is assumed that they will reach similar therapeutic effects or they are therapeutically equivalent." Under this proposed assumption, the performance of CQAs can serve as surrogate endpoints for clinical outcomes for providing totality-of-the-evidence for demonstrating biosimilarity between a proposed biosimilar product and the innovative biologic product (Table 2.3).

Analysis and Sample Size Requirement – In practice, since there are many CQAs that are relevant to clinical outcomes at various stages of the manufacturing process of a proposed biosimilar product, it is a concern regarding minimum requirement for sample size, especially when test or reference lots are not available. If only a few test or reference lots are available, the validity of the analysis of the test results could be a concern. In addition, in the interest of controlling the overall Type I error rate at a pre-specified level of significance, it is also a concern how to address the issue of multiplicity when only a limited number of test results are available. Detailed information regarding sample size requirement for analytical similarity assessment will be provided and discussed in Chapter 7.

TABLE 2.3

Total Sample Sizes Needed with Unscaled (ABE) and Scaled Average Bioequivalence (SABE)[a]

Method	A B E				S A B E			
Power	80%		90%		80%		90%	
GMR[b]	1.0	1.1	1.0	11	1.0	1.1	1.0	1.1
CV(%)[c]								
30%	16	34	20	46	15	27	18	38
35%	21	45	26	62	16	24	20	34
40%	27	57	33	79	16	22	20	30

[a] 4-period, 2-sequence investigations are considered. The sample sizes for SABE, with the expectations of FDA, are from Tothfalusi et al. (2009).
[b] GMR: Ratio of the geometric means of the two drug products, assumed *a priori* for the study design.
[c] CV: Coefficient of within-subject variation.

Remarks – In practice, a commonly asked question is this: does it matter if we fail to pass analytical similarity assessment for some CQAs but meet bio-similarity criteria for clinical outcomes? This issue is debatable. According to the definition of biosimilarity as described in the BPCI Act, we need to show that there are no clinically meaningful differences in terms of safety, purity, and potency. By definition, the proposed biosimilar product fails to provide totality-of-the-evidence for demonstrating biosimilarity in purity (i.e., some CQAs fail to pass analytical similarity assessment). In this case, however, one may argue that predictive models for those CQAs are not fully established and validated. In other words, dis-similarity in those CQAs may not trans-late to dis-similarity in clinical outcomes. Thus, it is necessary that regula-tory guidance be developed in order to resolve this controversial issue.

2.4 Global Harmonization

According to the regulatory requirements of different regions described in the previous section, there seems to be no significant difference in the gen-eral concept and basic principles in these guidelines. There are five well-recognized principles with regard to the assessment of biosimilar products: (i) the generic approach is not appropriate for biosimilars; (ii) biosimilar products should be similar to the reference in terms of quality, safety, effi-cacy; (iii) a step-wise comparability approach is required that indicates the similarity of the SBP to RBP in terms of quality; this is a prerequisite for reduction of submitted non-clinical and clinical data; (iv) the assessment of biosimilarity is based on a case-by-case approach for different classes of products; (v) the importance of pharmacovigilance is stressed.

However, differences have been noted in the scope of the guidelines, the choice of the reference product, and the data required for product approval. The concept of a "similar biological medicinal product" in the EU is applicable to a broad spectrum of products ranging from biotechnology-derived thera-peutic proteins to vaccines, blood-derived products, monoclonal antibodies, gen and cell-therapy, and so on. However, the scopes of other organization or countries are limited to recombinant protein drug products. Concerning the choice of the reference product, the EU and Japan require that the reference product should be previously licensed in their own jurisdiction, while other countries do not have this requirement. A detailed comparison of the guide-lines of the WHO, EU, Canada, Korea and Japan for the biosimilar products is summarized in Table 2.4.

In order to facilitate the global harmonization of evaluation of the follow-on biologic, the first workshop on implementing WHO *Guidelines on Evaluating Similar Biotherapeutic Products* into the regulatory and manufacturer's prac-tice at the global level was held on 24–26 August 2010 in Seoul, Republic of

TABLE 2.4

Comparison of Requirement for Evaluation of SBPs between Different Regions

	WHO	Canada	South Korea	EU	Japan
Term	SBPs	SEBs	Biosimilars	Biosimilars	Follow-on Biologics
Scope	Recombinant protein drugs			Mainly recombinant protein drugs	Recombinant protein drugs
Efficacy	Double blind or observer-blind; Equivalence or non-inferiority design		Equivalence design	Comparability margins should be pre-specified and justified	
Reference product	Authorized in a jurisdiction with well-established regulatory framework			Authorized in EU	Authorized in Japan
Stability	• Accelerated degradation studies • Studies under various stress conditions				Not necessary
Purity	Process-related and product-related impurities				
Manufacture	• Same standards required by the NRA for originator products • Full chemistry and manufacture data package				
Physico-chemical	• Primary and higher-order structure • Post-translational modifications				

(Continued)

TABLE 2.4 (CONTINUED)

Comparison of Requirement for Evaluation of SBPs between Different Regions

	WHO	Canada	South Korea	EU	Japan
Biological activity	• Qualitative measure of the function • Quantitative measure (e.g., enzyme assays or binding assays)				
Non-clinical studies	• *In vitro* (e.g., receptor-binding, cell-based assays) • *In vivo* (pharmacodynamics activity, at least one repeat dose toxicity study, antibody measurements, local tolerance)				
PK study design and criteria	• Single dose, steady-state studies, or repeated determination of PK • Cross-over or parallel • Include absorption and elimination characteristics • Traditional 80–125% equivalence range is used				
PD	Pharmacodynamic markers should be selected and comparative PK/PD studies may be appropriate				
Safety	Pre-licensing safety data and risk management plan				
Principles	• Generic approach is not appropriate for follow-on biologic • Follow-on biologic should be similar to the reference in terms of quality, safety, efficacy • Step-wise comparability approach: similarity of the SBP to RBP in terms of quality is a prerequisite for reduction of non-clinical and clinical data required for approval. • Case by case approach for different classes of products • Pharmacovigilance is stressed				

Korea. The workshop featured speakers from regulatory agencies from various countries, clinical and scientific experts, and representatives from the biopharmaceutical industry and WHO.

It was recognized in the workshop that some progress towards implementation and development of guidance documents in various countries had been made. For instance, the biosimilar guidance of Singapore and Malaysia are amended mainly based on the EU's biosimilar guidelines, while Brazil and Cuba choose the WHO and Canadian guidelines as the basis for developing regulations. However, there are also many challenges which need to be addressed for global harmonization of the regulatory framework for licensure of biotherapeutics. For example, the manufacturing of SBPs in the Arab region is not well controlled due to the lack of expertise in the assessment of biotechnology products and inexperience with regulatory processes. Besides, large emerging economies such as China and India are currently lagging behind in terms of their regulations and need to act rapidly in developing appropriate regulations for biosimilar product approval.

In summary, the status of SBPs and implementation of the WHO guidelines is highly diverse worldwide, and a harmonized approach for SBPs worldwide is unlikely to occur rapidly. While some countries have developed guidelines or are developing guidelines, some countries are taking a relaxed view and are not committed on the approach to adopt for approval for SBPs. Accordingly, in order to promote the global harmonization, national regulatory authorities should take an active role in building capacity for regulatory evaluation of biotherapeutics; the existing guidelines should be revised, because considerable experience has been gained through scientific advice, marketing authorization applications and workshops. The WHO should continue monitoring progress with the implementation of the guidelines on the evaluation of SBPs into regulatory and manufacturers' practices.

2.5 Conclusion Remarks

For the assessment of bioequivalence of generic drug products, with identified active ingredient(s) to the innovative drug products, a regulatory approval pathway through the conduct of bioequivalence studies is possible only under the Fundamental Bioequivalence Assumption. For the assessment of biosimilar products, a similar Fundamental Biosimilarity Assumption is necessarily established. It should be noted that biosimilar products are not identical but similar to the innovative products.

As described in Section 2.3, most regulatory requirements for the approval of biosimilar products are similar but slightly different in the definitions of biosimilarity, the scope of the guidelines, the choice of the reference product, and the data required for product approval. Little or no discussions regarding the criteria and/or the degree of biosimilarity and/or interchange-ability are given. In practice, a common question is how similar is very similar. Of particular interest to the sponsors is the question of how many studies are required for the regulatory approval of biosimilar products. Besides, the degree of similarity may have an impact on the interchangeability of biosimilar products. As indicated in the BPCI Act, biosimilar products are considered interchangeable provided that they can produce the same therapeutic effect in any given patient. This, however, is not possible to achieve. Alternatively, we would suggest that biosimilar products are considered interchangeable provided that they can produce the same therapeutic effect in any given patient with certain statistical assurances.

In summary, there are still many unsolved scientific issues regarding criteria, design, and analysis for the assessment of biosimilarity and/or inter-changeability of biosimilar (interchangeable) products. Detailed regulatory guidance for global harmonization is needed whenever possible.

3

CMC Requirements for Biological Products

3.1 Introduction

As indicated in the previous chapters, there are some fundamental differences between biologics (large-molecule) and generic (small-molecule) drugs. Small-molecule drugs are made from chemical synthesis, which is usually *not* sensitive to process changes, while biological products are made from living cells or organisms, which are very sensitive to process and environmental changes. A small change in manufacturing conditions could result in a drastic change in clinical outcomes. In practice, even minor modifications of the manufacturing process can cause variations in important properties of a biological product. Thus, quality, safety, purity, and potency of a biologic product can be determined by its manufacturing process. Biologics, which possess sophisticated three-dimensional structures and contain mixtures of protein isoforms, are 100-fold or 1,000-fold larger than small-molecule drugs. A biological product is a heterogeneous mixture and the current analytical methods cannot characterize these complex molecules sufficiently to confirm structural equivalence with the reference biologics.

Since biological products are manufactured in living systems that are inexact by their nature, they are influenced by the method-of-manufacturing system. Along with the manufacturing system, biological products are susceptible to environmental factors. As a result, the biological products are capable of causing a unique set of pathologies due to their origin during the manufacturing process. Thus, it is important to make sure that biological products are manufactured using a consistent and reproducible process which is in compliance with applicable regulations to ensure the quality, safety, purity, and potency (efficacy) of the biological products.

In the United States (US), regulatory requirements for biological products are codified in Chapter 21 Section 600 of Codes of Federal Regulations (CFR) (21 CFR 600). The CMC requirements for biosimilars in the European Union (EU) are those described in the ICH Common Technical Document (CTD) Quality Module 3 with supplemental information demonstrating comparability or similarity on quality attributes to the reference medicine product. At the present time, the US suggests submission following CTD format for drug

TABLE 3.1

CTD Module 3 Format

Section	Description
3.2.S.1	General Information
3.2.S.2	Manufacture
3.2.S.3	Characterization
3.2.S.4	Control
3.2.S.5	Reference Standards
3.2.S.6	Container Closure System
3.2.S.7	Stability

substance, which consists of general information, manufacture, characterization, control, reference standards, container closure system, and stability (see Table 3.1).

In the next section, chemistry, manufacturing, and controls (CMC) requirements for biological products are briefly outlined. Sections 3.3 and 3.4 discuss requirements for manufacturing process validation and quality control/assurance, including the concept of quality by design in BLA submissions. Section 3.5 deals with the design and analysis of stability studies for biological products. Section 3.6 provides some concluding remarks.

3.2 CMC Development

The CMC development for biological drug products starts with establishment of the expression system. A cell-line will be selected among bacterial, yeast, and mammalian host strains, and then the correct DNA sequence will be inserted. Elaborate cell-screening and selection methods are then used to establish a master cell bank. Extensive characterization on the master cell bank needs to be carried out to provide microbiological purity or sterility and identity (CBER/FDA, 1993). Bulk protein production involves developing robust and scalable fermentation and purification processes.

3.2.1 Fermentation and Purification Process

The goals for fermentation are to increase the expression level of a deficiency without compromising the correct amino acid sequence and post-translational modification. Achieving high expression requires optimizing culture medium and growth conditions, and efficient extraction and recovery procedures. The correct amino acid sequence and post-translation modification will need to be verified. Solubilization and refolding of insoluble proteins are sometimes necessary for proteins which have a tendency to aggregate under the processing conditions. Differences in the cell bank and production processes may create impurities that are different from the innovator's product. The purification process needs to remove impurities such as host-cell proteins, DNA, medium constituents, viruses, and metabolic by-products as much as possible. It is important for biosimilar manufacturers to accept appropriate yield losses to achieve high purity, because any increase in yield at the expense of purity is unacceptable and can have clinical consequences. The final product is produced by going through formulation, sterile-filtration, and fill/finish into the final containers. Selection of formulation components starts from basic buffer species for proper pH control and salt for isotonicity adjustment. Surfactants may be needed to prevent proteins from being absorbed onto container surfaces or water-air interface or other hydrophobic surfaces. Stabilizers are required to inhibit aggregation, oxidation, deamidations and other degradations. The container and closure system can be glass vials, rubber stoppers, and aluminum seals or pre-filled syringes or IV bags. The container and closure integrity need to be verified by sterility or dye-leak tests.

3.2.2 Drug Substance and Product Characterization

Biologics are not pure substances. They are heterogeneous mixtures. Each batch of a biologic product for clinical or commercial use needs to be produced in compliance with current Good Manufacturing Practice (cGMP) and is typically tested by a panel of assays (Table 3.2) to ensure the product meets pre-defined specifications for quality, purity, potency, strength, identity, and safety. The product purity is often measured by multiple assays, which measure different product-related variants (biologically active) or product related impurities (biologically inactive). Biologics are parenteral drugs and filled into the final containers through the aseptic process, so that microbiological control is critical. It is advisable to set up product specifications for a biosimilar within the variation of the

TABLE 3.2

Analytical Methods to Characterize Good Drug Characteristics of Biological Products

Type	Assays
Quality	Appearance, particulates, pH, osmolality
Purity	SDS-PAGE, SEC-HPLC, IEX-HPLC, RP-HPLC
Potency	*In vitro* or *in vivo* bioactivity assays
Strength	Protein concentration by A280
Identity	Western blot, Peptide mapping, Isoelectric focusing
Safety	Endotoxin, sterility, residual DNA, host-cell proteins

reference biologic product. Product characterization can be performed on selected batches for primary sequence, high order structures, isoform profiles, heterogeneity, product variants and impurities, and process-impurity profiles. Physicochemical characterization tests include IEF, CE, HIC, LCMS, carbohydrate analysis, N & C terminal sequencing, amino acid analysis, analytical ultracentrifugation, CD, and DSC (Chirino et al., 2004; Kendrick et al., 2009).

General Description – Biosimilar manufacturers will have no access to the manufacturing process and product specifications of the innovator's products because these are proprietary knowledge. To develop a biosimilar, a biosimilar manufacturer will need to first identify a marketed biologic product to serve as the reference biologic product. Then a detailed characterization of the reference biologic product will be performed. The information obtained from the characterization of the reference biologic product will be used to direct the process development of the biosimilar product and comparative testing to demonstrate similarity between the biosimilar product and the reference biologic product. A biosimilar will be manufactured from a completely new process, which may be based on a different host/vector system with different process steps, facilities, and equipment.

Drug Substance and Product Characterization – The drug substance and drug product should be positive for identity and have specified criteria for purity, potency, and microbial contamination. Acceptance criteria for release and stability attributes should be established. Results from release and stability testing should be provided in the IND. Raw data supporting drug substance characterization should be provided in the IND. The following good drug characteristics should be characterized.

Safety – Ensured by the specified limits for bioburden and endotoxin, miscellaneous process-related contaminants, which are usually characterized by LAL test, rabbit pyrogen test, or bacterial culture methods. The final drug product for injection should be sterile and within specified limits for endotoxin. Immunogenicity should be screened and monitored. Successfully reduced in MAb by replacing murine with human sequences.

Purity – Assesses capability of purification process to remove process-related impurities (e.g., endogenous viruses, host-cell proteins, DNA, leachables, anti-foam, antibiotics, toxins, solvents, and heavy metals), product-related impurities (e.g., aggregates, breakdown products, or product variants due to oxidation, deamidation, denaturation, or loss of C-term Lys in MAbs) and product substances and drug product (product variants that are active). Methods for characterization of purity include, but are limited to, reversed-phase HPLC, Peptide mapping, MS, SDS-PAGE, Western analysis, capillary electrophoresis, SEC, AUC, FFF, light scattering, Ion Exchange Chromatography, and carbohydrate analysis (capillary electrophoresis, HPAEC = high-pH anion-exchange chromatography, IEF for sialic acid). Product and process-related impurities and product-related substances should be within specified limits.

Identity – Unique for protein of interest, especially relevant for closely related proteins manufactured in the same facility. Identity is usually characterized by N-terminal sequencing, peptide mapping, or immunoassays (ELISA, Western blotting).

Potency – Required to assess biological activity of the product. Assay should be relevant for protein mechanism of action. For MAb or Fc fusion proteins, a binding assay may be sufficient for early development, but a functional assay relevant for the mechanism of action should be developed. If mechanism of action unknown, multiple bioactivities plus elucidating higher-order structure may be required. Potency of drug substance is usually characterized by animal-based assays, cell-based assays, reporter gene, or biochemical (e.g., enzyme activity). Assay of drug product should be relevant for protein mechanism of action.

Strength – Protein content, which can be characterized by RIA, ELISA, UV absorbance, or Bradford.

Stability – Drug substance and drug product stability should be demonstrated with appropriate stability-indicating assays. The drug product should maintain stability for the duration of the clinical trial.

3.2.3 Reference Standards and Container Closure System

Characterization of the reference standard is usually performed using part of the lot used for non-clinical studies. The established reference standard is then used to release the clinical lot. As development progresses, if the lot is too old or there is an insufficient amount of the previous lot, new reference standards may be required to account for changes in manufacturing. In this case, protocol for generating and/or qualifying new reference standards must be developed to incorporate new methods as new specifications evolve. Portions of each reference standard lot must be retained for future use or needs. During the development, reference materials which reflect degradation pathways critical in product quality control are needed as assay development, controls, and validation.

To fulfill regulatory requirements for licenses, extractable and leachable studies are often conducted. Extractables migrate from a contained closure system and/or other packaging components in a DP vehicle or solvent under exaggerated conditions, while leachables migrate spontaneously from a container closure system and/or other packaging components under normal conditions of use and storage. Extractables are helpful in predicting potential leachables and in selecting the appropriate container closure system. Leachables are often a subset of extrables or derived by their chemical modification. Sources of leachables in the product include syringes/prefilled syringes, ampoules, vials, bottles, IV bags, storage bags for product intermediates, closures (screw caps, rubber stoppers), and container liners (e.g., tube liners). Processing equipment usually includes stainless steel storage tanks/bioreactors, tubing, gaskets, valves, rings, filters, and purification resins.

As indicated by Markovic (2007), leachables could have an impact on safety and product quality. For example, when there is a change from HSA formulation to a polysorbate with an unchanged container closure system (pre-filled syringes with the uncoated rubber stoppers). The observation of serious adverse event leads to the hypothesis that leachables acted as adjuvants triggering immunogenicity. For another example, when there is a change from a lyophilized to a liquid formulation (divalent cation leached from the rubber stopper), which might cause activation of metalloprotease (a process-related impurity co-eluted with the API), this could impact product degradation at the N-terminal site (stability study).

Commonly errors for extractable and leachable studies include the absence of data on extractables and leachables from the container closure and absence of assessment of the impact of the extractables/leachables data on product specification and methods (potential to seed microaggregates, potential to alter the immunogenicity profile).

3.2.4 Practical Issues

As indicated in Chow (2013), commonly seen issues in the process of product characterization include, but are not limited to, (i) establishing specification prior to understanding the product, (ii) insufficient knowledge of the relationship between protein structure and potential safety/toxicity, (iii) lack of the examination of the impurity profile, and (iv) inadequate process validation/control. It is suggested that these issues be resolved prior to the conduct of analytical similarity assessment for demonstrating biosimilarity in critical quality attributes identified at various stages of the manufacturing process.

3.3 Manufacturing Process Validation

3.3.1 Manufacturing Process

A typical manufacturing process for biological product consists of expression vector (plasmid) cell banking system including master cell bank (MCB), working cell bank (WCB) and end of production cells (EOP), drug substance manufacturing and release, and drug product formulation and release (see also Figure 3.1).

As an example, consider a manufacturing process for therapeutic biologic protein products. Expression vectors are used for (i) transfer of genes from one organism to another, (ii) production of large amounts of protein, (iii) description of origin of the construct, (iv) plasmid mapping (e.g., restriction sites, integration sites, promoter, copy number) and stability, and (v) sequencing of gene of interest (Markovic, 2007). A working cell bank (WCB) is derived from the master cell bank (MCB) and is used to initiate a production batch. Table 3.3 provides a list of characterization of cell banks required for CMC.

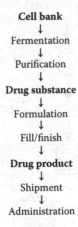

Cell bank
↓
Fermentation
↓
Purification
↓
Drug substance
↓
Formulation
↓
Fill/finish
↓
Drug product
↓
Shipment
↓
Administration

FIGURE 3.1
Example of a typical flowchart of manufacturing production.

TABLE 3.3

Characterization of Cell Banks

Test	MCB	WCB	EPC
Viability	X	X	X
Identity	X	X	
Purity	X	X	X
Stability	X		X
Karyology	X	X	
Tumorienicity	X		X
Sterility	X	X	X
Mycoplasma	X	X	X
Adventitious viruses	X		X
Species-specific	X		
Retrovirus	X		X

Abbreviations: EOP, End of production cells; MCB, Master cell bank; WCB, Working cell bank.

It should be noted that sources of adventitious agents include cell substrate (e.g., endogenous viruses and exogenous microbial contamination), raw materials (e.g., cell culture reagents such as animal and non-animal derived), and environment (e.g., water, air, and humans technicians). As indicated in Figure 3.1, fermentation and purification are two key components of the manufacturing process. The fermentation process and purification process are exhibited in Figures 3.2 and 3.3, respectively.

The goal of the manufacturing process is to produce sufficient quantities of quality product in a controlled and reproducible manner. Manufacturing

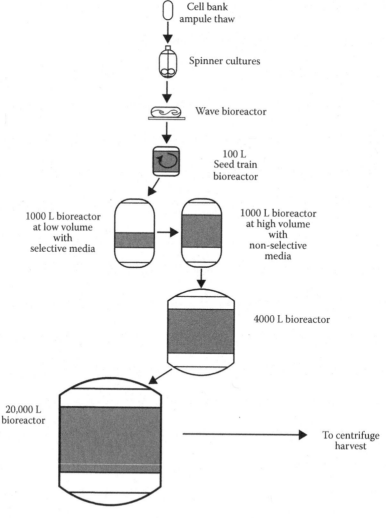

FIGURE 3.2
Example of fermentation process.

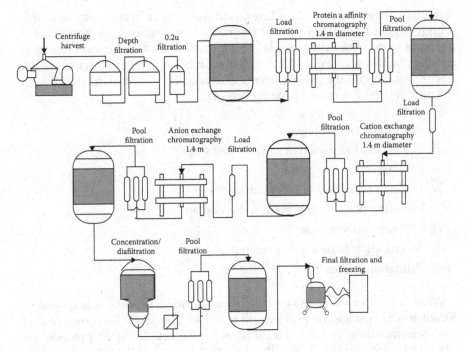

FIGURE 3.3
Example of purification process.

processes are dynamic. Changes can be process-related (e.g., modification of process, increase in scale, or change in location). Changes can be method related (e.g., improvement of analytical method and replacement of one or more analytical methods). The manufacturing process should be thoroughly investigated and understood. Quality should be designed into the process, rather than tested into the product via the analytical methods. When evaluating changes, one should focus on (i) biological products are complex mixtures, (ii) control starts with the cell banks and all raw materials, and (iii) the product is at risk from raw materials, operators, and the environment. No change can be assumed to be neutral. Release criteria alone are insufficient to fully evaluate the impact of changes.

3.3.2 Process Validation

As indicated in Chow and Liu (1995), the primary objective of process validation is to provide documented evidence that a manufacturing process does reliably what it purports to do. To accomplish this prospectively, a validation protocol is necessarily developed. A validation protocol should include the characterization of the product, the manufacturing procedure, and sampling plans, acceptance criteria, and testing procedures to be performed at identified critical stages of the manufacturing process. The statistically based sampling plans, testing procedures, and acceptance criteria ensure, with a high degree of confidence, that the manufacturing process does what it purports to do. Since the manufacturing process for a biological product is complex and involves several critical stages, it is important to discuss the following issues with the project scientists to acquire a good understanding of the product and the manufacturing process:

i. Identified critical stages

ii. Equipment to be used at each critical stage

iii. Possible issues/problems

iv. Testing procedures to be performed

v. Sampling plan, testing plan, and acceptance criteria

vi. Pertinent information

vii. Specification to be used as reference

viii. Validation summary

When a problem is observed in a manufacturing process, it is crucial to locate at which stage the problem occurred so that it can be corrected and the manufacturing process can do what it purports to do. In practice, the manufacturing process is usually evaluated by constant monitoring of its critical stages. Therefore, for the validation of a manufacturing process, it is recommended that project scientists be consulted to identify the critical stages of the manufacturing process. At each stage of the manufacturing process, it is also helpful to have knowledge of the equipment to be used and its components. The equipment may affect the conformance with compendia specifications for product quality. The validation protocol should establish statistically based sampling plans, testing plans, and acceptance criteria. Sampling plans and acceptance criteria are usually chosen such that (i) there is a high probability of meeting the acceptance criteria if the batch at a given stage is *acceptable* (this probability is 1 minus the producer's risk) and (ii) there is small probability of meeting the acceptance criteria if the batch at a given stage is *unacceptable* (this probability is the so-called consumer's risk).

3.3.3 Practical Issues

Note that commonly seen errors/omissions regarding manufacturing process include (i) manufacturing and/or testing locations not registered, (ii) manufacturing not planned during application review cycle, (iii) inadequate raw material control, (iv) failure to perform or provide results of process understanding studies and impurity clearance studies, (v) failure to validate commercial process, intermediate hold times, resin and membrane reuse cycles, buffer hold times, reprocessing if included, and shipping of intermediates/drug substance, and (vi) failure to demonstrate consistency of manufacture (e.g., according to a protocol with pre-specified criteria, using validated test methods, as three successive successful runs), and (vii) failure to demonstrate comparability between processes during development, and (viii) lack of retain samples to close gaps.

3.4 Quality Control and Assurance

The manufacture of biological products involves biological processes and materials, such as cultivation of cells or extraction of material from living organisms. These biological processes may display inherent variability, so that the range and nature of by-products are variable. Moreover, the materials used in these cultivation processes provide good substrates for growth of microbial contaminants. Thus, quality control/assurance of biological products usually involves biological analytical techniques which have a greater variability than that of chemical drug products. In-process controls therefore take on a greater importance in the manufacture of biological medicinal products.

3.4.1 General Principles

For quality control of biological products, it is suggested that the following general principles should be followed:

In-process Controls – In-process controls play an important role in ensuring the consistency of the quality of biological products. Those controls which are crucial for quality (e.g., virus removal) but cannot be carried out on the finished product, should be performed at an appropriate stage of production. Thus, critical stages of the manufacturing process need to be identified, accompanied with a sampling plan, acceptance criteria, and testing procedures.

Sample Retention – It may be necessary to retain samples of intermediate products in sufficient quantities and under appropriate storage conditions to allow the repetition or confirmation of a batch control.

Quality Control Requirement – Where continuous culture is used, special consideration should be given to the quality control requirements arising from this type of production method.

Statistical Process for QC – Continuous monitoring of certain production processes is necessary, for example, fermentation. Such data should form part of the batch record. Statistical process for QC will help in identifying problems or issues early, and appropriate actions can be taken to correct the problems or resolve the issues.

3.4.2 Quality by Design

Quality by design (QbD) is a concept that quality could be planned, and that most quality crises and problems relate to the way in which quality was planned. QbD has attracted much attention in biopharmaceutical research and development since the US FDA kicked off the initiative *Pharmaceutical Current Good Manufacturing Practices (cGMPs) for the 21st Century.* This initiative was intended to address a number of issues that are commonly encountered in biopharmaceutical development. These issues include low efficiencies in bringing innovative medicines to the market, waste of materials, and rework of finished products. Unlike traditional approaches, the concept of QbD is to use new advanced technology and quality risk management (QRM) techniques in the product and process development (Table 3.4). In practice, the concept of QbD starts with the understanding of the product and process of pharmaceutical development, including (i) understanding the mechanism of action (MOA) of the product, (ii) its safety profile, (iii) the variability of raw materials, (iv) the bulk and finished products, (v) the analytical method variations, (vi) the relationships between process parameters and the product's critical quality attributes (CQAs), and (vii) most importantly, the impact of specific CQAs on clinical responses and overall product quality. The understanding of these issues will enable product quality to be built by design.

As indicated in in ICH Q8 (R2), key steps to the implementation of QbD include (i) defining quality target product profile (QTPP), (ii) determining critical quality attributes (CQAs) through risk assessment, (iii) linking raw material

TABLE 3.4

Comparison of Traditional Approach and QbD Approach

Aspects	Traditional	Quality by Design
Pharmaceutical Development	Empirical; Univariate experiments	Systematic; Multivariate experiments
Manufacturing Process	Fixed	Adjustable
Process Control	In-process testing for go/no-go; Offline analysis w/slow response	PAT utilized for feedback and feed forward at realtime
Product Specification	Primary means of quality control; Based on batch data	Overall quality control strategy; Based on desired product performance (safety and efficacy)
Control Strategy	Based on intermediate and end product testing	Risk-based; Controls shifted upstream; Real-time release upstream
Lifecycle Management	Reactive to problems; Scale-up and post-approval changes	Continual improvement enabled within design space

Source: Winkle, H.N. (2007). Quality by design (QbD). Keynote presentation at the 2007 PDA/FDA Joint Regulatory Conference, September 24–28, 2007, Washington DC.
Abbreviation: PAT, Process Analytical Technology.

attributes (MAs) and process parameters (PPs) to CQAs, (iv) developing design space for key MAs and PPs, (v) developing and implementing effective control strategies, and (vi) managing the product life cycle through continued process improvement (ICH, 2006). The implementation of QbD, however, is often challenging to the sponsor due to the large number of variables to consider, interactions among the variables, uncertainties in measurements, and missing data involved in product and process development (Yang, 2016).

Quality Target Product Profile – The quality target product profile (QTPP) defines a new drug product in terms of its desired quality characteristics. ICH defines QTPP as a prospective summary of the quality characteristics of a drug product that ideally will be achieved to ensure that the desired quality, taking into account safety and efficacy of the drug product (ICH, 2006). ICH Q8 (R2) also lists some considerations for the QTPP such as (i) intended use in clinical setting, (ii) route of administration, (iii) dosage form, (iv) delivery systems, (v) dosage strength(s), (vi) container closure system, (vii) therapeutic moiety release or delivery and attributes affecting pharmacokinetic characteristics (e.g., dissolution, or aerodynamic performance) appropriate to the drug product dosage form being developed, and (viii) drug product quality criteria (e.g., sterility, purity, stability, and drug release) appropriate for the intended marketed product.

Critical Quality Attributes – The identification of critical quality attributes (CQAs) allows for the evaluation and control of product characteristics that have impact on quality, safety, and efficacy. A CQA is defined as a physical, chemical, biological, or microbiological property or characteristic that should be within an appropriate limit, range, or distribution to ensure the desired product quality (ICH, 2006). Thus, CQAs are indicators collectively whether the manufacturing process delivers product that meets its QTPP. The identification of CQAs is usually accomplished through a criticality assessment that evaluates the risk associated with each attribute. Criticality and/ or risk assessment are often done via laboratory testing, non-clinical evaluation, and clinical experience related to the quality attribute under evaluation as well as data from published literature. More discussions regarding the assessment of criticality and/or risk assessment for identified CQAs relevant to clinical outcomes based on either mechanism of action and/or pharmacokinetics (PK) and pharmacodynamics (PD) under appropriate statistical models are discussed further in Chapter 5.

Product and Process Design – Since changes in CQAs might have a significant impact on product safety and efficacy, it is critical to establish the relationships between CQAs and clinical performances (e.g., a predictive model for clinical outcomes). Under a well-established predictive model, acceptable ranges for CQAs may be obtained. This information provides a foundation for developing robust quality control strategies for the manufacturing process. In practice, it can be difficult to link specific critical quality attributes to clinical outcomes such as safety and efficacy. However, various models have been proposed in the literature to understand the relationships between CQAs and product safety and efficacy (see, e.g., Yang, 2013, 2016). More details can be found in Chapter 5.

Design Space – Design space is a key concept in the implementation of QbD. According to ICH Q8 (R2), a design space is referred to as the multi-dimensional combination and interaction of input variables (e.g., material attributes) and process parameters that have demonstrated to provide assurance of quality (ICH, 2006). A typical example for a design would be for a cell culture system, which may have ranges for temperature, pH, feed volume, and culture duration that ensure quality product. Design space is intimately related to quality risk management (QRM) principles. By linking manufacturing variations with the variability of CQAs, it sets the stage for developing effective manufacturing controls. Yang (2016) pointed out that a design space also has regulatory implications on post-approval changes. Note that working within the design space is not considered as a change, while movement out of the design space is considered a change, which would normally initiate a regulatory post-approval change process. Thus, a well-developed design space may enable a manufacturer to continuously improve the manufacturing process by adapting to novel technologies without incurring additional risk and creating more regulatory hurdles (Yang, 2016).

In recent years, FDA has considered QbD as a vehicle for the transformation of how drugs are discovered, developed, and commercially manufactured. In the past few years, the FDA has implemented the concepts of QbD into its pre-market processes. For evaluation of design space, there are five key elements: (i) specifications, (ii) design of the experiment (DOE) or experiments including the factors and the levels of the factors selected, (iii) model, the relationship between factors, interaction and the response, (iv) 95% confidence interval or region as well as the projected interval or region, and (v) design space based on the estimated and projected confidence intervals or regions.

Statistical Methods for Design Space – Several statistical methods have been proposed to determine design space through multivariate regression analysis. The two most commonly considered approaches are probably overlapping mean response surfaces and the composite desirability function. Note that a composite desirability function is a measure of how well a combination of factor settings optimizes the response. A design space can be determined using these approaches to be either a multivariate region in which the mean response values of CQAs are within specifications, or a region in which the composite desirability function exceeds a pre-specified threshold. These traditional methods do not take into account correlations among CQAs, nor do they account for variability in the model prediction (Perterson, 2008). As a result, it is suggested that an appropriately constructed design space should be able to account for measurement uncertainties, uncertainty about the parameters of the statistical models, and correlations among the measurements, in order to provide high assurance of product quality (Peterson, 2008, 2009). In recent years, significant advances have been made in developing design spaces using Bayesian multivariate analysis techniques (see, e.g., Perterson, 2004; Peterson and Yahyah, 2009; Peterson et al., 2009a, b; Stockdale and Cheng, 2009; Peterson and Lief, 2010; LeBrun, 2012; LeBrun et al. 2015). It should be noted that in practice, a design space is usually developed based on experiments and data from small scale processes.

Continued Process Verification – Continued process verification (CPV) is the next step in the lifecycle of the product/process, following successful process performance qualification. The goal of CPV is to demonstrate that the process continues to consistently produce product that meets its quality standard. In addition, CPV provides opportunities to either confirm that the control strategies are appropriate or identify other sources of variations that might require modification of specifications, in-process controls, or process improvements. Central to the fulfilment of these objectives is a system or systems for detecting unplanned departures from the process as designed, through monitoring and review of data collected from post-approval batches. It is also critical that the data collection and analysis is performed in accordance with cGMP requirements, and that proper corrective and preventive action(s) be taken when deviations are detected. The data should include relevant product quality attributes and process parameters, incoming materials or components measurements, and in-process tests. The use of statistical tools such as statistical process control (SPC) and process analytical technologies may potentially provide real-time process monitoring.

Traditionally, univariate SPC charts such as Shewhart, cumulative sum (CUSUM) charts, moving average (MA), and the exponentially weighted moving average (EWMA) charts have been used to monitor process performance over time (Montgomery, 2008). However, a major drawback of these univariate methods is that they do not account for interdependence among the variables being monitored. Because large and complex data sets including process parameters, quality attributes, properties of raw materials, and environmental factors of manufacturing areas are likely to include many and complex correlations, using independent univariate control charts may fail to identify out-of-control observations. The process data of two quality attributes (y_1, y_2) are plotted, with the confidence region for (y_1, y_2) indicated by the ellipse and the respective univariate control limits by the dotted lines. It is clear that the out of control batch would have been misidentified if only the univariate controls charts were used. Multivariate analysis techniques such as principal components analysis (PCA) and partial least squares (PLS) are effective statistical tools dealing with interdependence among variables. Through reducing the dimensionality of multivariate data, these methods make monitoring of complex data manageable. They have often been used by many companies for monitoring the performance of manufacturing processes and analytical methods, qualifying batches of raw materials, and demonstrating comparability pre- and post-changes.

Example – For illustration purpose, consider the following regression model proposed by Peterson (2008) and Stockdale and Cheng (2009) to describe the measured responses (Y) of the CQAs:

$$Y = Bz(x) + e \qquad (3.1)$$

where B is a $p \times q$ matrix of regression coefficients, $z(x)$ is a $q \times 1$ vector function of x and e is a $p \times 1$ vector of measurement errors having a multivariate normal distribution with mean **0** and covariance-variance matrix $\sum$. It is assumed that $z(x)$ is the same for each CQA though the method described in this section can be extended to the seemingly unrelated regressions (SUR) model (Peterson, 2006), in which each response Y_i ($i = 1, \ldots p$) has a different function $z_i(x)$. Under such circumstances, the SUR model provides greater flexibility and accuracy in modeling the CQAs (Peterson, 2007). A SUR model takes the form:

$$Y_j = z_j(x)'\beta_j + e_j, \qquad j = 1, \ldots p.$$

The SUR model includes the standard multivariate regression model as a special case where $z_j(x) = z(x)$. A design space based on the SUR model and posterior predictive probability can be similarly obtained.

Now, consider the example concerning the liquid chromatography method discussed in LeBlond (2014). The design space is {x: Pr[Y $\in$ A|x, *data*] $\geq$ 0.9}. To calculate the posterior predictive probability, minimally informative priors b_0, b_3, and $b_6 \sim N(0, 0.0001)$, b_1 and $b_2 \sim N(0, 0.001)$, and b_4, b_5, and $b_7 \sim N(0, 0.01)$ are used for the eight regression coefficients in model (1), based on examining the results of a frequentist analysis of the data, and a scalar inverse Wishart prior is used for the variance-covariance matrix. The variance-covariance matrix can be expressed as

$$\sum = \text{Diag}(\xi)Q\text{Diag}(\xi),$$

where Diag(ξ) is a diagonal matrix with diagonal elements $\xi = (\xi_1, \xi_2, \xi_3, \xi_4)$ and Q is a correlation matric given by

$$Q = \begin{pmatrix} 1 & \rho_{12} & \rho_{13} & \rho_{14} \\ \rho_{21} & 1 & \rho_{23} & \rho_{24} \\ \rho_{31} & \rho_{32} & 1 & \rho_{34} \\ \rho_{41} & \rho_{42} & \rho_{43} & 1 \end{pmatrix}$$

The parameters $\xi = (\xi_1, \xi_2, \xi_3, \xi_4)$ are chosen as $U(0, S_i)$ $(i = 1, \ldots, 4)$ with S_i being the frequentist estimate of the 99% upper confidence limit, of the square root of ith diagonal element of Σ, and Q is distributed according to $IW(I, 5)$, where I is the 4×4 identity matrix. Such a design space provides over 99% confidence that movement within the design space has no impact on the assay's ability to meet its specification.

3.5 Stability Analysis

For drug substance and drug product, real-time and accelerated (stress) stability data with several time points under upright and inverted conditions are necessarily collected to establish the expiration dating period of the drug substance and drug product. Stress studies (e.g., UV, exaggerated light, temperature and pH) are useful to elucidate product degradation pathways and for defining acceptance criteria. Limited time stability studies may be acceptable for short-term stability study. In practice, stability data generated from engineering lots is also acceptable. It should be noted that failure to demonstrate product stability could result in a potential hold issue.

For assessment of stability, the following tests are suggested: (i) safety (e.g., bioburden/sterility), (ii) purity (including product and process-related impurities and product-related substances), (iii) sialic acid (if appropriate), (iv) potency, (v) protein content/strength, (vi) pH, and (vii) appearance leachables (separate study, not part of routine stability testing) should be performed at a minimum.

In BLA submission, summary report of stability study usually includes, all stability data, supporting data from the clinical program, forced degradation data (to support the choice of stability indicating panel), data assessment to support expiry (shelf-life), stability protocol for commercial lots, data from an ICH compliant stability program (with a minimum of six months data under intended storage conditions) and conformance lots at commercial scale, and description and validation data for methods used only for assessing stability

Common errors include an insufficient number of lots, insufficient stability data, stability containers not representative of the drug substance container-closure system, absence of forced degradation data to identify stability indicating assays, and the absence of stability protocols.

Stability Guideline – Between 1993 and 2003, the ICH published a number of guidelines on stability for drug substances and drug products. These guidelines are listed in Table 3.5. As can be seen from Table 3.5, Q1A R2 is a revision of the parent guidelines published in 1993, which define the stability data package for registration of a new molecular entity as drug substance/drug product. Q1B makes recommendations on photostability testing,

TABLE 3.5

ICH Guidelines on Stability

Q1A	Stablility testing of new drug substances and products (R2 -2003)
Q1B	Stability testing of new drug substances and products (1996)
Q1C	Stability testing for new dosage forms (1996)
Q1D	Bracketing and matrixing designs for stability testing for new drugs substance and products (2002)
Q1E	Evaluation of Stability data (2003)
Q5C	Stability testing of biotechnological/biological products (1995)

while Q1C gives some recommendations on new dosage forms for authorized medicinal products. Q1D provides specific principles for the bracketing and matrixing in the study designs. Q1E suggests how to establish the shelf-life or retest period based on the stability studies performed. The Q5C is the main reference for biological medicinal substances and products. However, the principles defined in Q1 guidelines are also applicable.

Table 3.6 lists guidance on stability by the EU EMA. As can be seen from Table 3.6, CPMP/QWP/609/96 provides declaration of storage conditions. CPMP/QWP/2934/99 focuses on in-use stability testing, while CPMP/QWP/159/96 discusses maximum shelf-life for sterile products after first opening or following reconstitution. It should be noted that although regulatory requirements for stability testing for biologicals from the EU EMA are slightly different from those of ICH, they are similar enough for harmonization of regulatory requirements for stability testing of biosimilars. Thus, in this chapter, we will focus on the ICH Q5C stability guideline on biologicals.

Scope – The ICH Q5C stability guideline was published as an annex to the Tripartite ICH *Guideline for Stability of New Drug Substance and Products*. ICH Q5C intends to give guidance to applicants regarding the type of stability studies to be provided in support of marketing authorization applications for biological medicinal products. The ICH Q5C applies to well-characterized proteins and polypeptides, their derivatives and products of which they are components, and which are isolated from tissues, body fluids, cell cultures, or produced using rDNA technology. Table 3.7 lists medicinal products covered by ICH Q5C.

Batch Selection – As indicated by the ICH Q5C, stability evaluation should be done on active substance (bulk material), intermediates, and medicinal

TABLE 3.6

EU EMA Stability Guidelines on Biologicals

CPMP/QWP/609/96	Declaration of storage condition
CPMP/QWP/2934/99	In-use stability testing
CPMP/QWP/159/96	Maximum shelf-life for sterile products after first opening or following reconstitution

TABLE 3.7

Coverage of ICH Q5C Stability Guideline

Cover	Does Not Cover
Cytokines (IFN, IL, CSF, TNF)	Antibiotics
EPO	Allergenic extracts
Plasminogen activators	Heparins
Blood products	Vitamins
Growth hormones	Whole blood
Insulins	Cellular/Blood components products
Monoclonal antibodies	
Vaccines	

product (final container product). For stability data of drug substances, the ICH Q5C requires at least three batches representative of manufacturing scale of production be tested. Representative data are referred to as representative of (i) the quality of batches used in pre-clinical and clinical studies, (ii) manufacturing process and storage conditions, and (iii) containers/closures. If shelf-life is claimed to be greater than six months, a minimum of six months of stability data at the time of submission should be submitted. On the other hand, if shelf-life is claimed to be less than six months, the minimum amount of stability data in the initial submission should be determined on a case-by-case basis. Data from pilot-plant-scale batches of active substance produced at a reduced scale of fermentation and purification may be provided at the time the dossier is submitted to the regulatory agencies with a commitment to place the first three manufacturing scale batches into a long-term stability program after approval.

In practice, stability data for intermediates may be critical to the production of finished product. Thus, hold time and storage steps should be identified. The ICH Q5C suggests the manufacturer should generate in-house data and process limits that assure their stability within the bounds of developed processes. Along this line, an appropriate validation and/or stability study should be performed.

For stability data of the final drug product, similarly, the ICH Q5C requires at least three batches representative of manufacturing scale of production be tested. Drug product batches should be derived from different batches of drug substance. If shelf-life is claimed to be greater than six months, a minimum of six months stability data at the time of submission should be submitted. On the other hand, if shelf-life is claimed to be less than six months, the minimum amount of stability data in the initial submission should be determined on a case-by-case basis. Shelf-life should be derived from representative real time/real-conditions data. Data can be provided during the review and evaluation process. In here, representative data is referred to as representative of (i) the quality of batches used in pre-clinical and clinical studies, (ii) manufacturing process and storage conditions, and (iii) use of final containers/closures. Note that shelf-life will be based upon real time/ real-storage conditions data submitted for review. Pilot scale batches may be submitted with a commitment to place the first three manufacturing scale batches in a long-term stability program.

Study Design – Regarding study design and sample selection criteria, the ICH Q5C recommends a bracketing design or a matrixing design be used (see, e.g., Helboe, 1992; Nordbrock, 1992; DeWoody and Raghavarao, 1997; Pong and Raghavarao, 2000; Chow, 2007). Samples can then be selected for the stability program on the basis of a matrixing system and/or by bracketing. A bracketing design is a design that only samples on the extremes of certain design factors which are tested at all time points. Stability of any intermediate levels is considered represented by the stability of the extremes. Bracketing is generally not applicable for drug substances. Bracketing can be applied to studies with multiple strength of identical or closely-related formulations. In this case, only samples on the extremes of certain design factors (e.g., strength, container size, fill) are tested at all time points. Bracketing design can also be applied to studies with the same container closure system whether either the fill volume and/or the container size change.

A matrix design is a statistical design of a stability study that allows different fractions of samples be tested at different sampling time points (see, e.g., Nordbrock, 1992; Chow, 2007). Each subset of samples represents the stability of all samples at a given time point. Differences in the samples should be identified as covering different batched, different strengths, and different sizes of the same container closure system. A matrixing design should be balanced such that each combination of factors is tested to the same extent over the duration of the studies. It should be noted that all samples should be tested at the last time point before submission of application. For illustration purposes, consider the following examples concerning matrixing in a long-term stability study with one storage condition: (i) one-half reduction eliminates one in every two time points, and (ii) one-third design eliminates one in every three time points.

Storage Conditions – The ICH Q5C also defines storage conditions such as humidity, temperature, accelerated/stress conditions, light, container/closure, and stability after reconstitution of freeze-dried product. The ICH Q5C indicates that products are generally distributed in containers protecting against humidity. If demonstrated that container (storage conditions) provide sufficient protection against high and low humidity, relative humidities can be omitted. If humidity-protecting containers are not used, appropriate data should be provided. While most biologicals need precisely defined storage temperatures, real-time/real-temperature studies are confined to the proposed storage temperature. Requirement for light should be evaluated on a case-by-case basis.

For accelerated and stress conditions, shelf-life is established based on real time/real temperature data. In practice, accelerated studies can not only be supportive to established shelf-life, but also to provide information on

TABLE 3.8

Storage Conditions for Accelerated, Stress, and Long-Term Stability Studies

Stability Demonstration			
Long Term	Accelerated (6 Months)	Storage Statement	Additional Statement[a]
+25 ± 2°C/60%RH or +30 ± 2°C/65%RH	+40 ± 2°C/75%RH	No special storage conditions	Do not refrigerate or freeze
+25 ± 2°C/60%RH or +30 ± 2°C/65%RH	–	Do not store above +30 °C or 25 °C	Do not refrigerate or freeze
+5 ± 3°C	–	Store at +2 ±8°C	Do not freeze
<0°C	–	Store at -XX°C	

[a] where relevant.

post-development changes and validation of stability-indicating tests. For accelerated testing, testing conditions are normally one station higher than real storage conditions, which will help in elucidating the degradation profile (see Table 3.8). For stress testing, it can not only determine the best product stability indicators, but also reveal patterns of degradation. It is representative of accidental exposures to other conditions.

Testing Frequency – Shelf lives for biological products usually vary. ICH stability guidelines are based on a six-months to five-year shelf-life for most biological products. The recommended testing intervals for long term studies in pre-licensing are illustrated in Table 3.9.

To indicates that stability will be done at zero months, followed by every three months for the first year. Stability test will be done every six months for the second year and yearly thereafter.

Expiration Dating Period – As indicated in the 1987 FDA stability guideline and the 1993 ICH stability guideline, the expiration dating period of a drug product can be determined as the time at which the average drug characteristic remains within an approved specification (e.g., USP/NF, 2000) after manufacture (FDA, 1987; ICH, 1993; Chow, 2007). FDA suggests that an expiration dating period of a drug product be determined as the time point at which the 95% lower confidence bound of the mean drug characteristic intersects the approved lower specification of the drug product.

TABLE 3.9

Test Code Definitions

Code	Test Times After Time 0						
T0	3	6	9	12	18	24	36
T1	3		9	12		24	36
T2	3	6		12	18		36
T3		6	9		18	24	36

Note: Batches are tested at time 0.

The use of the one-sided 95% lower confidence bound of the mean degradation of the drug product is to assure that the drug product will remain within the approved specification for the identity, strength, quality, and purity prior to the expiration date.

FDA's approach for determination of the expiration dating period of a given batch of a drug product is briefly described below. For a given batch, let y_j be the assay result at time x_j, j = 1,...,n. The following simple linear model is usually assumed:

$$y_j = \alpha + \beta x_j + e_j, \quad j = 1,...n, \tag{3.2}$$

where α and β are unknown parameters, $x_j's$ are deterministic time points (storage times) selected in the stability study, and $e_j's$ are measurement errors independently and identically distributed as a normal (Gaussian) random variable with mean 0 and variance σ^2. According to the method suggested by the FDA, for a fixed time point, the 95% lower confidence bound for $\alpha + \beta x$ is given by

$$L(x) = \hat{\alpha} + \hat{\beta}x - \hat{\sigma}t_{n-2}\sqrt{\frac{1}{n} + \frac{(x-\bar{x})^z}{S_{XX}}}, \tag{3.3}$$

where t_{n-2} is the 95th percentile of the t-distribution with n-2 degrees of freedom, $\bar{x}$ is the average of $x_j's$,

$$\hat{\sigma}^2 = \frac{1}{n-2}\left(S_{yy} - S_{xy}^2/S_{xx}\right),$$

in which

$$S_{yy} = \sum_{j=1}^{n}(y_j - \bar{y})^2,$$

$$S_{xx} = \sum_{j=1}^{n}(x_j - \bar{x})^2,$$

$$S_{xy} = \sum_{j=1}^{n}(x_j - \bar{x})(y_j - \bar{y}),$$

and $\bar{y}$ is the average of $y_j's$.

In practice, if $L(x)$ is greater than the lower product specification, we claim that the product meets the product specification up to time x.

3.6 Concluding Remarks

A typical BLA includes (i) form FDA 356h (cover sheet), (ii) applicant information, (iii) product/manufacturing information, (iv) pre-clinical studies, (v) clinical studies, and (vi) labeling. Thus, CMC, pre-clinical, and clinical are three critical pieces in biosimilars development. A full-scale CMC development is required including expression system, culture, purification, formulation, analytics, and packaging. The EU has issued biosimilars guidelines based on comparative testing against a reference biologic drug (the original approved biologic). For approval of biosimilars in the US, we expect the CMC package will be similar to that in the EU which contains a full quality dossier with a comparability program including detailed product characterization comparison and reduced pre-clinical and clinical requirements.

CMC requirements for biological products have received increasing attention from the regulatory agencies worldwide. The following lists potential CMC hold issues for phase 1 IND:

  i. Comparability between preclinical and clinical lots not demonstrated
  ii. Insufficient characterization of cell banks (e.g., adventitious agents testing, identity, etc.)
  iii. Inadequate product characterization with regards to purity, identity, potency, and safety
  iv. Lack of final product release testing
  v. Lacking or inappropriate specifications for release and stability testing
  vi. Lacking or inadequate potency assay
  vii. Data supporting product stability have not been shown for the planned duration of clinical studies
  viii. Lack or inappropriate immunogenicity assays for high risk products
  ix. Lack of evidence for final drug product sterility

There has been significant progress in developing QbD approaches for pharmaceutical products since the US FDA kicked off the QbD initiative *Pharmaceutical Current Good Manufacturing Practices (cGMPs) for the 21st Century* in 2002. Pharmaceutical product and process development, due to the high level of complexity, would benefit from greater use of available data and tools for risk assessment and development of design space and control strategies. How to effectively synthesize information from various sources, quantify manufacturing variability and risk, identify CQAs, determine design space, and devise effective control strategies is the key to implementing a QbD approach. All of these rely on statistical methodologies for experimental design and data analysis. Statistical advances, particularly in the area of Bayesian modeling and computation have brought about new opportunities for applying QbD principles in biopharmaceutical development.

During early development of biosimilar products, accelerated stability studies can help provide key data and information on the effect of short-term exposure to environmental conditions. Short-term stability studies are typically performed over a six-month period. Long term stability studies (12-months and longer) allow evaluation of the quality of the drug both during and beyond its projected shelf-life. Matrix designs described in this chapter are generally applicable to many situations and can result in significant savings, with the 1/3 matrix on time readily acceptable for stable products. There are two basic approaches when analyzing data from a matrixing design. There are several methods used to evaluate and to compare potential designs.

4

Analytical Method Validation

4.1 Introduction

For analytical similarity assessment of functional and structural characterization of critical quality attributes, extensive analytical data are often collected to support (i) a demonstration that a proposed biosimilar product and
the United States (US)-licensed reference product are highly similar; (ii) a
demonstration that the proposed biosimilar product can be manufactured
in a well-controlled and consistent manner, leading to a product that is sufficient to meet appropriate quality standards; and (iii) a justification of the relevance of the comparative data generated using a non- US-licensed reference
product (e.g., a EU-approved reference product) to support a demonstration
of biosimilarity of the proposed biosimilar product to a US-licensed reference product if comparative data (e.g., pharmacokinetics/pharmacodynamics
and/or clinical data) between the proposed biosimilar product and the non-
US-licensed reference product are available.

In practice, the analytical similarity assessment for the proposed biosimilar product consists of a comparison of the proposed biosimilar product to
the US-licensed product for the purpose of demonstrating that the proposed
biosimilar product is *highly similar* to the US-licensed product, notwithstanding minor differences in clinically inactive components. Pairwise comparisons of the proposed biosimilar product to a non-US-licensed product (e.g.,
an EU-approved product) and the EU-approved product to the US-licensed
product are performed for the purpose of establishing the analytical portion
of the scientific bridge necessary to support the use of the data derived from
the clinical studies that used the EU-approved product as the comparator.

For a given critical quality attribute, an analytical method selected to
perform similarity between a proposed biosimilar product and a reference
product is usually selected based on product knowledge regarding structure, function, heterogeneity, and stability of the reference product. The analytical similarity assessment is usually grouped into categories that evaluate
biological activity, primary structure, higher-order structure, particles and
aggregates, product related substances and impurities, thermal stability
and forced degradation, general process, and process-related impurities.

For example, for characterization of structure, liquid chromatography-tandem mass spectrometry (LC-MS/MS) is often used to analyze amino acid by reduced peptide, mapping with ultraviolet (UV), and liquid chromatography. On the other hand, enzyme-linked immunosorbent assay (ELISA) is often used to study binding assays. The analytical methods used must be appropriately developed, validated, and deemed suitable for the intended use.

Validation of an analytical method plays an important role at the stage of laboratory development in manufacturing process of biosimilar product development. In practice, it is suggested that validation should be done according to some performance characteristics as described in the United States Pharmacopeia and National Formulary (USP/NF, 2000). A validated analytical method is expected to achieve a certain degree of accuracy, reliability, repeatability, and reproducibility. The evaluation of reliability, repeatability, and reproducibility of an analytical method is important in studies which are intended for detecting a scientifically meaningful difference in certain responses. In practice, reliability and repeatability/reproducibility are related to various sources of variability such as intra-analyst (or within day) variability, inter-analyst (or between day) variability, and variability due to some interactions such as analyst-by-day interactions. For analytical method validation, regulatory requirements specify the assay characteristics that are subject to validation and suggest experimental strategies to study these properties. It is suggested that these requirements should be followed in conjunction with an efficient validation plan with sound statistical experimental design in order to ensure the assay results are reliable, repeatable, and reproducible.

In the next section, regulatory requirement for analytical development and validation are discussed. Section 4.3 reviews some useful validation study designs, commonly considered validation parameters (performance characteristics), and the corresponding acceptance criteria. Statistical methods for analysis of validation data are given in Section 4.4. Section 4.5 introduces a newly proposed method for quality control/assurance of a given quality attribute during the manufacturing process in terms of reliability, repeatability, and reproducibility. A couple of examples regarding analytical method validation of LC-MS/MS and ELISA are given in Section 4.6. Some concluding remarks are given in the last section of this chapter.

4.2 Regulatory Requirements

4.2.1 FDA Guidance on Analytical Procedures and Methods Validation

In July 2015, FDA published a guidance on *Analytical Procedures and Methods Validation for Drugs and Biologics* (FDA, 2015d). This guidance replaces the

2000 draft guidance for industry on *Analytical Procedures and Methods Validation* and the 1987 *Guidelines for Submitting Samples and Analytical Data for Methods Validation*. It provides recommendations on how the applicant can submit analytical procedures and methods validation data to support the documentation of the identity, strength, quality, purity, and potency of drug substances and drug products. It intends to assist the applicant to assemble information and present data to support analytical methodologies in the regulatory submissions. The recommendations apply to drug substances and drug products covered in new drug applications (NDAs), abbreviated new drug applications (ANDAs), biologics license applications (BLAs), and supplements to these applications. This guidance complements the International Conference on Harmonization (ICH) guidance Q2(R1) *Validation of Analytical Procedures: Text and Methodology* for developing and validating analytical methods.

When an analytical procedure is approved/licensed as part of the NDA, ANDA, or BLA, it becomes the FDA-approved analytical procedure for the approved product. This analytical procedure may originate from FDA recognized sources (e.g., a compendial procedure from the United States Pharmacopeia/National Formulary (USP/NF)) or a validated procedure you submitted that was determined to be acceptable by FDA. To apply an analytical method to a different drug product, appropriate validation or verification studies for compendial procedures with the matrix of the new product should be considered.

An analytical procedure is developed to test a defined characteristic of the drug substance or drug product against established acceptance criteria for that characteristic. Early in the development of a new analytical procedure, the choice of analytical instrumentation and methodology should be selected based on the intended purpose and scope of the analytical method. Parameters that may be evaluated during method development are specificity, linearity, limits of detection (LOD) and limits of quantitation (LOQ), range, accuracy, and precision. During the early stages of method development, the robustness of methods should be evaluated because this characteristic can help to decide which method one should submit for approval. Analytical procedures in the early stages of development are initially developed based on a combination of mechanistic understanding of the basic methodology and prior experience. Experimental data from early procedures can be used to guide further development. You should submit development data within the method validation section if they support the validation of the method. To fully understand the effect of changes in method parameters on an analytical procedure, one should adopt a systematic approach for a method robustness study (e.g., a design of experiments with method parameters). One should begin with an initial risk assessment and follow with multivariate experiments. Such approaches allow understanding of factorial parameter effects on method performance. Evaluation of a method's performance may include analyses of samples obtained from various stages of the manufacturing process from in-process to the finished product. Knowledge gained during these studies on the sources of method variation can help you assess the method performance.

Analytical method validation is the process of demonstrating that an analytical procedure is suitable for its intended purpose. The methodology and objective of the analytical procedures should be clearly defined and understood before initiating validation studies. This understanding is obtained from scientifically-based method development and optimization studies. Validation data must be generated under a protocol approved by the sponsor, following current good manufacturing practices, with the description of methodology of each validation characteristic and predetermined and justified acceptance criteria, using qualified instrumentation. Protocols for both drug substance and product analytes or mixture of analytes in respective matrices should be developed and executed. You should include details of the validation studies and results with your application.

4.2.2 ICH Guidance on Assay Validation

ICH Q2R1 (1996) indicates that the objective of validation of an analytical procedure is to demonstrate that it is suitable for its intended purpose. A tabular summation of the characteristics applicable to identification, control of impurities, and assay procedures is included. Other analytical procedures may be considered in future additions to this document (ICH Q2R1, 1996).

As indicated in ICH Q2R1, there are four types of analytical procedures that need to be validated. These four types of analytical procedures include (i) identification tests, (ii) quantitative tests for impurities content, (iii) limit tests for the control of impurities, and (iv) quantitative tests of the active moiety in samples of drug substance or drug product or other selected component(s) in the drug product.

Identification tests are intended to ensure the identity of an analyte in a sample. This is normally achieved by comparison of a property of the sample (e.g., spectrum, chromatographic behavior, chemical reactivity, etc.) to that of a reference standard. Testing for impurities can be either a quantitative test or a limit test for the impurity in a sample. Either test is intended to accurately reflect the purity characteristics of the sample. Different validation characteristics are required for a quantitative test than for a limit test. Assay procedures are intended to measure the analyte present in a given sample. In the context of this document (ICH Q2R1, 1996), the assay represents a quantitative measurement of the major component(s) in the drug substance. For the drug product, similar validation characteristics also apply when assaying for the active or other selected component(s). The same validation characteristics may also apply to assays associated with other analytical procedures (e.g., dissolution).

Although there are many other analytical procedures, such as dissolution testing for drug products or particle size determination for drug substances, these have not been addressed in the initial text on validation of analytical procedures. Validation of these additional analytical procedures are equally important to those listed herein and may be addressed similarly (ICH Q2R1, 1996).

4.2.3 United States Pharmacopeia and National Formulary (USP/NF)

In its recent guidance entitled *Analytical Procedures and Methods Validation for Drugs and Biologics*, FDA indicates that the analytical procedure may originate from FDA recognized sources such as a compendial procedure from the United States Pharmacopeia/National Formulary (USP/NF) or a validated procedure submitted that was determined to be acceptable by FDA.

As indicated by Chapter 1225 of USP/NF, validation of an analytical procedure is the process by which it is established, by laboratory studies, that the performance characteristics of the procedure meet the requirements for the intended analytical applications. Typical analytical performance characteristics that should be considered in the validation of the types of procedures described in this document. Because opinions may differ with respect to terminology and use, each of the performance characteristics is defined in the next section of this chapter, along with a delineation of a typical method or methods by which it may be measured. The definitions refer to *test results*. The description of the analytical procedure should define what the test results for the procedure are. The test method should specify that one or a number of individual measurements be made, and their average, or another appropriate function (such as the median or the standard deviation), be reported as the test result. It may also require standard corrections to be applied, such as correction of gas volumes to standard temperature and pressure. Thus, a test result can be a result calculated from several observed values. In the simple case, the test result is the observed value itself. A test result also can be, but need not be, the final, reportable value that would be compared to the acceptance criteria of a specification. Validation of physical property methods may involve the assessment of chemometric models. However, the typical analytical characteristics used in method validation can be applied to the methods derived from the use of the chemometric models.

The effects of processing conditions and potential for segregation of materials should be considered when obtaining a representative sample to be used for validation of procedures. In the case of compendial procedures, revalidation may be necessary in the following cases: a submission to the USP of a revised analytical procedure; or the use of an established general procedure with a new product or raw material (see below in Data Elements Required for Validation). The ICH documents give guidance on the necessity for revalidation in the following circumstances: changes in the synthesis of the drug substance; changes in the composition of the drug product; and changes in the analytical procedure. Chapter 1225 is intended to provide information that is appropriate to validate a wide range of compendial analytical procedures. For some compendial procedures the fundamental principles of validation may extend beyond characteristics suggested in Chapter 1225. For these procedures the user is referred to the individual compendial chapter for those specific analytical validation characteristics and any specific validation requirements.

4.3 Analytical Method Validation

4.3.1 Validation Performance Characteristics

As indicated in ICH Q2R1, typical validation characteristics include accuracy, precision (repeatability and intermediate precision), specificity, detection limit, quantitation limit, linearity, and range, which are briefly described below.

The accuracy of an analytical procedure expresses the closeness of agreement between the value which is accepted either as a conventional true value or an accepted reference value and the value found. The precision of an analytical procedure expresses the closeness of agreement (degree of scatter) between a series of measurements obtained from multiple sampling of the same homogeneous sample under the prescribed conditions. Precision may be considered at three levels: repeatability, intermediate precision and reproducibility. This Specificity is the ability to assess unequivocally the analyte in the presence of components which may be expected to be present. Typically these might include impurities, degradants, matrix, etc.

The detection limit of an individual analytical procedure is the lowest amount of analyte in a sample which can be detected but not necessarily quantitated as an exact value, while the quantitation limit of an individual analytical procedure is the lowest amount of analyte in a sample which can be quantitatively determined with suitable precision and accuracy.

The linearity of an analytical procedure is referred to as its ability (within a given range) to obtain test results which are directly proportional to the concentration (amount) of analyte in the sample. On the other hand, the range of an analytical procedure is the interval between the upper and lower concentration (amounts) of analyte in the sample (including these concentrations) for which it has been demonstrated that the analytical procedure has a suitable level of precision, accuracy and linearity.

4.3.2 Study Design

In practice, it is convenient for the purposes of planning a validation experiment to dichotomize assay validation parameters into two categories: parameters related to the accuracy of the analytical method and parameters related to variability. In this way, the analyst will choose the test sample set that will be carried forward into the validation, in accord with accuracy parameters, and devise a replication plan for the validation to satisfy the estimation of the variability parameters.

The validation parameters related to the accuracy of an analytical method are accuracy, linearity, and specificity. Accuracy is usually established by spiking known quantities of analyte into the sample matrix and demonstrating that these can be completely recovered. Conventional practice is to spike using five levels of the analyte: 50%, 75%, 100%, 125%, and 150% of the declared content of analyte in the drug. Note that this is more stringent than the ICH Q2R1 guidance, which requires a minimum of nine determinations over a minimum of three concentration levels. The experimenter is not restricted to these levels but should plan to spike through a region that embraces the range of expected measurements from samples that will be tested during development and manufacture. Accuracy can also be determined relative to a referee method; in the case of complex mixtures, such as combination vaccines, accuracy can be judged relative to a monovalent control. Linearity can be established from the spiking experiment. For example, if a sample is tested in two-fold dilution, the measured concentrations in those samples should yield a two-fold series. It, however, should be note that a dilution series is usually employed when the assessment of linearity cannot be achieved through spiking (i.e., when the purified analyte does not exist, such as in vaccines). When the samples are available, specificity can be established through testing of the analyte-free matrix.

The assay validation parameters related to variability are precision, repeatability, ruggedness, limit of detection (LOD), limit of quantitation (LOQ), and range. The precision of an assay expresses the degree of closeness between a series of measurements obtained from multiple sampling of the same homogeneous sample under the prescribed conditions. Precision is frequently referred to as inter-run (or between run) variability, where a run of an assay represents the independent preparation of assay reagents, tests samples, and a standard curve. The repeatability of an assay reflects the precision under the same operating conditions over a short interval of time, and is frequently called intra-run variability. Repeatability and precision can be studied simultaneously, using the levels of a sample required to establish accuracy and linearity, in a consolidated validation study design such as that depicted in Table 4.1. Here precision is associated with the multiple experimental runs, while repeatability is associated with the run-by-level interaction. In many cases, the experimenter may wish to test true within-run replicates rather than employ this subtle statistical artifact. The runs in this design can be strategically allocated to ruggedness parameters, such as laboratories, operators, and reagent lots, using a combination of nested and factorial experimental design strategies. Thus operators might be nested within laboratory, while reagent lot can be crossed with operator.

The robustness of an analytical method is referred to as a measure of its capacity to remain unaffected by small, but deliberate, variations in method

TABLE 4.1

Strategic Validation Design Employing Five Levels of Analyte
Tested in n Independent Runs of the Assay

Run	Level				
	1	2	3	4	5
1	y_{11}	y_{12}	y_{13}	y_{14}	y_{15}
$\vdots$	$\vdots$	$\vdots$	$\vdots$	$\vdots$	$\vdots$
n	y_{n1}	y_{n2}	y_{n3}	y_{n4}	y_{n5}

parameters, and might be more suitably established during the development of the assay. Note that while ruggedness parameters represent uncontrollable factors affecting the analytical method, robustness parameters can be varied and should be controlled when it has been observed that they have an effect on assay measurement.

The LOD and LOQ of an assay can be obtained from replication of the standard curve during the implementation of the consolidated validation experiment. The LOD of an analytical procedure is defined as the lowest amount of analyte in a sample that can be detected but not necessarily quantitated as an exact value, while the LOQ is the lowest amount of analyte in a sample that can be quantitatively determined with suitable precision and accuracy. The LOQ need not be restricted to a lower bound on the assay, but might be extended to an upper bound of a standard curve, where the fit becomes flat (for example, when using a four-parameter logistic regression equation to fit the standard curve). Finally, the composite of the ranges identified with acceptable accuracy and precision is called the range of the assay.

4.3.3 Choice of Validation Performance Characteristics

The choice of parameters that will be explored during an assay validation is determined by the intended use as well as by the practical nature of the analytical method. For example, a biochemical assay using a standard curve to establish drug content might require the exploration of all of these parameters if the assay is to be used to determine low as well as high levels of the analyte (such as an assay used to determine drug level in clinical samples or an assay that will be used to measure the content of an unstable analyte). If the assay is used, however, to determine content in a stable preparation, the LOD and LOQ need not be established. On the other hand, an assay for an impurity requires adequate sensitivity to detect and/or quantify the analyte; thus the LOD and LOQ become the primary focus of the validation of an impurity assay.

Many of these assay validation parameters have limited meaning in the context of biological assay and various other potency assays. There is no means to explore the accuracy and linearity of assays in animals or tissue culture, where the scale of the assay is defined by the assay; thus validation experiments for this sort of analytical method are usually restricted to a study of the specificity and ruggedness of the procedure. As discussed previously, the accuracy of a potency assay may be limited to establishing linearity with dilution when the purified analyte is unavailable to conduct a true spike-recovery experiment.

4.3.4 Acceptance Criteria

As previously defined, the goal of the validation experiment is to establish that the analytical method is fit for use. Thus for an impurity assay, for example, the goal of the validation experiment should be to show that the sensitivity of the analytical procedure (as measured by the LOD for a limit assay and by the LOQ for a quantitative assay) is adequate to detect a meaningful level of the impurity. Assays for content and potency are usually used to establish that a drug conforms to specifications that have been determined either through clinical trials with the drug or from process/product performance. It is important to point out that the process capability limits should include the manufacturing distribution of the product characteristic under study, the change in that characteristic because of instability under recommended storage conditions, as well as the effects on the measurement of the characteristic because of the performance attributes of the assay. In the end, the analytical method must be capable of reliably discriminating satisfactory from unsatisfactory product against these limits. The portion of the process range that is a result of measurement, including measurement bias as well as measurement variability, serves as the foundation for setting acceptance criteria for assay validation parameters.

Validation parameters related to accuracy are rated on the basis of recovery or bias (bias = 100 - recovery, usually expressed as a percentage), while validation parameters related to random variability are appraised on the basis of variability (usually expressed as % RSD), where RSD is relative standard deviation. These can be combined with the process variability to establish the capability of the process (CP) of the measured characteristic (Montgomery, 1991):

$$CP = \frac{\text{Specification range}}{6 \times \text{Product variation}}$$
$$= \frac{\text{Specification range}}{6 \times \sqrt{\sigma_{product}^2 + \sigma_{assay}^2 + bias^2}}.$$

The capability of the process is related to the percentage of measurements that are likely to fall outside of the specifications. Thus, acceptance criteria on the amount of bias and analytical variability can be established based upon the knowledge of the product variability and the desired process capability.

When the limits have not been established for a particular product characteristic, such as during the early development of a drug or biologic, acceptance criteria for an assay attribute might be specified on the basis of a typical
expectation for the analytical method as well as on the nature of the measurement in the particular sample. Thus, for example, high-performance liquid
chromatography (HPLC) for the active compound in the final formulation of
a drug might be typically capable of yielding measurement equal to or less
than 10% RSD, while an immunoassay used to measure antigen content in
a vaccine might only be capable of achieving up to 20% RSD. Measurement
of a residual or an impurity by either method, on the other hand, may only
achieve up to 50% RSD, owing to the variable nature of measurement at low
analyte concentrations.

4.4 Analysis of Validation Data

4.4.1 Assessment of Accuracy, Linearity, and Specificity

Accuracy and Linearity – Analysis of analytical performance characteristics
related to *accuracy* and *linearity* can be parameterized as the following simple
linear model (see also Schofield, 2008)

$$x = \alpha + \beta\mu$$
$$= \mu + [\alpha + (\beta - 1)\mu], \tag{4.1}$$

where μ is the known analyte content and x is the measured amount. Thus,
the accuracy can be expressed as $x = \mu$ at a single concentration, or as $x = \alpha +$
$\beta\mu$ across a series of concentrations, where $\alpha = 0$ and $\beta = 1$. As it can be seen
from (4.1), parameter α is related to the accuracy of the analytical method
and represents the constant bias in an assay measurement, which is usually reported in the units of measurement of the assay. On the other hand,
$\beta - 1$ is related to the linearity of the analytical method and represents the
proportional bias in an assay measurement, which is usually reported in the
units of measurement in the assay per unit increase in that measurement. In
practice, data from a spiking experiment are often be utilized to estimate α
and $\beta - 1$. The estimates can then be compared with the acceptance criteria
established for these parameters for achieving desired statistical inference
(performance).

To illustrate the concepts of accuracy (either absolute or relative bias) and
linearity described above, consider the example given in Table 2 of Schofield
(2008), which is reproduced in Table 4.2. As it can be seen from Table 4.2, a
validation study has been performed in which samples have been prepared

TABLE 4.2

Example of Results from a Validation Experiment in Which Five Levels of an Analyte Were Tested in Six Runs of the Assay

[C]	1	2	3	4	5	6	Avg.
3	3.3	3.4	3.2	3.1	3.4	3.3	3.3
4	4.2	4.2	4.2	4.4	4.0	4.3	4.2
5	5.0	4.9	4.9	5.0	4.8	4.9	4.9
6	5.8	5.8	5.6	5.9	5.9	5.7	5.8
7	6.4	6.7	6.7	6.8	6.6	6.4	6.6

with five levels of an analyte (denoted by [C] in mg), and their content is determined in six runs of an assay. The assay results are also displayed in Figure 4.1. Analysis of assay results from Table 4.2 yield

$$\hat{\alpha} = -0.04\,(-0.08, 0.01) \text{ and } \hat{\beta} - 1 = -0.19\,(-0.30, -0.15)$$

As a result, we may conclude that the proportional bias is significant based on the fact that the confidence interval excludes 0. However, the conclusion of non-linearity should take magnitude of the bias into consideration, especially when the bias is in excess of a pre-specified acceptance limit. For example, there is a 0.2 mg decrease per unit increase in concentration. It could be as much as −0.3 mg per unit decrease in concentration based on the confidence interval. If the range in the concentration of the drug is typically 4–6 mg (i.e., a 2 mg range), this would predict that the bias as a result of non-linearity is 0.4 mg (0.6 mg in the confidence interval). This can be judged to be of consequence or not in testing product. For example, if the specification limits on the analyte are 5 ± 1 mg/dose, much of this range will be consumed

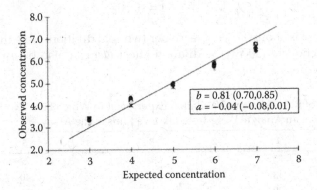

FIGURE 4.1

Validation results from an assay demonstrating proportional bias, with low-concentration samples yielding high measurements and high-concentration samples yielding low measurements.

by the proportional bias, potentially resulting in an undesirable failure rate in the measurement of that analyte.

In some situations where the purified analyte is unavailable for spiking and the levels of the analyte in the validation have been attained through dilution, then the data generated from that series can be alternatively evaluated using the following model

$$x = \alpha\mu^{\beta}, \tag{4.2}$$

where μ may represent either the expected concentration upon dilution or the actual dilution. Model (4.2) can be linearized by taking the logs as follows

$$\log(x) = \log(\alpha) + \beta \log(\mu). \tag{4.3}$$

Similarly, the proportional bias can be estimated as $2^{\beta-1} - 1$ if μ is in units of concentration and as $2^{\beta+1} - 1$ if μ is in units of dilution assuming that the dilution increment is two-fold. Note that the proportional bias is also known as dilution effect which is expressed as percent bias per dilution increment. For illustration purpose, let us consider the example given in Table 3 of Schofield (2008), which is reproduced in Table 4.3. In this example, a dilution series of a sample is performed in each of the five runs of an assay. The plot of assay results is given in Figure 4.2.

Under the linearized model (4.3), an estimate of the slope can be obtained as follows

$$\hat{\beta} = 1.01\,(-1.04, -0.98),$$

and the corresponding dilution effect can be estimated as

$$2^{-1.01+1} - 1 = -0.007.$$

Thus, there is about 0.7% decrease per two-fold dilution. Note that as indicated by Schofield (2008), the dilution effect (*de*) can also be estimated as

TABLE 4.3

Example of Results from a Validation Experiment in Which Five Two-Fold Dilutions of an Analyte Were Tested in Five Runs of the Assay

Dilution	1	2	3	4	5	Titer
1	75	90	93	79	72	81
2	43	45	46	35	40	83
4	20	23	22	18	21	83
8	11	10	12	10	10	85
16	4	5	6	4	5	78

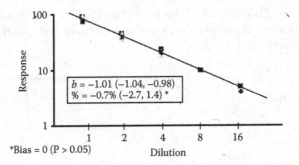

FIGURE 4.2
Results from a validation series demonstrating satisfactory linearity, i.e., approximately a two-fold decrease in response with two-fold dilution.

a function of the estimated slopes for the standard and the test sample as follows

$$de = 2^{1-\hat{\beta}_T/\hat{\beta}_S},$$

(4.4)

where $\hat{\beta}_S$ and $\hat{\beta}_T$ are the estimated slopes for the standard and the test sample, respectively. It can be verified that the standard error of $\hat{\beta}_T/\hat{\beta}_S$ is given by

$$SE\left(\frac{\hat{\beta}_T}{\hat{\beta}_S}\right) = \sqrt{\frac{1}{\hat{\beta}_S^2}\left[Var(\hat{\beta}_T) + \left(\frac{\hat{\beta}_T}{\hat{\beta}_S}\right)^2 Var(\hat{\beta}_S)\right]}.$$

As a result, estimate (4.4) is considered a more reliable estimate of the dilution effect as compared to the earlier estimate by taking variability in the measurements into consideration.

Another approach for assessment of accuracy and linearity of an analytical method is to simply compare an *experimental* analytical method with a validated *referee* method. In this case, paired measurements in the two assays can be made on a panel of samples, which can be used to establish the linearity and the accuracy of the experimental method relative to the referee method. Since the paired measurements are a sample from a multivariate population, the linearity of the experimental method to the referee method can be assessed using principal component analysis (see Morrison, 1967). As a result, a *concordance slope*, denoted by β_C, can be estimated from the elements of the first characteristic vector of the sample covariance matrix (a_{21}, a_{11}), i.e.,

$$\hat{\beta}_C = \frac{a_{21}}{a_{11}},$$

(4.5)

where a_{i1} is the ith element of the first characteristic vector of the sample covariance matrix. Similarly, the standard error of the estimated concordance slope can be obtained as

$$SE(\hat{\beta}_C) = \sqrt{\left[\frac{Var(a_{21})}{a_{21}^2} + \frac{Var(a_{11})}{a_{11}^2} - 2\frac{Cov(a_{21}, a_{11})}{a_{21}a_{11}} \right] \hat{\beta}_C^2},$$

where

$$Var(a_{i1}) = \left[\frac{l_1 l_2}{n(l_2 - l_1)^2} \right] a_{i1}^2,$$

and

$$Cov(a_{21}, a_{11}) = \left[\frac{l_1 l_2}{n(l_2 - l_1)^2} \right] a_{12} a_{22},$$

in which l_i, $i = 1,2$ are the corresponding characteristics roots. Thus, the concordance slope can be used to measure the discordance of the experimental analytical method relative to the referee method as follows:

$$d = 2^{\beta_C - 1}, \tag{4.6}$$

and its corresponding confidence interval can be obtained similarly. Note that the concordance slope approach is similar to the dilution effect discussed earlier, which expresses the percentage difference in reported potency per two-fold increase in result as measured in the referee method.

Specificity – The specificity of an analytical method is usually evaluated by comparing assay results from a placebo sample (i.e., a sample containing all constituents except the analyte of interest) relative to background. A statistical evaluation may include either a comparison of instrument measurements obtained for this placebo with background measurements or from measurements made on the placebo alone. Schofield (2008) indicated that in either case, an increase over the detection level of the assay would indicate that some constituent of the sample (in addition to analyte) is contributing to the measurement in the sample. The assessment can be made by comparing the upper bound of the confidence interval to determine whether it falls below some pre-specified level of the assay.

4.4.2 The Assessment of Assay Parameters Related to Variability

Assay parameters related to variability include, but are not limited to, reliability (precision), repeatability, reproducibility, LOD, LOQ, and range, which will be briefly described below.

Reliability, Repeatability, and Reproducibility – Reliability or precision (usually assessed by inter-run or between-run variability) and repeatability (usually assessed by intra-run or within-run variability) can generally be established by the analysis of variance (ANOVA) on the results obtained from the data generated for evaluation of accuracy and linearity (see, e.g., Winer, 1971). Reproducibility, on the other hand, involves other sources of variability such as variability due to the use of different laboratories, different operators (analysts), and reagent lots. These sources of variability need to be incorporated into the pattern of runs performed during the validation to estimate the effects of these sources of variability in order to assess the reproducibility of the assay results.

For illustration purpose, the analysis of variance table of the assay results given in Table 4.2 is summarized in Table 4.4. As it can be seen from Table 4.4, the within-run variability and between run variability can be estimated as

$$\hat{\sigma}^2_{within-run} = \hat{\sigma}^2_W = MSE,$$

$$\hat{\sigma}^2_{between-run} = \hat{\sigma}^2_B = \frac{MSR - MSE}{5},$$

and

$$\hat{\sigma}^2_{total} = \hat{\sigma}^2_T = \frac{\hat{\sigma}^2_B}{r} + \frac{\hat{\sigma}^2_W}{nr},$$

where MSE is mean square error, MSR is mean square for runs, r is the number of independent runs performed on the sample, and n is the number of replicates within a run. Note that MSE is in fact a composite of pure error and the interaction of run and level. In practice, although pure error can be separated from the interaction by performing replicates at each level, the within-run variability is usually reported as the composite of pure error and the level-by-run interaction. Note that the total variability is the variance of $\bar{y}$. Thus, based on the total variability, a confidence interval can be obtained (see, e.g., Burdick and Graybill, 1992) as follows

TABLE 4.4

Analysis of Variance Table Showing Expected Mean Squares as Functions of Intra-Run and Inter-Run Components of Variability

Effect	df	MS	Estimated Mean Square	F
Level	4	Mean square for levels (MSL)	$\hat{\sigma}_W^2 + Q(L)$	
Run	5	MSR	$\hat{\sigma}_W^2 + 5\hat{\sigma}_B^2$	MSR/MSE
Error	20	MSE	$\hat{\sigma}_W^2$	

$$\hat{\sigma}_{total}^2 = \hat{\sigma}_T^2 = \frac{\hat{\sigma}_B^2}{r} + \frac{\hat{\sigma}_W^2}{nr}$$

$$= \left(\frac{1}{Jr}\right)MSR + \left(\frac{n-J}{Jnr}\right)MSE$$

$$= c_1 MSR + c_2 MSE$$

$$= \sum_q c_q S_q,$$

where J is the number of replicates of a sample tested in each run of the assay validation experiment ($J = 5$ in Table 4.4). Thus, the $(1 - 2\alpha) \times 100\%$ confidence interval for the total variability can be constructed as follows:

$$\left(L = \hat{\sigma}_T^2 - \sqrt{\sum_q G_q^2 c_q^2 S_q^4}, \quad U = \hat{\sigma}_T^2 + \sqrt{H_q^2 c_q^2 S_q^4} \right), \tag{4.7}$$

where

$$G_q = \frac{1}{F_{\alpha,dfq,\infty}}, H_q = \frac{1}{F_{1-\alpha,dfq,\infty}},$$

and $F_{\alpha,dfq,\infty}$ and $F_{1-\alpha,dfq,\infty}$ are the percentiles of the F distribution.

LOD and LOQ – The LOQ (limit of detection) and the LOQ (limit of quantitation) are usually determined using the standard deviation of the responses (i.e., $\hat{\sigma}$) and the slope of the calibration curve (i.e., $\hat{\beta}$). For example,

$$LOD = 3.3 \,\hat{\sigma}/\hat{\beta}, \tag{4.8}$$

where the coefficient of 3.3 is derived from the 95th percentile of the standard normal distribution, i.e., $3.3 \approx 2 \times 1.645$. The purpose is to limit both the

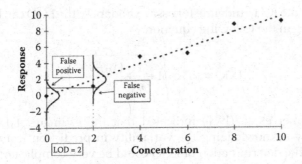

FIGURE 4.3
Determination of the LOD using the linear fit of the standard curve plus restriction on the proportions of false positives and false negatives.

false-positive and false-negative error rates to the equal to or less than 5%. To provide a better understanding, the concept of LOD is illustrated in Figure 4.3. On the other hand, LOQ is set to be

$$LOQ = 10\,\hat{\sigma}/\hat{\beta}, \qquad (4.9)$$

where the coefficient of 10 corresponds to the restriction of no more than 10% of relative standard deviation (RSD) in the measurement variability. In other words, for a given $\bar{x}$, we have

$$\frac{RSD}{100} = \frac{\hat{\sigma}}{\bar{x}} < 0.10.$$

Note that if one wishes to have no more than 20%, the LOQ becomes $LOQ = 20\,\hat{\sigma}/\hat{\beta}$. The concept of LOQ is illustrated in Figure 4.4. Note that as

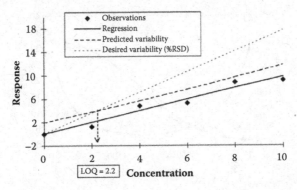

FIGURE 4.4
Determination of the LOQ using the linear fit of the standard curve and a restriction on the variability (%RSD) in the assay.

indicated in the ICH guideline for assay validation, the LOQ can be obtained by solving x_Q in the following equation:

$$LOQ = 10\left\{ \hat{\sigma}\sqrt{1 + \frac{1}{n} + \frac{(x_Q - \bar{x})^2}{SXX}} \right\}. \tag{4.10}$$

Oppenheimer et al. (1983) indicated that the estimate of LOQ can be improved by acknowledging the variability in prediction from the curve. In practice, the determination of LOQ could be very complicated if the calibration function is non-linear, which may involve four-parameter logistic regression equation and the first-order Taylor expansion of the solution equation for x_Q (see, e.g., Morgan, 1992). The determination of LOQ for this equation is illustrated in Figure 4.5.

Range – Another validation parameter is the range of assay results. When the measurement of an analyte is obtained from multiple runs of an assay, these results can be examined by noting the range, which is defined as either log(max)-log(min) or max-min depending upon whether the measurements are scaled proportionally or absolutely. Range is commonly considered as a tool for analytical method quality control of bias and variability, which are often used to bridge to the assay validation and to warrant the continued reliability of the analytical method. For this purpose, a bound on the range is often derived as follows

$$\text{Range} \le \hat{\sigma}_{run}(d_2 + kd_3), \tag{4.11}$$

where $\hat{\sigma}_{run}$ represents run-to-run variability, d_2 and d_3 are the factors used to scale the standard deviation to a range, and k is a factor indicating the desired degree of control (e.g., $k = 2$ indicates approximately 95% confidence).

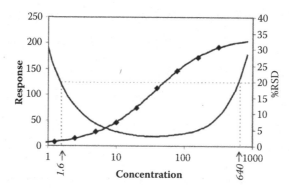

FIGURE 4.5
Quantifiable range determined as limits within which assay variability (%RSD) is less than or equal to 20%.

In this case, extra variability in replicates can be detected utilizing standard statistical processes for quality control.

In the case where extra variability is detected, additional measurements can be obtained and a result for the sample can be determined from the remainder of the measurements made after the maximum and minimum results have been eliminated from the calculation. When extra variability has been identified in instrument replicates of a standard concentration or a dilution of a test sample, that concentration of the standard or dilution of the test sample might be eliminated from subsequent calculations.

Measurements on control sample(s) help identify shifts in scale and therefore bias in a run of an assay. Information in multiple control samples can typically be reduced to one or two meaningful scores, such as an average and/or a slope. These can help isolate problems and provide information regarding the cause of a deviation in a run of an assay. Suppose, for example, that three controls of various levels are used to monitor shifts in sensitivity in the assay. The average can be used to detect absolute shifts, because of some underlying source of constant variability, such as inappropriate preparation of standards or the introduction of significant background. The slope can be used to detect proportional shifts as a result of some underlying source of proportional variability.

Suitability criteria on the performance of column peaks can detect degeneration of a column, while characteristics of the standard curve, such as the slope or EC_{50} (EC_{50} corresponds to C in a four-parameter logistic regression) help predict an assay with atypical or substandard performance. Care must be taken, however, to avoid either meaningless or redundant controls. The increased false-positive rate inherent in the multiplicity of a control strategy must be considered, and a carefully contrived quality control scheme should acknowledge this.

4.5 Evaluation of Reliability, Repeatability, and Reproducibility

As mentioned in the Introduction section, a validated analytical method is expected to achieve a certain degree of accuracy, reliability, and reproducibility However, it is a concern whether the test results obtained from a validated testing procedure are repeatable (with similar test samples) and/or reproducible (under similar but slightly different experimental conditions). Salah, Chow, and Song (2017) proposed a method a method to assess reliability, repeatability, and reproducibility of an analytical method by monitoring relevant variability control charts under a mixed effects nested design. The proposed method was motivated based on the concept of empirical power (reproducibility) proposed by Shao and Chow (2002) to determine acceptance limits of variability control charts and consequently to ensure that there is a high probability of repeatability and reproducibility. Salah, Chow, and Song's proposed method is briefly outlined in the following subsequent subsections.

4.5.1 Study Design and Statistical Model

In practice, study designs that are commonly considered include crossed design (e.g., two-way crossed design with and/or without interactions), nested design (e.g., one design factor is nested within another design factor), crossed-then-nested design, and nested-then-crossed design (see, e.g., JMP, 2012). In this section, consider a three factor mixed effects nested design. Let A, B, and C denote factor A with $i = 1, \ldots, a$ levels, factor B with $j = 1, \ldots, b$ levels, and factor C with $k = 1, \ldots, c$ levels. In the three factors mixed-effects nested design, factor B is nested within factor A and factor C is nested within factor B. Under this study design, the following statistical model is useful in describing the data collected from the study design using the SAS software

$$y_{ijkl} = \mu + \alpha_i + \beta_{(i)j} + \gamma_{(ij)k} + e_{ijkl},$$
$$i = 1, \ldots, a; \ j = 1, \ldots, b; \ k = 1, \ldots, c; \ l = 1, \ldots, n, \tag{4.12}$$

where y_{ijkl} is the lth test result observed at the kth level of factor C, which is nested within the jth level of factor B, which is in turn nested within the ith level of factor A; μ is the overall mean; α_i is the ith fixed effect of factor A; $\beta_{(i)j}$ represents the jth random effect that nested in the ith level of factor A; $\gamma_{(ij)k}$ is the kth random effect that nested within the ith level of factor A and the jth level of factor B; e_{ijkl} is the random error in observing y_{ijkl}. Under model (4.12), analysis of variance (ANOVA) table is given in Table 4.5.

TABLE 4.5

ANOVA of the Three Factor Nested Design

Factor	df	SS	MS	E(MS)
A	a-1	SSa	SSa/(a-1)	$\sigma_e^2 + abc\sigma_A^2 + bc\sigma_B^2 + n\sigma_C^2$
B	a(b-1)	SSb	SSb/a(b-1)	$\sigma_e^2 + bc\sigma_B^2 + n\sigma_C^2$
C	ab(c-1)	SSc	SSc/ab(c-1)	$\sigma_e^2 + n\sigma_C^2$
Error	abc(n-1)	SSe	SSe/abc(n-1)	σ_e^2
Total	abcn-1			

Abbreviations: df = degree of freedom; E(MS) = expected mean sum of squares; MS = mean sum of squares; SS = sum of squares.

As it can be seen from Table 4.5, variance components such as σ_e^2, σ_A^2, σ_B^2, and σ_C^2 can be estimated from the ANOVA table as follows

$$\hat{\sigma}_C^2 = \frac{1}{n}\left[\frac{SSc}{ab(c-1)} - \frac{SSe}{abc(n-1)}\right],$$

$$\hat{\sigma}_B^2 = \frac{1}{bc}\left[\frac{SSb}{a(b-1)} - \frac{SSc}{ab(c-1)}\right],$$

$$\hat{\sigma}_A^2 = \frac{1}{abc}\left[\frac{SSa}{a-1} - \frac{SSb}{a(b-1)}\right],$$

$$\hat{\sigma}_e^2 = \frac{SSe}{abc(n-1)},$$

where

$$SSa = bcn\sum_{i=1}^{a}(\bar{y}_i\ldots - \bar{y}\ldots.)^2,$$

$$SSb = cn\sum_{i=1}^{a}\sum_{j=1}^{b}(\bar{y}_{ij}.. - \bar{y}_i\ldots)^2,$$

$$SSc = n\sum_{i=1}^{a}\sum_{j=1}^{b}\sum_{k=1}^{c}(\bar{y}_{ijk}. - \bar{y}_{ij}..)^2,$$

$$SSe = \sum_{i=1}^{a}\sum_{j=1}^{b}\sum_{k=1}^{c}\sum_{l=1}^{n}(\bar{y}_{ijkl} - \bar{y}_{ijk}.)^2,$$

where

$$\bar{y}\ldots. = \frac{1}{abcn}\sum_{i=1}^{a}\sum_{j=1}^{b}\sum_{k=1}^{c}\sum_{l=1}^{n}y_{ijkl},$$

$$\bar{y}_i\ldots = \frac{1}{bcn}\sum_{j=1}^{b}\sum_{k=1}^{c}\sum_{l=1}^{n}y_{ijkl},$$

$$\bar{y}_{ij}.. = \frac{1}{cn}\sum_{k=1}^{c}\sum_{l=1}^{n}y_{ijkl},$$

$$\bar{y}_{ijk}. = \frac{1}{n}\sum_{l=1}^{n}y_{ijkl}.$$

Restricted maximum likelihood (REML) can be used to estimate variance components. A boundary constraint is put on variance component estimates to avoid negative values.

4.5.2 Variability Monitoring

For monitoring reliability, repeatability, and reproducibility, of a given instrument (machine), percentage of variation for each source of variability is calculated using as a control of the process variability (see, e.g., JMP, 2012). In absence of process variability, tolerance limits may be defined according to historical observed variabilities. If the observed variability is within this, then we claim that process is in control (i.e., the measurement system is reliable and repeatable/reproducible). If the variability is outside the tolerance interval, we claim that the measurement system is out of control and corrective action should be made to bring the variability back within this interval.

In practice, the selection of a tolerance interval according to historical data has been criticized lack of statistical justification. In addition, the availability of process variability to play the role of a control is rare. Alternatively, we may consider selecting an acceptance limit based on the following concept of reproducibility proposed by Shao and Chow (2002). As an example, suppose we are interested in selecting an acceptance limit for monitoring variability associated with factor C, i.e., σ_C^2. We then calculate the following empirical power

$$P_C = P\left\{\text{significant result} \mid \hat{\sigma}_C^2 \equiv \sigma_C^2\right\}.$$

In other words, we calculate the power that $\hat{\sigma}_C^2 < \sigma_{C0}^2$ by assuming that the observed $\hat{\sigma}_C^2$ is true σ_C^2. As a result, the acceptance limit σ_{C0}^2 can be chosen in order to have a desired empirical power (i.e., reproducibility) P_{C0}. Following a similar idea, acceptance limits for other variance components such as σ_A^2, σ_B^2, and σ_e^2 can be selected.

4.5.3 Sample Size for Comparing Variabilities

For testing equality in total variabilities between products, i.e., a test product (T) and a reference product (R), the following hypotheses are often considered

$$H_0 : \sigma_T^2 = \sigma_R^2 \quad \text{vs.} \quad H_a : \sigma_T^2 \neq \sigma_R^2, \tag{4.13}$$

where σ_T^2 and σ_R^2 are the total variabilities associated with the test product and the reference product, respectively. Under the null hypothesis of (4.13), it can be verified that the test statistic

$$T = \frac{\hat{\sigma}_T^2}{\hat{\sigma}_R^2} \tag{4.14}$$

is distributed as an F random variable with $n_T - 1$ and $n_R - 1$ degrees of freedom. Thus, we would reject H_0 at the α level of significance if

$$T > F_{\alpha/2, n_T-1, n_R-1} \text{ or } T < F_{1-\alpha/2, n_T-1, n_R-1}$$

Under the alternative hypothesis, without loss of generality, we assume that $\sigma_T^2 < \sigma_R^2$, i.e., $\sigma_T^2 + \sigma_0^2 = \sigma_R^2$, where σ_0^2 is considered a scientifically/clinically meaningful difference. As a result, the power of the test (4.14) can be evaluated at $\sigma_T^2 - \sigma_R^2 = \sigma_0^2$ under the alternative hypothesis, which is given by

$$Power = P\left\{ T < F_{1-\alpha/2, n_T-1, n_R-1} \right\} = P\left\{ F_{n_T, n_R} > \frac{\sigma_T^2}{\sigma_R^2} F_{\alpha/2, n_T-1, n_R-1} \right\}.$$

Under the assumption that $n_T = n_R = n$ and fixed σ_T^2 and σ_R^2, the sample size needed for achieving a desired power of $1 - \beta$ can be obtained by solving the following equation

$$\frac{\sigma_T^2}{\sigma_R^2} = \frac{F_{1-\beta, n-1, n-1}}{F_{\alpha/2, n-1, n-1}} \tag{4.15}$$

Similarly, equations for sample size calculation for testing non-inferiority/superiority and equivalence can be obtained. The results are summarized in Table 4.6.

TABLE 4.6

Sample Size Required for Comparing Variabilities

Test For	Sample Size Equation
Equality	$\dfrac{\sigma_T^2}{\sigma_R^2} = \dfrac{F_{1-\beta, n-1, n-1}}{F_{\alpha/2, n-1, n-1}}$
Non-inferiority/ Superiority	$\dfrac{\sigma_T^2}{\delta^2 \sigma_R^2} = \dfrac{F_{1-\beta, n-1, n-1}}{F_{\alpha, n-1, n-1}}$
Equivalence	$\dfrac{\delta^2 \sigma_T^2}{\sigma_R^2} = \dfrac{F_{\beta/2, n-1, n-1}}{F_{1-\alpha, n-1, n-1}}$

Note: $n_T = n_R = n$; δ is non-inferiority margin or equivalence limit.

4.5.4 An Example

Suppose a sponsor is interested in conducting a study for evaluation of reliability and repeatability/reproducibility of a testing procedure for testing hair fibers post treatment. The study objective is to determine the effect of a test treatment on hair fiber coloration, which is measured by a machine. The machine consists of two robots. Each robot can handle four plates and each plate consists of two wells. In each well, two measures are done. Thus, the study design is considered a nested design. In other words, PLATE is nested within ROBOT and WELL is nested within ROBOT and PLATE. As a result, the study design is a typical three factor (factor A is ROBOT, factor B is PLATE, and factor C is WELL) nested design as described in model (4.12). In this case, the following SAS model can be used to analyze the data collected under the study design:

$$Response = Overall\ Mean + Robot + Plat(Robot) + Well(Robot, Plate) + Error \quad (4.16)$$

Thus, there are four variance components: inter-robot (i.e., robot-to-robot) variability, inter-plate (i.e., plate-to-plate) within robot variability, inter-well (i.e., well-to-well) within plate (within the same robot) variability, and random error. Denote by σ_R^2, σ_p^2, σ_w^2, and σ_e^2 robot-to-robot variability, plate-to-plate (within robot) variability, well-to-well (within plate by the same robot) variability, and random error, respectively. Thus, σ_e^2 is a measure of precision (reliability) of the test procedure, σ_w^2 is an indicator of repeatability of the test procedure, and $\sigma_R^2 + \sigma_p^2$ determines whether the test result is reproducible. Let p_1, p_2, and p_3 be the desirable reliability, repeatability, and reproducibility of the test procedure, respectively. Also, let σ_{01}^2, σ_{02}^2, and σ_{03}^2 be the upper limit for achieving the desirable reliability, repeatability, and reproducibility respectively. Thus, we have

$$p_i = P\left\{ \text{significant result} \mid \hat{\sigma}_i^2 = \sigma_{0i}^2 \right\} \quad (4.17)$$

where $i = 1$ (reliability), 2 (repeatability), and 3 (reproducibility), $\hat{\sigma}_1^2 = \sigma_e^2$, $\hat{\sigma}_2^2 = \hat{\sigma}_w^2$, and $\hat{\sigma}_3^2 = \hat{\sigma}_R^2 + \hat{\sigma}_p^2$, respectively. Under model (4.16), estimates of variance components can be obtained. Thus, for a given data set $\left(\text{i.e., } \hat{\sigma}_i^2 \right)$ and pre-specified p_i, we can solve σ_{0i}^2 by using the above equation (see, e.g., Shao and Chow, 2002). Note that under model (4.16), commonly considered estimates for variance components are estimates obtained from the analysis of variance (ANOVA).

Under the three factors nested design, test results for hair fibre are given in Table 4.7. The hair fibre data were analyzed using SAS PROC GLM procedure under model (4.16). The results are summarized in Table 4.8. Following expected values of mean squares given in Table 1, the restricted maximum

TABLE 4.7

Hair Fibre Example Data

Plate	Well	Robot No. 1		Plate	Well	Robot No. 2	
P305	A	29.83	28.70	P264	A	30.10	27.98
	B	29.05	30.77		B	29.01	28.35
P345	A	28.12	30.00	P335	A	33.57	30.59
	B	28.24	28.40		B	28.92	29.24
P517	A	29.64	30.44	P490	A	30.89	28.52
	B	29.75	29.52		B	28.51	29.98
P656	A	29.32	31.63	P502	A	30.89	30.12
	B	29.54	31.61		B	29.79	31.28

TABLE 4.8

ANOVA Table for Hair Fiber Data

Source	DF	Sum Squares	Mean Square	F Value	Pr > F
Robot	1	0.316	0.316	0.24	0.633
Plate	6	15.312	2.552	1.92	0.140
Well(Robot*Plate)	8	10.480	1.310	0.98	0.483
Error	16	21.306	1.332		

likelihood (REML) estimates for variance components of $\hat{\sigma}_1^2 = \sigma_e^2$, $\hat{\sigma}_2^2 = \hat{\sigma}_w^2$, and $\hat{\sigma}_3^2 = \hat{\sigma}_R^2 + \hat{\sigma}_p^2$ can be obtained as follows

$$\hat{\sigma}_1 = 0.5737, \hat{\sigma}_3 = 0.0013, \text{ and } \hat{\sigma}_3 = 0.7016.$$

Assuming that the observed variance components due to ROBOT, PLATE nested within ROBOT, and WELL nested within ROBOT and PLATE are the true values, for a given sample size, the limits that guarantee there is a desired reliability, repeatability, and reproducibility (say 90%) for detecting a clinically meaningful difference Δ can be obtained by solving the empirical power function for the corresponding variance components (4.17). Table 4.9 gives the

TABLE 4.9

Acceptance Limits for Assurance of Reliability, Repeatability, and Reproducibility under the Three-Factor Nested Design for the Hair Fiber Study

Δ	Reliability (σ_{20})	Repeatability (σ_{20})	Reproducibility (σ_{30})
3%	1.355	1.355	2.119
4%	2.409	2.409	3.768
5%	3.764	3.764	5.887
Observed	0.574	0.001	0.702

Note: under the three-factor nested design, the degrees of freedom for evaluation of reliability, repeatability, and reproducibility are 16, 8, and 7, respectively.

acceptance limits for assurance of reliability, repeatability, and reproducibility under the three-factor nested design for the hair fibre study.

For evaluation of reliability and repeatability, if the study objective is to detect a clinically meaningful difference of 3.0%, then the observed variability should be less than 1.355 (i.e., $\hat{\sigma}_1 = \hat{\sigma}_e \leq \sigma_{20} = 1.355$) in order to have a 90% reliability. In the hair fiber example, we observed $\hat{\sigma}_1 = \hat{\sigma}_e = 0.574$ which is not lower than σ_{20} with 80% power. This indicates that the reliability for detecting a 3.0% difference is lower than 90%. Since $\hat{\sigma}_1 = \hat{\sigma}_e = 574$ is less than $\hat{\sigma}_{20} = 2.409$ with 80% power, there is at least 90% reliability for detecting a difference of 4% with an observed variance significantly lower than 2.409.

For evaluation of reproducibility, if the study objective is to detect a clinically meaningful difference of 4.0%, then the observed variability should be less than 3.768 (i.e., $\hat{\sigma}_3 = \hat{\sigma}_P + \sigma_R \leq \sigma_{30} = 3.768$) in order to have a 90% reliability.

Note that JMP (2012) suggested the use of various variability charts to monitoring quality and reliability of an instrument for testing critical quality attributes. The variability charts can be further used to evaluate repeatability and reproducibility by examining whether the observed variabilities fall within some pre-specified acceptance limits for repeatability and reproducibility or whether these variabilities are small enough compared to the measured process variability.

To ensure there are high (desirable) probabilities for repeatability and reproducibility, it is suggested that the acceptance limits be chosen based on the concepts of repeatability and reproducibility probabilities as described in Shao and Chow (2002). In other words, the acceptance limits are chosen in such a way that there is high (desirable) probability of repeatability and reproducibility if the observed variabilities are less than the acceptance limits.

It should be noted that the acceptance limits based on the repeatability/reproducibility probability are random variables. The proposed acceptance limits for evaluation of repeatability and reproducibility are derived based on the observed variance components due to design factors which depend upon sample size under the combinations of the study factors. The properties and performances of these acceptance limits in monitoring (evaluating) reliability, repeatability, and reproducibility require further research.

4.6 Concluding Remarks

A carefully conceived assay validation should follow assay development, and precede routine implementation of the method for product characterization. The purpose of the validation is to demonstrate that the assay is fundamentally reliable, and possesses operating characteristics that make it suitable for use. The statistical estimates collected during the validation can also be used to design a test plan, that provides the maximum reliability for the minimum effort in the laboratory. Most importantly, validation of the assay does not end with the documentation of the validation experiment but should continue with assay quality control. The proper choice of control samples and control parameters helps assure the continued validity of the procedure during routine use in the laboratory.

Assay validation plays an important role in biosimilar product development as many biosimilar drug products are known to be sensitive to environmental factors such as light and temperature at various stages of manufacturing process. A small change or variation of critical quality attribute may translate to a significant change in clinical outcomes (in terms of safety and efficacy) of the final product. Assay validation is key to the success of analytical similarity assessment. Adequate assay validation cannot only provide accurate and reliable assessment of biosimilarity between a proposed biosimilar product and a US-licensed reference product, but also quality control and assurance of the raw material, in-process material, and end product of biosimilar products.

5

Critical Quality Attributes

5.1 Background

ICH Q8 (R2) defines a critical quality attribute (CQA) as a physical, chemical, biological, or microbiological property or characteristic that should be within an appropriate limit, range, or distribution to ensure the desired product quality (ICH, 2006). Thus, CQAs identified at various stages of a manufacturing process collectively indicate whether the manufacturing process delivers product that meets its quality target product profile (QTPP). In practice, the identification of CQAs allows for the evaluation and control of product characteristics that have impact on quality, purity, safety and efficacy. The identification of CQAs is usually performed through the conduct of a criticality assessment that evaluates the risk and benefit associated with each attribute. The study of CQAs has led to the initiative of quality by design (QbD) concept for (i) meeting desired product quality, (ii) achieving further improvement in product quality, and (iii) enhancing quality control and assurance in the manufacturing process.

In its guidance on *Scientific Considerations in Demonstrating Biosimilarity to a Reference Product*, the FDA recommends a stepwise approach for obtaining totality-of-the-evidence for demonstrating biosimilarity between a proposed biosimilar product and an innovative biologic drug product (FDA, 2015a). The stepwise approach starts with the assessment of analytical similarity of CQAs for structural and functional characterization in the manufacturing process of biosimilar products that are relevant to clinical outcomes and consequently may have an impact on assessment of biosimilarity. For biosimilar product development, it is well recognized that small changes in identified CQAs might have a significant impact on product safety, purity, and efficacy. Thus, one concern is how changes in identified CQAs translate to changes in clinical outcomes? To address this question, it is critical to establish the relationships between CQAs and clinical outcomes (safety and efficacy). The information can help in determining the criticality of the CQAs relevant to the clinical outcomes. In practice, however, it can be difficult to link a specific CQA to clinical outcomes (safety and efficacy). FDA suggests that the identification of CQAs and the determination of criticality of the identified CQAs that are relevant to clinical outcomes should be done based on mechanism of action (MOA) and/or pharmacokinetics (PK) and pharmacodynamics (PD) under an appropriate statistical model.

In practice, there is often a large number of CQAs that may be relevant to clinical outcomes. Thus, it is almost impossible to assess analytical similarity for all of these CQAs individually. As a result, FDA suggests that the sponsors identify CQAs that are relevant to clinical outcomes and classify them into three tiers depending upon their criticality or risk ranking, i.e., most relevant (Tier 1), mild to moderately relevant (Tier 2), and least relevant (Tier 3) to clinical outcomes. To assist the sponsors, FDA also proposes some statistical approaches for the assessment of analytical similarity for CQAs from different tiers. For example, FDA recommends equivalence tests for CQAs from Tier 1, a quality range approach for CQAs from Tier 2, and descriptive raw data and graphical presentation for CQAs from Tier 3 (see, e.g., Cristl, 2015; Tsong, 2015; Chow, 2014; Chow, 2015).

In the next section, several statistical methods for identification of CQAs via the study of the link between CQAs and clinical outcome are briefly described. Stepwise approach for demonstrating biosimilarity is reviewed in Section 5.3. Under a well-established relationship between CQAs and clinical outcomes, a couple of proposed criteria for classification of identified CQAs based on its criticality and/or risking raking are discussed in Section 5.4. Some concluding remarks are given in Section 5.5.

5.2 Identification of CQAs

To identify CQAs in a manufacturing process of biosimilar drug products, the link (or relations) between CQAs and clinical outcomes in terms of safety, purity, and potency (efficacy) are necessarily established. FDA suggested that CQAs should be linked with mechanism of action (MOA) and/ or pharmacokinetics (PK) and pharmacodynamics (PD) that relate to clinical outcomes of the product under study. For this purpose, several models have been proposed in the literature in order to understand the relationships between CQAs and product safety, purity, and efficacy.

5.2.1 Link between CQAs and Clinical Outcomes

The purpose of the study of the link between CQAs and clinical outcomes is multi-fold. First, it is to establish a predictive model between clinical outcomes (response variable) and CQAs (explanatory variables or predictors). In this case, we will be able to assess of the impact on clinical outcomes with changes in CQAs under the established predictive model. Second, under the established predictive model, acceptable ranges of changes in CQAs can be derived for given similarity margins in clinical outcomes in terms of safety and efficacy. Third, the acceptable ranges (limits) of the CQAs can be used for developing in-house specifications and/or release targets for the purposes of quality control and assurance in the manufacturing process of the proposed biosimilar product.

For example, based on PK/PD, Yang (2013) proposed a method to study a functional relationship between three correlated CQAs of a monoclonal antibody and a PK surrogate efficacy marker, i.e., area under the drug concentration curve (AUC). A lower (upper) limit on the AUC which is considered acceptable to regulatory agency, can be translated into an acceptable limit on changes in these CQAs that are allowed for assurance of efficacy (safety) of the final product. Thus, Yang's proposed model not only allows for the assessment of the impact of changes in these CQAs on the product safety/efficacy, but also enables the establishment of clinically relevant joint acceptance ranges for the CQAs.

For another example, Schenerman et al. (2009) considered an impurity safety factor (ISF) to link the toxicity of an impurity to the maximum acceptable level in the product based on the amount of impurity through the following function:

$$ISF = LD_{50} \div r, \qquad (5.1)$$

where LD_{50} is the amount of an impurity that results in lethality in 50% of animals tested, and r is the maximum amount of an impurity in the target dose of the product. As it can be seen from (5.1), the higher the ISF, the lower the safety risk associated with the impurity is. As a result, a lower limit on the ISF which is considered acceptable to a regulatory agency can be translated into an acceptable limit on the maximum amount of impurity that is allowed in the final target dose of the proposed biosimilar product.

5.2.2 Statistical Design and Methods

Statistical design of experiments – For quality control and assurance in a manufacturing process, a traditional approach is to consider a so-called one-factor-at-a-time (OFAT) study. In other words, the OFAT approach considers one CQA at a time. This approach is inefficient and time consuming because (i) there is often a large number of CQAs that might be related to clinical outcomes and (ii) not all the process parameters ad material attributes have a significant impact on CQAs. Thus, a statistical design experiment is necessarily implemented to ensure efficiency for obtaining the necessary information regarding the effects of process parameters and input material attributes on the CQAs in order to establish a design space for achieving spirit of quality by design as described in Chapter 3.

It should be noted that prior to the characterization studies, a risk analysis is usually performed to select experimental factors that have high or medium impact on CQAs. This information in conjunction with other historical data and manufacturing experience and can serve as the basis for establishing a design space, product specifications, and manufacturing control strategies.

Statistical Methods – For determining design space, several statistical methods based on multivariate regression analysis have been proposed in the literature. In this section, we will focus on two traditional approaches: overlapping mean response surfaces and the composite desirability function. Using these approaches, a design space can be determined to be either a multivariate region in which the mean response values of CQAs are within specifications, or a region in which the composite desirability function exceeds a pre-specified threshold. These traditional methods, however, do not take into account correlations among CQAs, nor the variability associated with prediction in the predictive model. In practice, to provide high assurance of product quality, it is suggested that an appropriately constructed design space should be able to account for measurement uncertainties, uncertainty about the parameters of the statistical models, and correlations among the measurements. In recent years, Bayesian multivariate analysis techniques have been proposed for design space development based on experiments and data obtained from small-scale processes. However, how to update such design spaces based on data from pilot and full-scale data still remains unresolved and needs further research. In what follows, the methods of overlapping mean response surfaces and Bayesian approach are briefly described.

Overlapping Mean Response Surfaces – Let $Y = (Y_1, ..., Y_p)'$ be a $p \times 1$ vector of measures of p CQAs, A the set of joint acceptable ranges of the CQAs, and x a $k \times 1$ vector of process parameters and other controllable inputs, such as material attributes. Suppose that the relationship between Y and x can be characterized through the model

$$Y = f(x; \theta) + \varepsilon$$

where $f(x; \theta)$ is a mathematical function, θ is model parameters, and ε is measurement errors. Let $\hat{\theta}$ be the vector of estimates of the model parameters. Therefore, the estimate of expected mean values at x are given by $\hat{E}[Y \mid x] = f(x; \hat{\theta})$. The design space based on the overlapping mean approach is defined as

$$\left\{ x : \hat{E}[Y \mid x] \in A \right\}.$$

Consider an example regarding the performance of a liquid chromatography method (Perterson, 2004, 2009). Data were collected from a Box-Behnkin experiment, which has three factors, $X = (x_1, x_2, x_3) = $ (% isopropyl, temperature, pH), and four responses, $Y = (y_1, y_2, y_3) = $ (resolution, run time, signal to noise ratio, tailing). For this assay, the acceptance region A for the responses is given as

$$A = \left\{ y : y_1 \geq 1.8, y_2 \leq 15, y_3 \geq 300, 0.75 \leq y_4 \leq 0.85 \right\}.$$

We may fit the data using a simple regression model as follows:

$$Y = z(x)\theta + \varepsilon$$

where

$$z(x) = (1, x_1, x_2, x_3, x_1^2, x_2^2, x_3^2, x_1 x_2),$$

$$\theta = \begin{pmatrix} \theta_0^{(1)} & \cdots & \theta_0^{(4)} \\ \vdots & \ddots & \vdots \\ \theta_7^{(1)} & \cdots & \theta_7^{(4)} \end{pmatrix},$$

and ε is normally distributed with mean 0 and covariance matrix Σ Peterson (2004) indicated that there are several serious drawbacks of this approach. First, the method does not account for uncertainties in the model parameter estimates. Second, it ignores potential correlations among the responses. Third, it does not quantify the level of assurance for meeting specifications when the parameters are inside the design space (see also Yang, 2013).

Bayesian Approach – To overcome the drawbacks of the overlapping mean response surfaces, alternatively, Peterson (2008) suggested a Bayesian approach should be used to establish a design space. A Bayesian design space and various extensions were also studied by Lebrun (2012). In general terms, a Bayesian design space is defined as

$$\left\{ x : \Pr[Y \in A \,|\, x, data] \geq R \right\} \tag{5.2}$$

where Y, A, and x are defined as before, "Pr" stands for posterior predictive probability, *data* is the data set from a controlled experiment, which includes measured values of the CQA(s) and various settings of the process parameters, and R is a pre-selected level of reliability that must be met.

As it can be seen from (5.2), the posterior predictive probability accounts for the uncertainty in the model parameters and correlation among the response variables. Thus, it overcomes the drawbacks of the overlapping mean response surface method. In addition, the Bayesian method can be easily extended to accommodate many different types of experiments, such as split-plot, as well as experiments involving mixed effects. The Bayesian design space can be determined by estimating the probability $\Pr[Y \in A | x, data]$ over a grid of x values. The posterior predictive probability can be estimated either through a closed-form solution or by Markov Chain Monte Carlo (MCMC) simulations.

NOTE: that under the Bayesian framework, several regression models can be used to describe the measured responses of the CQAs (Y). For example, the following model can be used

$$Y = Bz(x) + e \tag{5.3}$$

where B is a $p \times q$ matrix of regression coefficients, z(x) is a $q \times 1$ vector function of x, and e is a $p \times 1$ vector of measurement errors having a multivariate normal distribution with mean 0 and covariance-variance matrix Σ.

NOTE: that in (5.3), it is assumed that z(x), which is the same for each CQA though the method described in this section, can be extended to the seemingly unrelated regressions (SUR) model (Peterson, 2006), in which each response $Y_i (i = 1, ...,p)$ has a different function $z_i(x)$. Under such circumstances, the SUR model provides greater flexibility and accuracy in modeling the CQAs (Peterson, 2007). A SUR model takes the form:

$$Y_j = z_j(x)'\beta_j + e_j, \quad j = 1,...,p$$

The SUR model includes the standard multivariate regression model as a special case where $z_j(x) = z(x)$. A design space based on the SUR model and posterior predictive probability can be similarly obtained.

For the Bayesian approach, the selection of prior information regarding process/method parameters, CQAs, and so on is a critical which has an impact on the resultant Bayesian inference. Peterson (2008) discussed various options in which prior information can be elicited for a design space model. For example, an informative prior distribution may be established from experiments done at the pilot scale, which is conducted for developing a design space for future large-scale manufacturing. In practice, however, data obtained from pilot small-scale experiments may not be sufficient for constructing informative priors. Peterson (2008) proposed the potential use of three additional sources of information for specifying informative prior distributions.

Under the design space, posterior predictive probability can be obtained as follows. Let $Y_{obs} = (Y_1, \dots, Y_n)'$ be an $n \times p$ matrix consisting of n observations of the $1 \times p$ response vector Y. Let $Z_{exp} = (Z_1, \dots, Z_2)'$ be an $n \times q$ matrix with $Z_i = z(x_i)$, $i = 1, \dots, n$, and let x_i be the ith condition of the controllable input variables and process parameters. Assuming that a non-informative prior is used to describe the parameters (B, Σ):

$$p(B, \Sigma) \propto | \Sigma |^{-(p+1)/2}. \tag{5.4}$$

It can be verified that the posterior predictive distribution of a future observation $\tilde{Y}$ in this case is a multivariate-t with degrees of freedom $v = n - p - q + 1$:

$$\tilde{Y} \,|\, x, data \sim t_v\left(\hat{B}z(x), H\right) \tag{5.5}$$

where

$$H = \left[1 + z(x)'D^{-1}z(x)\right]\hat{\Sigma}],$$

$$D = \sum_{i=1}^{n} Z_i Z_i',$$

and $\hat{B}, \hat{\Sigma}$ are the least squares estimates of B and Σ given by

$$\hat{B} = (Z_{exp}' Z_{exp})^{-1} Z_{exp}' Y_{obs}$$

$$\hat{\Sigma} = [Y_{obs} - (\hat{B}Z)']'[Y_{obs} - (\hat{B}Z)']/v \quad \text{with } v = n - p - q + 1.$$

Thus, the posterior predictive distribution, which is a multivariate t-distribution, can be simulated as follows:

Step 1. Draw W from the multivariate normal distribution $N(0, H)$;

Step 2. Draw U from the chi-square distribution χ_v^2;

Step 3. Calculate $Y_j = \sqrt{v}W_j/\sqrt{U} + \hat{\mu}_j$, for $j = 1, \dots, p$ where Y_j, W_j, and $\hat{\mu}_j$ are the jth elements of Y, W, and $\hat{B}z(x)$, respectively.

As indicated by Peterson (2009), the posterior predictive probability can be approximated using Monte Carlo simulation as follows:

$$p(x) = \Pr\left[Y \in A \,|\, x, data\right]$$

$$\approx \frac{1}{N} \sum_{s=1}^{N} I(Y^{(s)} \in A),$$

where $Y^{(s)}$, $s = 1, \ldots, N$, are independent random multivariate-t variables simulated from the above-mentioned procedure, and $I(\cdot)$ is an indicator function taking values of either 0 or 1. The design space can be obtained by estimating the posterior predictive probability $p(x)$ over a grid of x values and comparing it to the reliability threshold R.

Alternatively, as discussed previously, informative prior distributions may be used in the derivation of the posterior predictive probability. For example, one may use conjugate prior distributions for $p(B|\Sigma)$ and $p\Sigma$:

$$B \,|\, \Sigma \sim N_{p \times q}(B_0, \Sigma, \Sigma_0),$$
$$\Sigma \sim W_1^{-1}(\Omega, v_0),$$

where B_0 is the mean vector and $\Sigma \, \Sigma_0$ are the covariance matrices of the columns and rows of B, respectively; Σ follows an inverse-Wishart distribution with Ω being an *a priori* response scale matrix, and v_0 the degrees of freedom.

Lebrun (2012) notes that the posterior predictive probability in this case is also a multivariate t-distribution. Thus, using the Monte Carlo simulation procedure previously described, the Bayesian design space can be constructed.

5.3 Stepwise Approach for Demonstrating Biosimilarity

As defined in the BPCI Act, a biosimilar product is a product that is *highly similar* to the reference product notwithstanding minor differences in clinically inactive components and there are no clinically meaningful differences in terms of safety, purity, and potency. Based on the definition of the BPCI Act, biosimilarity requires that there are no *clinically meaningful differences* in terms of safety, purity, and potency. Safety could include pharmacokinetics and pharmacodynamics (PK/PD), safety and tolerability, and immunogenicity studies. Purity includes all critical quality attributes during the manufacturing process. Potency is referred to as efficacy studies. As indicated earlier, in the 2015 FDA guidance on scientific considerations, FDA recommends that a stepwise approach be considered for providing the totality-of-the-evidence to demonstrating biosimilarity of a proposed biosimilar product as compared to a reference product (FDA, 2015a).

The stepwise approach is briefly summarized by a pyramid illustrated in Figure 5.1. The stepwise approach starts with analytical studies for structural and functional characterization. The stepwise approach continues with animal studies for toxicity, clinical pharmacology studies such as PK/PD

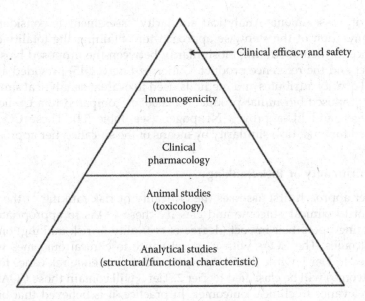

Clinical efficacy and safety

Immunogenicity

Clinical pharmacology

Animal studies (toxicology)

Analytical studies (structural/functional characteristic)

FIGURE 5.1
Stepwise approach for biosimilar product development.

studies, followed by investigations of immunogenicity, and clinical studies for safety/tolerability and efficacy.

The sponsors are encouraged to consult with medical/statistical reviewers of FDA with the proposed plan or strategy of the stepwise approach for regulatory agreement and acceptance. This is to make sure that the information provided is sufficient to fulfill the FDA's requirement for providing totality-of-the-evidence for the demonstration of biosimilarity of the proposed biosimilar product as compared to the reference product. As an example, more specifically, the analytical studies are to assess similarity in CQAs at various stages of the manufacturing process of the biosimilar product as compared to those of the reference product. To assist the sponsors to fulfill the regulatory requirement for providing totality-of-the-evidence of analytical similarity, the FDA suggests several approaches depending upon the criticality of the identified quality attributes relevant to the clinical outcomes.

5.4 Tier Assignment for Critical Quality Attributes

Tsong (2015) indicated that critical quality attributes (CQAs) are necessarily tested for the functional, structural, and physicochemical characterization of the proposed biosimilar product as compared to a reference product (either a US-licensed product or an EU-approved reference product) for analytical similarity assessment. Analytical similarity assessment is considered as the foundation of the stepwise approach for obtaining the totality-of-the-evidence for demonstrating biosimilarity between the proposed biosimilar product and the reference product. Gutierrez-Lugo (2015) provided a list of critical quality attributes and methods used to evaluate analytical similarity of the proposed biosimilar product (EP2006) as compared to a US-licensed Neupogen and EU-approved Neupogen (see Table 5.1). These CQAs are assessed for analytical similarity by means of the so-called tier approach.

5.4.1 Criticality or Risk Ranking

The tier approach first assesses the criticality or risk ranking of the CQAs relevant to clinical outcome and classify these CQAs to appropriate tiers depending upon their impact (degree of criticality or risk ranking) on clinical outcomes. The CQAs with most relevance to clinical outcomes will be assigned to Tier 1, while the CQAs with mild-to-moderate relevance to clinical outcomes will be classified to Tier 2. Tier 3 will contain those CQAs with least relevance to clinical outcomes. In practice, it is believed that biological activity assays are the best representation available to test the clinically relevant mechanism of action (MOA) and therefore should be assigned to Tier 1. Other CQAs which are tested in comparative physicochemical and functional assessment (outside of those relevant to MOA) are of potential relevance to similarity which are considered most appropriate for Tier 2 or Tier 3.

FDA, however, suggested a critical risk ranking of quality attributes with regard to their potential impact on activity, PK/PD, safety, and immunogenicity, with quality attributes being assigned to tiers commensurate with their risk. As a result, it is suggested that a statistical approach should be considered to serve as a decision tool for certain critical quality attributes that are relevant to the demonstration of similarity. In other words, it is suggested that an appropriate statistical model should be used not only to determine the relevance or association between CQAs and clinical outcomes but also to assess the criticality or risk ranking of the CQAs relevant to clinical outcome by establishing a predictive model. The established predictive model can then be used to determine the degree of criticality or risk ranking for assignment of the identified CQAs to appropriate tiers.

5.4.2 Statistical Model

According to the United States Pharmacopoeia (USP), *in vitro* and *in vivo* correlation (IVIVC) is referred to as the establishment of a relationship between a biological property, or a parameter derived from a biological property produced from a dosage form, and a physicochemical property of the same dosage form. Typically, AUC (area under a blood- or plasma-concentration time curve) or peak concentration (Cmax) is considered

TABLE 5.1

Example of Critical Quality Attributes for Neupogen (Filgrastim)

Quality Attribute	Methods
Primary structure	• N-terminal sequencing • Peptide mapping with UV and MS detection • Protein molecular mass by ESI MS • Protein molecular mass MALDI-TOF MS • DNA sequencing of construct cassette • Peptide mapping coupled with MS/MS
Bioactivity	• Proliferation of murine myelogenous leukemia cells (NFS-60)
Receptor binding	• Surface Plasmon Resonance
Protein content	• RP-HPLC
Clarity	• Nephelometry
Sub-visible particles	• Micro flow imaging
Higher order structure	• Far and Near UV circular dichroism • ^{1}H nuclear magnetic resonance • ^{1}H-^{15}N heteronuclear single quantum coherence spectroscopy • LC-MS (disulfide bond)
High molecular weight variants/ aggregates	• Size exclusion chromatography • Reduced and non-reduced SDS-PAGE
Oxidized species	• RP-HPLC • LC/MS
Covalent dimers	• LC/MS
Partially reduced species	• LC/MS
Sequence variants: His → Gln Asp → Glu Thr → Asp	• RP-HPLC • LC/MS
fMet1 species	• RP-HPLC • LC/MS
Succinimide species	• RP-HPLC • LC/MS
Phosphoglucunoylation	• LC/MS
Acetylated species	• LC/MS
N-terminal truncated variants	• LC-MS/MS
Norleucine species	• RP-HPLC • LC/MS
Deamidated species	• RP-HPLC • LC/MS • IEF • CEX

the parameter derived from the biological property, while the physico-chemical property is the *in vitro* dissolution profile. Basically, under the Fundamental Bioequivalence Assumption, the IVIVC is to use dissolution tests as a surrogate for human studies (i.e., when the drug absorption profiles in terms of AUC or Cmax are similar, it is assumed that they are therapeutically equivalent). In addition, one of the main roles of IVIVC is to assist in the quality control of functional and/or structural characteristics during manufacturing process.

For simplicity and illustration purpose, we will consider the case where the relationship between CQA and the clinical outcome is linear. The nonlinear case can be similarly treated. Let x and y be the response of a critical quality attribute and the clinical outcome, respectively. In practice, if the critical quality attribute is relevant to the clinical outcome, it is assumed that the clinical outcome can be predicted by the CQA accurately and reliably with some statistical assurance. One of the statistical criteria is to examine the degree of closeness (or the degree of relevance) between the observed response y and the predicted response $\hat{y}$ via an established statistical model. To study this, we will first study the association between x and y and build up a model. Then, validate the model based on some criteria. For simplicity, we assume that x and y can be described by the following linear model:

$$y = \beta_0 + \beta_1 x + \varepsilon, \tag{5.6}$$

where ε follows a normal distribution with a mean of 0 and variance of σ_e^2. Suppose that n pairs of observations $(x_1, y_1), \ldots, (x_n, y_n)$ are observed in a translation process. To define the notation, let

$$X^T = \begin{pmatrix} 1 & 1 & \cdots & 1 \\ x_1 & x_2 & \cdots & x_n \end{pmatrix}$$

and

$$Y^T = \begin{pmatrix} y_1 & y_2 & \cdots & y_n \end{pmatrix}.$$

Then, under model (5.6), the maximum likelihood estimates of the parameters β_0 and β_1 are:

$$\begin{pmatrix} \hat{\beta}_0 \\ \hat{\beta}_1 \end{pmatrix} = (X^T X)^{-1} X^T Y$$

with

$$\text{var}\begin{pmatrix} \hat{\beta}_0 \\ \hat{\beta}_1 \end{pmatrix} = (X^T X)^{-1} \sigma_e^2.$$

Furthermore, σ_e^2 can be estimated by the mean squared error (MSE), which is given by

$$\hat{\sigma}_e^2 = \frac{1}{n-2} \sum_{i=1}^{n} (y_i - \hat{y}_i)^2.$$

Thus, we have established the following relationship:

$$\hat{y} = \hat{\beta}_0 + \hat{\beta}_1 x. \tag{5.7}$$

For a given $x = x_0$, suppose that the corresponding observed value is given by y; however, using (5.7), the corresponding fitted value is $\hat{y} = \hat{\beta}_0 + \hat{\beta}_1 x_0$. Note that $E(\hat{y}) = \hat{\beta}_0 + \hat{\beta}_1 x_0 = \mu_0$ and

$$\text{var}(\hat{y}) = (1 \ \ x_0)(X^T X)^{-1} \begin{pmatrix} 1 \\ x_0 \end{pmatrix} \sigma_e^2 = c\sigma_e^2,$$

where

$$c = (1 \ \ x_0)(X^T X)^{-1} \begin{pmatrix} 1 \\ x_0 \end{pmatrix}.$$

Furthermore, $\hat{y}$ is normally distributed with mean μ_0 and variance $c\sigma_e^2$, i.e.,

$$\hat{y} \sim N(\mu_0, c\sigma_e^2).$$

We may validate the translation model by considering how close is an observed y to its predicted value $\hat{y}$ which is fitted to the regression model (5.7).

To assess the closeness, we propose the following two measures which are based either on the absolute difference or relative difference between y and $\hat{y}$:

Criterion I. $p_2 = P\{|y - \hat{y}| < \delta\}$,

Criterion II. $p_2 = P\left\{\left|\dfrac{y - \hat{y}}{y}\right| < \delta\right\}$.

In other words, it is desirable to have a high probability that the difference or the relative difference between y and $\hat{y}$, given by p_1 and p_2, respectively, is less than a clinically or scientifically meaningful difference δ. Then, for either $i = 1$ or 2, it is of interest to test the following hypotheses:

$$H_0 : p_i \leq p_0 \quad versus \quad H_a : p_i > p_0, \tag{5.8}$$

where p_0 is some pre-specified constant. The idea is to reject H_0 and in favor of H_a. In other words, we would like to reject the null hypothesis H_0 and conclude H_a, which implies that the established model is considered validated.

Measure of Closeness Based on the Absolute Difference – It should be noted that we have

$$(y - \hat{y}) \sim N\left(0, (1 + c)\sigma_e^2\right).$$

Therefore, p_1 can be estimated by

$$\hat{p}_1 = \Phi\left(\frac{\delta}{\sqrt{(1 + c)\hat{\sigma}_e^2}}\right) - \Phi\left(\frac{-\delta}{\sqrt{(1 + c)\hat{\sigma}_e^2}}\right).$$

Using the delta method through a Taylor expansion, for sufficiently large sample size n,

$$\mathrm{var}(\hat{p}_1) \approx \phi\left(\left(\frac{\delta}{\sqrt{(1 + c)\hat{\sigma}_e^2}}\right) - \phi\left(\frac{-\delta}{\sqrt{(1 + c)\hat{\sigma}_e^2}}\right)\right)^2 \frac{\delta}{2(1 - \delta)(n - 2)\sigma_e^2},$$

where $\phi(z)$ is the probability density function of a standard normal distribution. Furthermore, $var(\hat{p}_1)$ can be estimated by V_1, where V_1 is given by

$$V_1 = \frac{2\delta^2}{(1+c)(n-2)\hat{\sigma}_e^2} \phi^2 \left(\frac{\delta}{\sqrt{(1+c)\hat{\sigma}_e^2}} \right).$$

By Slutsky's Theorem, $\dfrac{\hat{p}_1 - p_0}{\sqrt{V_1}}$ can be approximated by a standard normal distribution. For the testing of the hypotheses $H_0 : p_1 \leq p_0$ versus $H_a : p_1 < p_0$, we would reject the null hypothesis H_0 if

$$\frac{\hat{p}_1 - p_0}{\sqrt{V_1}} > z_{1-\alpha},$$

where $z_{1-\alpha}$ is the $100(1-\alpha)$th percentile of a standard normal distribution.

Measure of Closeness Based on the Relative Difference – On the other hand, for evaluation of p_2, we note that y^2 and $\hat{y}^2$ follow a non-central χ_1^2 distribution with non-centrality parameter μ_0^2/σ_e^2 and $\mu_0^2/c\sigma_e^2$, respectively, where $\mu_0 = \hat{\beta}_0 + \hat{\beta}_1 x_0$. Hence, $c\hat{y}^2/y^2$ is doubly non-central F distributed with $v_1 = 1$ and $v_2 = 1$ degrees of freedom and non-centrality parameters $\lambda_1 = \mu_0^2/c\sigma_e^2$ and $\lambda_2 = \mu_0^2/\sigma_e^2$. By Johnson and Kotz (1970), a non-central F distribution can be approximated by

$$\frac{1+\lambda_1 v_1^{-1}}{1+\lambda_2 v_2^{-1}} F_{v,v'},$$

where $F_{v,v'}$ is a central F distribution with degrees of freedom

$$v \frac{(v_1+\lambda_1)^2}{v_1+2\lambda_1} = \frac{\left(1+\mu_0^2/c\sigma_e^2\right)^2}{1+2\mu_0^2/c\sigma_e^2}$$

and

$$v' \frac{(v_2+\lambda_2)^2}{v_2+2\lambda_2} = \frac{\left(1+\mu_0^2/\sigma_e^2\right)^2}{1+2\mu_0^2/\sigma_e^2}.$$

Thus,

$$p_2 = P\left\{\left|\frac{y - \hat{y}}{y}\right| < \delta\right\}$$

$$= P\left\{(1-\delta)^2 < c\left(\frac{\hat{y}}{y}\right)^2 < (1+\delta)^2\right\}$$

$$\simeq P\left\{\frac{(1-\delta)^2}{c} < \frac{1+\lambda_1}{1+\lambda_2}F_{v,v'} < \frac{(1-\delta)^2}{c}\right\}$$

$$= P\left\{\frac{(1-\delta)^2}{c}\frac{1+\lambda_2}{1+\lambda_1} < F_{v,v'} < \frac{1+\lambda_2}{1+\lambda_1}\frac{(1-\delta)^2}{c}\right\}$$

Thus, p_2 can be estimated by

$$\hat{p}_2 = P\left\{\frac{(1-\delta)^2}{c}\frac{1+\hat{\lambda}_2}{1+\hat{\lambda}_1} < F_{\hat{v},\hat{v}'} < \frac{1+\hat{\lambda}_2}{1+\hat{\lambda}_1}\frac{(1-\delta)^2}{c}\right\} = P\{u_1 < F_{\hat{v},\hat{v}'} < u_2\},$$

where

$$u_1 = \frac{(1+\hat{\lambda}_2)}{c(1+\hat{\lambda}_1)}(1-\delta)^2,$$

$$u_2 = \frac{(1+\hat{\lambda}_2)}{c(1+\hat{\lambda}_1)}(1+\delta)^2,$$

and $(\hat{\lambda}_1, \hat{\lambda}_2, \hat{v}, \hat{v}')$ are the corresponding maximum likelihood estimates of $(\lambda_1, \lambda_2, v, v')$.

For a sufficiently large sample size, by Slutsky's Theorem, $\hat{p}_2$ can be approximated by a normal distribution with mean p_2 and variance V_2, where

$$V_2 = \left(\frac{\partial \hat{p}_2}{\partial \beta_0}, \frac{\partial \hat{p}_2}{\partial \beta_1}, \frac{\partial \hat{p}_2}{\partial \sigma_e^2}\right)\begin{pmatrix} (X^T X)^{-1}\hat{\sigma}_e^2 & 0 \\ 0' & \dfrac{2\hat{\sigma}_e^4}{n-2} \end{pmatrix}\begin{pmatrix} \dfrac{\partial \hat{p}_2}{\partial \beta_0} \\ \dfrac{\partial \hat{p}_2}{\partial \beta_1} \\ \dfrac{\partial \hat{p}_2}{\partial \sigma_e^2} \end{pmatrix};$$

with

$$\frac{\partial \hat{p}_2}{\partial \beta_0} = \frac{2(c-1)\hat{\mu}_0}{c^2 \hat{\sigma}_e^2 (1+\hat{\lambda}_1)^2} \left[(1+\delta)^2 f(u_2) - (1-\delta)^2 f(u_1) \right]$$

$$\frac{\partial \hat{p}_2}{\partial \beta_1} = \frac{2(c-1)x_0 \hat{\mu}_0}{c^2 \hat{\sigma}_e^2 (1+\hat{\lambda}_1)^2} \left[(1+\delta)^2 f(u_2) - (1-\delta)^2 f(u_1) \right]$$

$$\frac{\partial \hat{p}_2}{\partial \sigma_e^2} = \frac{\hat{\lambda}_1 - \hat{\lambda}_2}{c^2 \hat{\sigma}_e^2 (1+\hat{\lambda}_1)^2} \left[(1+\delta)^2 f(u_2) - (1-\delta)^2 f(u_1) \right]$$

where $f(u)$ is the probability density function of an F distribution with degrees of freedom $\hat{v}$ and $\hat{v}'$. Thus, the hypotheses given in (5.8) for one-way translation based on probability of relative difference can be tested. In particular, H_0 is rejected if

$$Z = \frac{\hat{p}_2 - p_0}{\sqrt{V_2}} > z_{1-\alpha},$$

where $z_{1-\alpha}$ is the $100(1-\alpha)$th percentile of a standard normal distribution. Note that V_2 is an estimate of $\mathrm{var}(\hat{p}_2)$ which is obtained by simply replacing the parameters with their corresponding estimates of the parameters.

An Example – For the two measures proposed in the Section 5.3.2, p_1 is based on the absolute difference between y and $\hat{y}$. To test the hypothesis $H_0: p_1 \le p_0$ versus $H_a: p_1 > p_0$ or the given α, p_0, the set of data (x_i, y_i) and the selected observation (x_0, y_0), we have to compute the value of $\hat{p}_1$.

$$Z = \frac{\hat{p}_1 - p_0}{\sqrt{V_1}} > z_{1-\alpha},$$

If the null hypothesis is rejected. Note that the value of $\hat{p}_1$ depends on the value of δ. Furthermore, it can be shown that $\left(\hat{p}_1 - p_0 - z_{1-\alpha} \sqrt{V_1} \right) > 0$ is an increasing function of δ over $(0, +\infty)$. That is, $\left(\hat{p}_1 - p_0 - z_{1-\alpha} \sqrt{V_1} \right) > 0$ is equivalent to $\delta > \delta_0$. Thus, the hypothesis can be tested based on δ_0 instead of $\hat{p}_1$ as long as we can find the value of δ_0 for the given dataset (x_i, y_i) and the selected observation (x_0, y_0). An example is given for illustration.

TABLE 5.2

Data to Establish a Predictive Model

X	0.9	1.1	1.3	1.5	2.2	2.0	3.1	4.0	4.9	5.6
Y	0.9	0.8	1.9	2.1	2.3	4.1	5.6	6.5	8.8	9.2

Suppose that the (Table 5.2) are obtained from an IVIVC study, where x is a given dose level and y is the associated toxicity measure:

This set of data is fitted to Model (5.6). The estimates of the model parameters are given by

$$\hat{\beta}_0 = -0.704, \hat{\beta}_1 = 1.851, \hat{\sigma}^2 = 0.431.$$

Based on this model, given $x = x_0$, the fitted value is given by $\hat{y} = -0.704 + 1.851x_0$. Given $\alpha = 0.05$, $p_0 = 0.8$, $(x_0, y_0) = (1.0, 1.2)$ and $(5.2, 9.0)$, we obtain $\delta_0 = 1.27$ and 1.35, respectively. If the required difference is $\delta > \delta_0$, the null hypothesis will be rejected. We conclude that the probability that the difference between y and $\hat{y}$ is less than δ is larger than 0.8. Note that δ_0 changes for different selected observations (x_0, y_0).

5.4.3 Validity of the Translational Model

In this section, the validity of the translation model is assessed by the two proposed closeness measures p_1 and p_2, respectively. Without loss of generality, choose $\alpha = 0.05$ and $p_0 = 0.8$.

Case 1: Testing of H_0: $p_1 \leq p_0$ versus H_a: $p_1 > p_0$ – Using the above results, for $x_0 = 1.0$, δ is 1.112, since $|y_0 - \hat{y}| = |1.2 - 1.147| = 0.053$, which is less than $\delta = 1.112$, therefore H_0 is rejected. Similarly, for $x_0 = 5.2$, the corresponding δ is 1.178, then

$$|y_0 - \hat{y}| = |9.0 - 8.921| = 0.079,$$

which is again smaller than $\delta = 1.178$, thus H_0 is rejected.

Case 2: Testing of H_0: $p_2 \leq p_0$ versus H_a: $p_2 > p_0$ – Suppose that $\delta = 1$, for the given two values of x, estimates of p_2 and the corresponding values of the test statistic Z are given in the (Table 5.3).

TABLE 5.3

Summary of Test Results

x_0	y_0	$\hat{y}$	$\hat{p}_2$	Z	
1.0	1.2	1.147	0.870	1.183	Do not reject H_0
5.2	9.0	8.921	0.809	1.164	Do not reject H_0

5.4.4 Two-Way Translational Process

The above translational process is usually referred to as a *one-way transla-tion* in translational medicine. That is, the information observed at basic research discoveries is translated to the clinic. As indicated by Pizzo (2006), the translational process should be a *two-way translation*. In other words, we can exchange x and y in (5.6)

$$x = \gamma_0 + \gamma_1 y + \varepsilon$$

and come up with another predictive model $\hat{x} = \hat{\gamma}_0 + \hat{\gamma}_1 y$. Following the simi-lar ideas, using either one of the measures p_i, the validation of a two-way translational process can be summarized by the following steps:

Step 1: For a given set of data (x, y), established a predictive model, say, $y = f(x)$.

Step 2: Select the bound δ_{yi} for the difference between y and $\hat{y}$. Evaluate $\hat{p}_{yi} = P\{|y - \hat{y}| < \delta_{yi}\}$. Assess the one-way closeness between y and $\hat{y}$ by testing the hypotheses (5.8). Proceed to the next step if the one-way translation process is validated.

Step 3: Consider x as the dependent variable and y as the independent variable. Set up the regression model. Predict x at the selected obser-vation y_0, denoted by $\hat{x}$, based on the established model between x and y (i.e., $x = g(y)$), i.e. $\hat{x} = g(y) = \hat{\gamma}_0 + \hat{\gamma}_1 y$.

Step 4: Select the bound δ_{xi} for the difference between x and $\hat{x}$. Evaluate the closeness between x and $\hat{x}$ based on a test for the following hypotheses

$$H_0 : p_i \leq p_0 \quad \text{vs} \quad H_1 : p_i > p_0$$

where $p_i = P\left\{\left|\dfrac{y - \hat{y}}{y}\right| < \delta_{yi} \text{ and } \left|\dfrac{x - \hat{x}}{x}\right| < \delta_{xi}\right\}$.

The above test can be referred to as a test for two-way translation. If, in Step 4, H_0 is rejected in favor of H_1, this would imply that there is a two-way translation between x and y (i.e., the established predictive model is vali-dated). However, the evaluation of p involved the joint distribution of $\dfrac{x - \hat{x}}{x}$ and $\dfrac{y - \hat{y}}{y}$. An exact expression is not readily available. Thus, an alternative approach is to modify Step 4 of the above procedure and proceed with a con-ditional approach instead. In particular,

Step 4 (modified): Select the bound δ_{xi} for the difference between x and $\hat{x}$. Evaluate the closeness between x and $\hat{x}$ based on a test for the following hypotheses:

$$H_0 : p_{xi} \leq p_0 \quad \text{vs} \quad H_\alpha : p_{xi} > p_0 \tag{5.9}$$

where $p_{xi} = P\{|x - \hat{x}| < \delta_{xi}\}$.

NOTE: that the evaluation of p_{xi} is much easier and can be computed in a similar way by interchanging the role of x and y for the results given below.

An Example – Using the data set given in Table 5.2, we set up the regression model $x = \gamma_0 + \gamma_1 y + \varepsilon$ with y as the independent variable and x as the dependent variable. The estimates of the model parameters are $\hat{y}_0 = 0.468$, $\hat{y}_1 = 0.519$, and $\hat{\sigma}^2 = 0.121$. Based on this model, for the same α and p_0, given $(x_0, y_0) = (1.0, 1.2)$ and $(5.2, 9.0)$, the fitted values are given by $\hat{x} = 0.468 + 0.519 y_0$.

Case 1: Testing of H_0: $p_{x1} \leq p_0$ versus H_a: $p_{x1} > p_0$ – Using the above results, for $y_0 = 1.2$, δ is 0.587, since $|x_0 - \hat{x}| = |1.0 - 1.09| = 0.09$, which is less than $\delta x = 0.587$, therefore H_0 is rejected. Similarly, for $y_0 = 9.0$, the corresponding δ is 0.624, then $|x_0 - \hat{x}| = |5.2 - 5.139| = 0.061$, which is again smaller than $\delta = 0.624$, thus H_0 is rejected.

Case 2: Testing of H_0: $p_{x2} \leq p_0$ versus H_a: $p_{x2} > p_0$ – Suppose that $\delta = 1$, for the given two values of y, estimates of p_{x2} and the corresponding values of the test statistic Z are given in the (Table 5.4).

5.4.5 Remarks

For analytical similarity assessment, FDA suggested the identified critical quality attributes (CQAs) should be classified into three tiers depending upon their criticality or risk ranking relevant to clinical outcomes. That is, CQAs that are most relevant to clinical outcomes should be assigned to Tier 1. CQAs that are less or least relevant to clinical outcomes should be classified to Tier 2 or Tier 3, respectively. However, no criteria regarding the degree of criticality or risk ranking were mentioned in the FDA draft guidance on

TABLE 5.4

Summary of Test Results

x_0	y_0	$\hat{x}_0$	$\hat{p}_{x2}$	Z	
1.0	1.2	1.090	0.809	1.300	Do not reject H_0
5.2	9.0	5.139	0.845	16.53	Do not reject H_0

analytical similarity assessment (FDA, 2017b). In practice, it is then suggested that the probability of closeness between y and $\hat{y}$ once the predictive model between the CQA and clinical outcome has been established. For example, if the probability is greater than 80% then we may assign the CQA to Tier 1. On the other hand, if the probability is within 50% and 80%, we may consider the CQA is less relevant to clinical outcomes and then assign the CQA to Tier 2. If the probability of closeness is less than 50%, we may assign the CQA to Tier 3 which is considered least relevant to clinical outcomes.

In practice, however, data that can link the CQAs and clinical outcomes are usually not available. In this case, an identified CQA that is considered relevant to clinical outcomes may be mis-classified to wrong tiers. Thus, in practice, it is of interest to study the impact of mis-classification on the decision whether analytical similarity assessment has provided sufficient totality-of-the-evidence for demonstration of highly similarity between a proposed biosimilar product and the reference product.

5.5 Concluding Remarks

For identifying CQAs at various stages of the manufacturing process, most sponsors assign CQAs based on the mechanism of action (MOA) or pharmacokinetics (PK) which are believed to be relevant to clinical outcomes. It is a reasonable assumption that change in MOA or PK of a given quality attribute is predictive of clinical outcomes. However, the primary assumption that there is a well-established relationship between *in vitro* assays and *in vivo* testing (i.e., the *in vitro* assays and *in vivo* testing correlation; IVIVC) needs to be validated. Under the validated IVIVC relationship, the criticality (or risk ranking) can then be assessed based on the degree of the relationship. In practice, however, most sponsors provide clinical rationales for the assignment of the CQAs without using a statistical approach for the establishment of IVIVC. The assignment of the CQAs without using a statistical approach is considered subjective and hence somewhat misleading.

For a given quality attribute, FDA suggests a simple approach by testing one sample (randomly selected) from each of the lots. Basically, FDA's approach ignores lot-to-lot variability for the reference product. In practice, however, lot-to-lot variability inevitably exists even when the manufacturing process has been validated. In other words, we would expect that there are differences in mean and variability from lot-to-lot, i.e., $\mu_{Ri} \neq \mu_{Rj}$ and $\sigma^2_{Ri} \neq \sigma^2_{Rj}$ for $I \neq j$, $i,j = 1, \ldots K$. In this case, it is suggested that FDA's approach be modified (e.g., performing tests on multiple samples from each lot) in order to account for the within-lot and between-lot (lot-to-lot) variabilities for fair and reliable comparisons.

6

FDA Tiered Approach for Analytical Similarity Assessment

6.1 Background

Following the passage of the *Biologics Price Competition and Innovation* (BPCI) Act in 2009, the FDA circulated three guidances on the demonstration of biosimilarity of biosimilar products for public comments in February 2012. These draft guidances were subsequently finalized in April, 2015 (FDA, 2015a-c). These guidances are intended not only (i) to assist sponsors in demonstrating that a proposed therapeutic protein product is biosimilar to a reference product for the purpose of submitting a marketing application under section 351(k) of the Public Health Service (PHS) Act, but also (ii) to describe the FDA's current thinking on factors considered to demonstrate that a proposed protein product is highly similar to a reference product which was licensed under section 351(a) of the PHS Act. In the guidance on *Scientific Considerations in Demonstrating Biosimilarity to a Reference Product*, FDA introduces the concept of a stepwise approach for obtaining totality-of-the-evidence for the regulatory review and approval of biosimilar applications (FDA, 2015a).

The stepwise approach starts with analytical studies for assessment of analytical similarity in critical quality attributes (CQAs) in manufacturing process that are relevant to clinical outcomes between a proposed biosimilar product and an innovative biological product. Analytical similarity assessment focuses on structural and functional characterization in manufacturing process of the proposed biosimilar product as the foundation for achieving totality-of-the-evidence for demonstration of similarity between the proposed biosimilar product and the innovative biological product. In practice, there is often a large number of CQAs that may be relevant to clinical outcomes. Thus, it is almost impossible to assess analytical similarity for all of these CQAs individually. As a result, FDA suggests the sponsors first identify CQAs that are relevant to clinical outcomes based on mechanism of action (MOA) and/or pharmacokinetics (PK) and pharmacodynamics (PD) (if applicable). Then, classify the identified CQAs into three tiers depending upon their criticality or risk ranking, i.e., CQAs that are most relevant, mild to moderately relevant, and least relevant to clinical outcomes are classified into Tier 1, Tier 2, and Tier 3, respectively. To assist the sponsors, FDA also proposes some statistical approaches for the assessment of analytical similarity for CQAs from different tiers. For example, FDA recommends equivalence test for CQAs from Tier 1, a quality range approach for CQAs from Tier 2, and descriptive raw data and graphical presentation for CQAs from Tier 3 (see, e.g., Cristl, 2015; Tsong, 2015; Chow, 2014; Chow, 2015).

The extensive analytical data are intended to support (i) a demonstration that the proposed biosimilar product and US-licensed reference are highly similar; (ii) a demonstration that the proposed biosimilar product can be manufactured in a well-controlled and consistent manner, leading to a product that is sufficient to meet appropriate quality standards; and (iii) a justification of the relevance of the comparative data generated using a non-US licensed (e.g., EU-approved) reference product to support a demonstration of biosimilarity of the proposed biosimilar product to US-licensed reference. The purpose of this chapter is not only to provide a close look at these approaches by providing interpretation and/or statistical justification whenever possible, but also to discuss some challenging issues to the FDA's proposed approach (mainly on the equivalence test for Tier 1 CQAs). In addition, some recommendations and alternative methods are proposed.

In the next section, the stepwise approach for demonstrating biosimilarity as suggested by the FDA draft guidance is briefly outlined. Sections 6.3 and 6.4 provide brief descriptions of the Tier 1 equivalence test and the quality range approach for Tier 2 CQAs and the method of descriptive raw data and graphical comparison for Tier 3 CQAs, respectively. Some practical considerations are discussed in Section 6.5. Some concluding remarks are given in the last section of this chapter.

6.2 Stepwise Approach for Demonstrating Biosimilarity

As defined in the BPCI Act, a biosimilar product is a product that is *highly similar* to the reference product, notwithstanding minor differences in clinically inactive components, and which has no clinically meaningful differences in terms of safety, purity, and potency. Based on the definition of the BPCI Act, biosimilarity requires that there are no *clinically meaningful differences* in terms of *safety, purity* and *potency*. Safety could include pharmacokinetics and pharmacodynamics (PK/PD), safety and tolerability, and immunogenicity studies. Purity includes all critical quality attributes during manufacturing process. Potency is referred to as efficacy studies. As indicated earlier, in the 2015 FDA guidance on scientific considerations, FDA recommends that a stepwise approach be considered for providing the totality-of-the-evidence to demonstrating biosimilarity of a proposed biosimilar product as compared to a reference product (FDA, 2015a).

The stepwise approach is briefly summarized by a pyramid illustrated in Figure 5.1. The stepwise approach starts with analytical studies for structural and functional characterization. The stepwise approach continues with animal studies for toxicity, clinical pharmacology studies such as PK/PD studies, followed by investigations of immunogenicity, and clinical studies for safety/tolerability and efficacy.

The sponsors are encouraged to consult with medical/statistical reviewers of FDA with the proposed plan or strategy of the stepwise approach for regulatory agreement and acceptance. This is to make sure that the information provided is sufficient to fulfill the FDA's requirement for providing totality-of-the-evidence for the demonstration of biosimilarity of the proposed biosimilar product as compared to the reference product. As an example, more specifically, the analytical studies are to assess similarity in CQAs at various stages of the manufacturing process of the biosimilar product as compared to those of the reference product. To assist the sponsors in fulfilling the regulatory requirement for providing totality-of-the-evidence of analytical similarity, the FDA suggests several approaches depending upon the criticality of the identified quality attributes relevant to the clinical outcomes.

6.3 Tier 1 Equivalence Test

Analytical similarity assessment is referred to as the comparisons of functional and structural characterizations between a proposed biosimilar product and a reference product in terms of CQAs that are relevant to clinical outcomes. FDA suggests that the sponsors identify CQAs that are relevant to clinical outcomes and classify them into three tiers depending the criticality or risk ranking (e.g., most, mild to moderate, and least) relevant to clinical outcomes. At the same time, FDA also recommends some statistical approaches for the assessment of analytical similarity for CQAs from different tiers. FDA recommends an equivalence test for CQAs from Tier 1, quality range approach for CQAs from Tier 2, and descriptive raw data and graphical presentation for CQAs from Tier 3 (see, e.g., Cristl, 2015; Tsong, 2015; Chow, 2015). They are briefly outlined in the subsequent subsections.

Interval Hypotheses and Statistical Methods – For Tier 1, FDA recommends that an equivalency test be performed for the assessment of analytical similarity. As indicated by the FDA, a potential approach could be a similar approach to bioequivalence testing for generic drug products (FDA, 2003; Chow, 2015). In other words, for a given critical attribute, we may test for equivalence by the following interval (null) hypothesis:

$$H_0 : \mu_T - \mu_R \le -\delta \text{ or } \mu_T - \mu_R \ge \delta, \tag{6.1}$$

where $\delta > 0$ is the equivalence limit (or similarity margin), and μ_T and μ_R are the mean responses of the test (the proposed biosimilar) product and the reference product lots, respectively. Analytical equivalence (similarity) is concluded if the null hypothesis of non-equivalence (*dis*-similarity) is rejected. Note that Yu (2004) defined inequivalence as when the confidence interval falls entirely outside the equivalence limits. Similarly to the confidence interval approach for bioequivalence testing under the raw data model, analytical similarity would be accepted for a quality attribute if the $(1-2\alpha) \times 100\%$ two-sided confidence interval of the mean difference is within $(-\delta, \delta)$.

Under the null hypothesis (6.1), FDA indicates that the equivalence limit (similarity margin), δ, would be a function of the variability of the reference product, denoted by σ_R. It should be noted that each lot contributes one test value for each attribute being assessed. Thus, σ_R is the population standard deviation of the lot values of the reference product.

Testing One Sample Per Lot – For testing one sample from each reference lot for obtaining an estimate of σ_R, Wang and Chow (2015) evaluated statistical properties of the FDA's recommended method. Without loss of generality, suppose multiple test samples from each reference lot are available. Let x_{Rij} be the test value of the jth test sample from the ith reference lot, $i = 1,\ldots, k, j = 1,\ldots,n_i$, and it follows a normal distribution with mean μ_i and variance σ_i^2, where μ_i and σ_i^2 are also random variables. The expectations of μ_i and σ_i^2 are μ and σ_2 and the variances are σ_μ^2 and σ_σ^2. Then, the variance of x_{Rij} is given by

$$\sigma_R^2 = Var(x_{Rij})$$
$$= Var\left(E\left(x_{Rij} \mid \mu_i, \sigma_i^2\right)\right) + E\left(Var\left(x_{Rij} \mid \mu_i, \sigma_i^2\right)\right)$$
$$= Var(\mu_i) + E\left(\sigma_i^2\right) = \sigma_\mu^2 + \sigma^2.$$

Define $\bar{x}_{Ri\cdot} = \dfrac{1}{n_i}\sum_{j=1}^{n_i} x_{Rij}$, $\bar{\bar{x}}_{R\cdot\cdot} = \dfrac{1}{k}\sum_{i=1}^{k} \bar{x}_{Ri\cdot}$. Then

$$E(\overline{x}_{Ri\cdot}) = \frac{1}{n_i}\sum_{j=1}^{n_i}E\big(E(x_{Rij}\,|\,\mu_i)\big) = \frac{1}{n_i}\sum_{j=1}^{n_i}E(\mu_i) = \mu,$$

$$Var(\overline{x}_{Ri\cdot}) = Var\big(E\big(\overline{x}_{Ri\cdot}\,|\,\mu_i,\sigma_i^2\big)\big) + E\big(Var\big(\overline{x}_{Ri\cdot}\,|\,\mu_i,\sigma_i^2\big)\big)$$

$$= Var(\mu_i) + \frac{1}{n_i}E\big(\sigma_i^2\big) = \sigma_\mu^2 + \frac{1}{n_i}\sigma^2,$$

where

$$E(\overline{\overline{x}}_{R\cdot\cdot}) = \frac{1}{k}\sum_{i=1}^{k}E(\overline{x}_{Ri\cdot}) = \mu,$$

$$Var(\overline{\overline{x}}_{R\cdot\cdot}) = \frac{1}{k^2}\sum_{i=1}^{k}Var(\overline{x}_{Ri\cdot}) = \frac{1}{k}\sigma_\mu^2 + \frac{\sigma^2}{k^2}\left(\sum_{i-1}^{k}n_i\right).$$

If we assume that $n_1 = \cdots n_k = n$, then we have

$$\hat{\sigma}_R^2 = \frac{1}{nk-1}\sum_{i=1}^{k}\sum_{j=1}^{n_i}\big(x_{Rij} - \overline{\overline{x}}_{R\cdot\cdot}\big)^2,$$

$$E\big(\hat{\sigma}_R^2\big) = \frac{1}{nk-1}\sum_{i=1}^{k}\sum_{j=1}^{n_i}E\big(\big(x_{Rij}-\mu\big)^2 + \big(\overline{\overline{x}}_{R\cdot\cdot}-\mu\big)^2 - 2(x_{Rij}-\mu)(\overline{\overline{x}}_{R\cdot\cdot}-\mu)\big)$$

$$= \frac{1}{nk-1}\sum_{i=1}^{k}\sum_{j=1}^{n_i}\big(Var(x_{Rij}) - Var(\overline{\overline{x}}_{R\cdot\cdot})\big)$$

$$= \frac{1}{nk-1}\big(n(k-1)\sigma_\mu^2 + (nk-1)\sigma_\mu^2\big) = \sigma_\mu^2 + \sigma^2 - \frac{n-1}{nk-1}\sigma_\mu^2.$$

Thus, it can be seen that the FDA's approach is an unbiased estimate of σ_R. However, when $n > 1$, the FDA's recommended approach with multiple test samples per lot is a biased estimate of σ_R. In the interest of having an unbiased estimate with multiple test samples per lot, Wang and Chow (2015) proposed an alternative approach to correct the biasedness. As indicated by Wang and Chow (2015), multiple test samples per lot provide valuable information regarding the heterogeneity across lots, which is useful especially when extreme lots (i.e., lots with extremely low or high variability) are selected for the equivalence test.

Ideally, the reference variability, σ_R, should be estimated based on some sampled lots randomly selected from a pool of reference lots for the statistical equivalence test. In practice, it may be a challenge when there is a limited number of available lots. Thus, FDA suggests the that the sponsor provide a plan on how the reference variability, σ_R, will be estimated with a justification.

An Example of Calculation – To provide a better understanding of the process for Tier 1 equivalence tests, consider the data received from Brent Morse. Consider EU lots ($n_R = 4$) as the reference product and the US lots ($n_T = 10$) as the test product. Although there are replicates, for illustration purpose, we only consider a single test value per lot as recommended by the FDA. Suppose the test values follow normal distribution in both reference and test products. For critical quality attribute in Tier 1, as discussed above, Tier 1 equivalence tests can be performed by following steps:

Step 1. Calculate mean and variance for reference product, which are given by

$$\hat{\mu}_R = \frac{1}{n_R} \sum_{i=1}^{n_R} x_{Ri} \quad \text{where } n_R \text{ is the totle number of reference lots}$$

$$\hat{\sigma}_R^2 = \frac{1}{n_R - 1} \sum_{i=1}^{n_R} (x_{Ri} - \hat{\mu}_R)^2$$

Step 2. Establish equivalence acceptance criterion (EAC) as follows:

$$\delta = 1.5 \times \hat{\sigma}_R$$

Step 3. Obtain mean for test product and the pooled variance, which are given by

$$\hat{\mu}_T = \frac{1}{n_T} \sum_{i=1}^{n_T} x_{Ti} \quad \text{where } n_T \text{ is the totle number of test lots}$$

$$\hat{\sigma}^2 = \frac{1}{n_R + n_T - 2} \left(\sum_{i=1}^{n_R} (x_{Ri} - \hat{\mu}_R)^2 + \sum_{i=1}^{n_T} (x_{Ti} - \hat{\mu}_T)^2 \right)$$

Step 4. Construct 90% confident interval (CI) for $\mu_T - \mu_R$, which is given below

$$CI = \left(\hat{\mu}_R - \hat{\mu}_T - t_{0.95} * \hat{\sigma} * \sqrt{\frac{1}{n_R} + \frac{1}{n_T}}, \hat{\mu}_R - \hat{\mu}_T - t_{0.95} * \hat{\sigma} * \sqrt{\frac{1}{n_R} + \frac{1}{n_T}} \right),$$

where $t_{0.95}$ is the 95th percentile of T distribution with df $= n_R + n_T - 2$

Step 5. Decision-making

If the constructed 90% CI is totally within the EAC, then we claim that the test product has passed the Tier 1 equivalence test.

Note that the above calculation can be similarly performed if we assume that the observed data are log-normally distributed (see also Table 6.1).

An Example of Tier 1 Equivalence Test – Suppose there are k lots of a reference product (RP) and n lots of a test product (TP) available for analytical similarity assessment, where k > n. For a given CQA, a Tier 1 equivalence test can be summarized in the following steps:

Step 1. *Matching number of RP lots to TP lots* – Since k > n, there are more reference lots than test lots. The first step is then to match the number of RP lots to TP lots for a head-to-head comparison. To *match* RP lots to TP lots, FDA suggests *randomly* selecting n lots out of the k RP lots. If the n lots are not randomly selected from the k RP lots, justification needs to be provided to prevent *selection bias*.

Step 2. *Use the remaining independent RP lots for estimating* σ_R – After the matching, the remaining $k - n$ lots are then used to *estimate* σ_R in order to set up the equivalency acceptance criterion (EAC). It should be noted that if $k - n \leq 2$, it is suggested that all RP lots should be used to estimate σ_R.

Step 3. *Calculate the equivalency acceptance criterion (EAC): EAC = 1.5 $\times$ $\hat{\sigma}_R$* – Based on the estimate of σ_R, denoted by $\hat{\sigma}_R$, FDA recommends EAC be set as $1.5 \times \hat{\sigma}_R$, where $c = 1.5$ is considered a regulatory standard.

Step 4. *Based on c (regulatory standard), $\hat{\sigma}_R$, and $\Delta = \mu_T - \mu_R$, an appropriate sample size can be chosen for the analytical similarity assessment* – As an example, suppose that there are 21 RP lots and 7 TP lots. We first randomly select 7 out of the 21 RP lots to match the 7 TP lots. Suppose that based on the remaining 14 lots, an estimate of σ_R is given by $\hat{\sigma}_R = 1.039$. Also, suppose that the true difference between the biosimilar product and the reference product is proportional to σ_R, say $\Delta = \sigma_R/8$. Then, the following table with various sample sizes (the number of TP lots available) and the corresponding test size and statistical power for detecting the difference of $\sigma_R/8$ is helpful for the assessment of analytical assessment.

TABLE 6.1

Tier 1 Equivalence Test Based on Single Test Value for Each Lot

Description and Formulas	Calculation
Reference lots	
n_R: number of reference lots	$n_R = 4$
$x_{Ri}\ i = 1, \dots, n_R$	$x_{Ri} = 112.4, 119.1, 107.0, 110.3$
Test lots	
n_T: number of test lots	$n_T = 10$
$x_{Ti}\ i = 1, \dots, n_T$	$x_{Ti} = 103.6, 104.9, 142.6, 112.7, 130.3,$
	$113.3, 75.9, 109.5, 119.6, 118.1$
Establishment of EAC	
$EAC = (\hat{\mu}_T - \hat{\mu}_R) \pm 1.5 * \hat{\sigma}_R$	
$\hat{\mu}_R = \dfrac{1}{n_R} \sum_{i=1}^{n_R} x_{Ri}$	$\hat{\mu}_R = \dfrac{1}{4}(112.4 + 119.1 + 107.0 + 110.3) = 112.3$
$\hat{\mu}_T = \dfrac{1}{n_T} \sum_{i=1}^{n_T} x_{Ti}$	$\hat{\mu}_T = \dfrac{1}{10}(103.6 + \dots + 119.6 + 118.1) = 116.05$
$\hat{\sigma}_R^2 = \dfrac{1}{n_R - 1} \sum_{i=1}^{n_R} (x_{Ri} - \hat{\mu}_R)^2$	$\hat{\sigma}_R^2 = \dfrac{1}{4-1}\left[(112.4 - 112.3)^2 + \dots + (110.3 - 112.3)^2\right] = 26.11.$
	$\hat{\sigma}_R = \sqrt{\hat{\sigma}_R^2} = \sqrt{26.11} = 5.11.$
	$EAC = 3.75 \pm 1.5 * 5.11 = (-3.92, 11.42)$

(Continued)

TABLE 6.1 (CONTINUED)

Tier 1 Equivalence Test based on Single Test Value for Each Lot

Description and Formulas	Calculation
90% Confidence Interval for $\mu_T - \mu_R$	
Pooled variance	$\hat{\sigma}^2 = \dfrac{1}{4+10-2}(78.34 + 3109.43) = 265.65$
$\hat{\sigma}^2 = \dfrac{1}{n_R + n_T - 2}\left(\displaystyle\sum_{i=1}^{n_R}(x_{Ri} - \hat{\mu}_R)^2 + \sum_{i=1}^{n_T}(x_{Ti} - \hat{\mu}_T)^2\right)$	$\hat{\sigma} = 16.30$
$CI = \hat{\mu}_T - \hat{\mu}_R \pm t_{0.95} * \hat{\sigma} * \sqrt{\dfrac{1}{n_R} + \dfrac{1}{n_T}}$	$CI = 3.75 \pm (1.78) * (16.30) * (0.58)$ $= (-13.08, 20.58)$
where $t_{0.95}$ is the 95th percentile of t distribution with df = $n_R + n_T - 2$	where $t_{0.95}$ with df = $n_R + n_T - 2 = 4 + 10 - 2 = 12$ is 1.78.
Conclusion	
Test product fails to pass Tier 1 equivalence test. In other words, Test product is not biosimilar to the Reference product	90%CI (-13.08,20.58) does not fall within EAC = (-3.92, 11.42).
	This may be due to the worst lots with test values of 75.9 and 142.6.

TABLE 6.2

Assessment of Analytical Similarity for CQAs from Tier 1

Number of RP Lots	Number of TP Lots	Selection of c	Test Size (Confidence Interval)	Statistical Power at (1/8)*RP SD
6	6	1.5	9% (82% CI)	74%
7	7	1.5	8% (84% CI)	79%
8	8	1.5	7% (86% CI)	83%
9	9	1.5	6% (88% CI)	86%
10	10	1.5	5% (90% CI)	87%

As it can be seen from Table 6.2, there is 79% power for 84% CI of $\hat{\Delta} = \hat{\mu}_T - \hat{\mu}_R$ to fall within $\pm$EAC assuming that the number of lots = 7 and true difference between TP and RP is $\sigma_R/8$. It should be noted that this approach has inflated alpha from the 5% to 16%. Note that for a fixed regulatory standard c, the sponsor may appropriately select sample size (the number of lots) for achieving a desired power (for detecting a $\sigma_R/8$ difference) and significance level for analytical similarity assessment. As can be seen from the above, if one wishes to reduce the test size (i.e., α level) from 8% to 5%, 10 TP lots need to be tested. Testing 10 TP lots will give an 87% power for detecting a $\sigma_R/8$ difference.

Remarks – To assist the sponsor in performing Tier 1 equivalence tests, FDA circulated a draft guidance on analytical similarity assessment in September 2017. In the draft guidance, FDA recommends that a minimum of 10 lots should be used when performing a Tier 1 equivalence test in biosimilar product development.

6.4 Other Tiered Approaches

6.4.1 Quality Range Approach for Tier 2

For Tier 2, FDA suggests that analytical similarity be assessed on the basis of the concept of quality ranges, i.e., $\pm x\sigma$, where σ is the standard deviation of the reference product and x should be appropriately justified. Thus, the quality range of the reference product for a specific quality attribute is defined as $(\hat{\mu}_R - x\hat{\sigma}_R, \hat{\mu}_R + x\hat{\sigma}_R)$. Analytical similarity would be accepted for the quality attribute if a sufficient percentage of test lot values (e.g., 90%) falls within the quality range.

For a given critical attribute the quality range is set based on test results of available reference lots. If $x = 1.645$, we would expect 90% of the test results from reference lots to lie within the quality range. If x is chosen to be 1.96, we would expect that about 95% of test results of reference lots will fall within the quality range. As a result, the selection of x could impact the quality range and consequently the percentage of test lot values that will fall within the quality range. Thus, FDA indicates that the standard deviation multiplier (x) should be appropriately justified.

The quality range approach for comparing populations between a proposed biosimilar product and a reference product is a reasonable approach under the assumption that $\mu_T = \mu_R$ and $\sigma_T = \sigma_R$. Under this assumption, we expect that a high percentage (say 90%) of test values of the test product will fall within the quality range obtained based on the test values of the reference product. Thus, one of the major criticisms of the quality range approach is that it ignores the fact that there are differences in population mean and population standard deviation between the proposed biosimilar product and the reference product, i.e., $\mu_T \neq \mu_R$ and $\sigma_T \neq \sigma_R$. In practice, it is recognized that biosimilarity between a proposed biosimilar product and a reference product could be established even under the assumption that $\mu_T \neq \mu_R$ and $\sigma_T \neq \sigma_R$. Thus, under the assumption that $\mu_T = \mu_R$ and $\sigma_T = \sigma_R$, the quality range approach for analytical similarity assessment for CQAs from Tier 2 is considered more stringent as compared to equivalence testing for CQAs from Tier 1 (most relevant to clinical outcomes), regardless that they are mild-to-moderately relevant to clinical outcomes. This is because that equivalence testing allows a possible mean shift of $\sigma_R/8$, while the quality range approach does not. In what follows, several examples for the possible scenarios of (i) $\mu_T \approx \mu_R$ or there is a significant mean shift (either a shift to the right or a shift to the left), and (ii) $\sigma_T \approx \sigma_R$, $\sigma_T > \sigma_R$, or $\sigma_T < \sigma_R$.

Example 1 – First consider the case where $\mu_T \approx \mu_R$ and $\sigma_T \approx \sigma_R$. In this case, if we choose $x = 1.645$, we would expect 90% of the test results from the test lots to lie within the quality range obtained based on the test values of the reference lots. This case is illustrated in Figure 6.1.

FIGURE 6.1
Quality range approach when $\mu_T \approx \mu_R$ and $\sigma_T \approx \sigma_R$ (dark gray dots represent test values of reference lots, while light gray represents test values of test lots).

FIGURE 6.2
Quality range approach when $\mu_T \approx \mu_R$ and $\sigma_T > \sigma_R$ (dark gray dots represent test values of reference lots, while light gray represents test values of test lots).

Example 2 – When $\mu_T \approx \mu_R$ but $\sigma_T > \sigma_R$, if we choose $x = 1.645$, we would expect less than 90% of the test results from test lots to lie within the quality range obtained based on the test values of the reference lots. The percentage of test values from test lots decreases as $C = \sigma_T/\sigma_R > 1$ increases. This case is illustrated in Figure 6.2.

Example 3 – The case where $\mu_T > \mu_R$ but $\sigma_T \approx \sigma_R$ is illustrated in Figure 6.3. As can be seen from Figure 6.3, if we choose $x = 1.645$, we would expect less than 90% of the test results from test lots to lie within the quality range obtained based on the test values of the reference lots. The percentage of test values from test lots drop significantly if the difference between $\varepsilon = \mu_T - \mu_R$ increases (i.e., μ_T shifts away from μ_R).

Example 4 – In practice, it is not uncommon to encounter the case where $\mu_T > \mu_R$ and $\sigma_T > \sigma_R$, which is illustrated in Figure 6.4. As can be seen from Figure 6.4, if we choose $x = 1.645$, we would expect less than 90% of the test results from test lots to lie within the quality range obtained based on the test values of the reference lots. The percentage of test values from test lots could be very low, especially when both $C = \sigma_T/\sigma_R > 1$ and $\varepsilon = \mu_T - \mu_R$ increases.

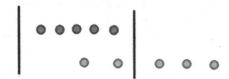

FIGURE 6.3
Quality range approach when $\mu_T > \mu_R$ and $\sigma_T \approx \sigma_R$ (dark gray dots represent test values of reference lots, while light gray represents test values of test lots).

FIGURE 6.4
Quality range approach for the case where $\mu_T > \mu_R$ and $\sigma_T > \sigma_R$ (dark gray dots represent test values of reference lots, while light gray represents test values of test lots).

Remarks – As discussed above, it is suggested that the quality range approach should be modified as follows;

$$(\hat{\mu}_R - x\hat{\sigma}_R - |\hat{\varepsilon}|, \hat{\mu}_R + x\hat{\sigma}_R + |\hat{\varepsilon}|),$$

where $|\hat{\varepsilon}| = |\hat{\mu}_T - \hat{\mu}_R|$ is the mean shift between the proposed biosimilar product and the reference product.

6.4.2 Raw Data and Graphical Comparison for Tier 3

For CQAs in Tier 3 with lowest risk ranking, FDA recommends an approach that uses raw data/graphical comparisons. The examination of similarity for CQAs in Tier 3 is by no means less stringent, which is acceptable because they have least impact on clinical outcomes in the sense that a notable dissimilarity will not affect clinical outcomes.

The method of raw data and graphical comparison is easy to implement and yet it is subjective. One of the major criticisms is that it is not clear how the approach can provide totality-of-the-evidence for demonstrating biosimilarity. For CQAs in Tier 1, they are least relevant to clinical outcomes and yet should carry less weight as compared to those CQAs from Tier 1 and Tier 2. There is little or no information regarding what results will be accepted by the method of data and graphical comparison. In practice, if significant differences in graphical comparisons of some CQAs are observed, should this observation raise a concern? In this case, if it is possible, the degree of criticality risk ranking of these CQAs should be assessed whenever possible.

To illustrate the use of the method of raw data and graphical comparison, similarly as above, we consider the following scenarios of (i) $\mu_T \approx \mu_R$ or there is a significant mean shift (either a shift to the right or a shift to the left), and (ii) $\sigma_T \approx \sigma_R$, $\sigma_T \gg \sigma_R$, or $\sigma_T \ll \sigma_R$.

An Example – The method of raw data and graphical comparison display test values (assay results) between Test lots and Reference lots for CQAs from Tier 3. Raw data and graphical comparison are rather subjective and lack of standards for comparison especially by knowing that (i) CQAs in Tier 3 are least relevant to clinical outcomes, and (ii) it is expected that $\mu_T \neq \mu_R$ and $\sigma_T \neq \sigma_R$ for biosimilar products. As a result, it is difficult to provide totality-of-the-evidence because it is not clear how much weight Tier 3 will carry. For illustration purposes, Figures 6.5–6.7 provide plots for the cases where (i) $\mu_T \approx \mu_R$ and $\sigma_T \neq \sigma_R$, (ii) $\mu_T \approx \mu_R$ and $\sigma_T \approx \sigma_R$, and (iii) $\mu_T \neq \mu_R$ and $\sigma_T \neq \sigma_R$, respectively. As can be seen from these figures, although graphical comparison may be different yet they are least relevant to clinical outcomes, it is not clear whether a significant difference in distribution of certain CQAs has raised a flag of safety or efficacy concern to demonstrating biosimilarity between the proposed biosimilar product and the reference product (either a US-licensed product or an EU-approved reference product).

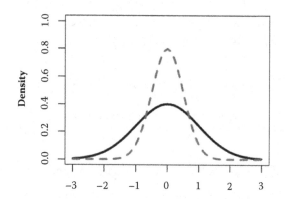

FIGURE 6.5
Graphical comparison for the case where $\mu_T \approx \mu_R$ and $\sigma_T \neq \sigma_R$.

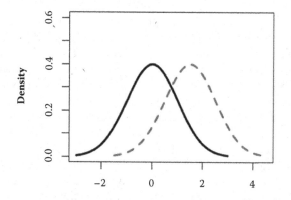

FIGURE 6.6
Graphical comparison for the case where $\mu_T \approx \mu_R$ and $\sigma_T \approx \sigma_R$.

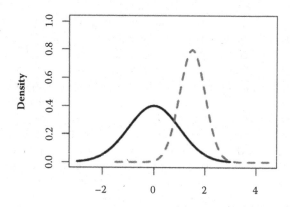

FIGURE 6.7
Graphical comparison for the case where $\mu_T \neq \mu_R$ and $\sigma_T \neq \sigma_R$.

6.5 Some Practical Considerations

The idea of FDA's proposed equivalence test for Tier 1 CQAs comes from the bioequivalence assessment for generic drugs which contain the same active ingredient(s) as the reference drug product. It may not be appropriate to apply the idea directly to the assessment of biosimilarity of biosimilar products. The FDA's proposed equivalence test is sensitive to (i) the primary assumptions made, (ii) the selection of c, and (iii) the estimation of σ_R Chow (2015) commented on these issues as follows.

Primary Assumptions – Basically, FDA's proposed equivalence test ignores (i) the lot-to-lot variability of both the reference product and the proposed biosimilar product, (ii) the difference between means, and (iii) the inflation/deflation in variability between the reference product and the proposed biosimilar product. Suppose that there are K reference lots which will be used to establish EAC for the equivalence test. FDA suggests that one sample is randomly selected from each lot. The standard deviation of the reference product σ_R can be estimated based on the K test results. Let x_i, $i = 1,...,K$ be the test result of the ith lot. x_i, $i = 1,...,K$ are assumed to be independently and identically distributed with mean μ_R and variance σ_R^2. In other words, we assume that $\mu_{Ri} = \mu_{Rj} = \mu_R$ and $\sigma_{Ri}^2 = \sigma_{Rj}^2 = \sigma_R^2$ for $i \neq j$, $i, j = 1,...,K$. Thus, the expected value of $E(\bar{x}) = \mu_R$ and $Var(\bar{x}) = \sigma_R^2/K$. In practice, it is well recognized that $\mu_{Ri} \neq \mu_{Rj}$ and $\sigma_{Ri}^2 \neq \sigma_{Rj}^2$ for $i \neq j$, where μ_{Ri} and σ_{Ri}^2 are the mean and variance of the ith lot of the reference product. A similar argument applies to the proposed biosimilar (test) product. As a result, the selection of reference lots for the estimation of σ_R is critical for the proposed approach.

In addition, FDA assumes that the difference in mean responses between the reference product and the proposed biosimilar product is proportional to the variability of the reference product. In other words, $\Delta = \mu_T - \mu_R$ (in log scale) $\propto \sigma_R$. FDA suggests that the power for detecting a clinically meaningful difference be evaluated at $\sigma_R/8$. Thus, under the assumption, the FDA's proposed equivalence testing is straightforward and easy to implement. However, Chow (2014) indicated that FDA's proposed testing procedure depends upon the selection of the regulatory standard $c = 1.5$, the anticipated difference $\Delta = \mu_T - \mu_R$, and the compromise between the test size (type I error) and statistical power (type II error) for detecting Δ (Chow, 2015).

Justification for the Selection of c – FDA indicates that a potential approach is to assume that the equivalence limit (similarity margin) is proportional to the reference product variability, i.e., $\delta = c * \sigma_R$. The constant c can be selected as the value that provides adequate power to show equivalence if there is only a small difference in the true mean between the biosimilar and reference products, when a moderate number of reference product and biosimilar lots is available for testing. FDA's recommended approach for the assessment of analytical similarity for a critical attribute is to choose $\delta = 1.5\ \sigma_R$ (i.e., $c = 1.5$) and then to select an appropriate sample size for achieving a desired power in order to establish similarity at the $\alpha = 5\%$ level of significance when the true underlying mean difference between the proposed biosimilar and reference product lots is equal to $\sigma_R/8$. FDA did not provide scientific/ statistical justification for the selection of $c = 1.5$ for equivalence acceptance criterion (EAC). Since FDA's proposed equivalence test was motivated by the bioequivalence assessment for generic drug products, the selection of $c = 1.5$ can be justified by the following steps:

Step 1. We start with $0.8 = \delta_L \leq \mu_T - \mu_R \leq \delta_U = 1.25$, where μ_T and μ_R are the reference mean and test mean (in log-scale), respectively.

Step 2. For drug products with large variabilities (i.e., highly variable drug products), FDA recommends the scaled average bioequivalence (SABE) criterion by adjusting the above bioequivalence limits for variability of the reference product (Haidar et al., 2008; Tothfalusi, Endrenyi, Garica Areta, 2009). This gives

$$0.8\sigma_R = \delta_L * \sigma_R \leq \mu_T - \mu_R \leq \delta_U * \sigma_R = 1.25 * \sigma_R$$

Step 3. FDA assumes that the difference between means is proportional to σ_R and allows a mean shift of $\dfrac{\sigma_R}{8} = 0.125$, which is the half-width of the margin. The worst possible scenario for the shift is that the true mean difference falls on $1.25 * \sigma_R$. In this case, FDA expands the margin by $0.25 * \sigma_R$. Thus, the upper margin of EAC becomes

$$1.25 * \sigma_R + 0.25 * \sigma_R = 1.5 * \sigma_R$$

Estimate of σ_R – FDA proposed that the equivalence test using available lot values be mainly based on the assumptions that (i) there is no lot-to-lot variability within the reference product and the test product, and (ii) the difference in mean responses is proportional to the variability of reference product. In practice, however, it is recognized that $\mu_{Ri} \neq \mu_{Rj}$ and $\sigma_{Ri}^2 \neq \sigma_{Rj}^2$ for $i \neq j$. The differences between lots and heterogeneity among lots are major challenges to the *validity* of the FDA's proposed approaches for both equivalence testing for CQAs in Tier 1 and the concept of quality range CQAs from Tier 2. Under the assumptions that $\mu_{Ri} \neq \mu_{Rj}$ and $\sigma_{Ri}^2 \neq \sigma_{Rj}^2$ for $i \neq j$, it is *not* clear what are the statistical properties/finite sample performances and corresponding impact on the assessment of analytical similarity and consequently on providing totality-of-the-evidence to demonstrate similarity.

Heterogeneity within/between Test and Reference Products – Let σ_R^2 and σ_T^2 be the variabilities associated with the reference product and the test product, respectively. Also, let n_R and n_T be the number of lots for analytical similarity assessment for the reference product and the test product, respectively. Thus, we have

$$\sigma_R^2 = \sigma_{WR}^2 + \sigma_{BR}^2 \text{ and } \sigma_T^2 = \sigma_{WT}^2 + \sigma_{BT}^2,$$

where σ_{WR}^2, σ_{BR}^2 and σ_{WT}^2, σ_{BT}^2 are the within-lot variability and between-lot (lot-to-lot) variability for the reference product and the test product, respectively. In practice, it is very likely that $\sigma_R^2 \neq \sigma_T^2$ and often $\sigma_{WR}^2 \neq \sigma_{WT}^2$ and $\sigma_{BR}^2 \neq \sigma_{BT}^2$ even $\sigma_R^2 \neq \sigma_T^2$. This has posted a major challenge to the FDA's proposed approaches for the assessment of analytical similarity for CQAs from both Tier 1 and Tier, especially when there is only one test sample from each lot from the reference product and the test product. FDA's proposal ignores lot-to-lot (between lot) variability, i.e., when $\sigma_{BR}^2 = 0$ or $\sigma_R^2 = \sigma_{WR}^2$. In other words, sample variance based on x_i, $i = 1,..., K$ from the reference product may underestimate the true σ_R^2, and consequently may not provide a fair and reliable assessment of analytical similarity for a given quality attribute.

In practice, it is well recognized that $\mu_{Ri} \neq \mu_{Rj}$ and $\sigma_{Ri}^2 \neq \sigma_{Rj}^2$ for $i \neq j$, where μ_{Ri} and σ_R^2 are the mean and variance of the ith lot of the reference product. A similar argument is applied to the proposed biosimilar (test) product. As a result, the selection of reference lots for the estimation of σ_R is critical for the proposed approach. The selection of reference lots has an impact on the estimation of σ_R and consequently on the EAC. Suppose there are K reference lots available and n lots will be tested for analytical similarity. FDA suggests using the remaining $K-n$ lots to establish EAC to avoid selection bias. It sounds a reasonable approach if $K \gg n$. In practice, however, there are few lots available. In this case, the FDA's proposed approach may not be feasible.

Sample Size Requirement – In practice, one of the major problems to a biosimilar sponsor is the availability of reference lots for analytical similarity testing. FDA suggests that an appropriate sample size (the number of lots from the reference product and from the test product) be used for achieving a desired power (say 80%) to establish similarity based on a two-sided test at the 5% level of significance assuming that the mean response of the test product differs from that of the reference product by $\sigma_R/8$.

Furthermore, since sample size is a function of α (type I error), β (type II error or 1 minus power), δ (treatment effect), and σ^2 (variability), it is a concern that we may have inflated the type I error rate for achieving a desired power to detect a clinically meaningful effect size (adjusted for variability) with a pre-selected small sample size (i.e., a small number of lots).

Some Thoughts on FDA Recommended Approaches – Suppose that there are K reference lots to establish EAC for the equivalence test for Tier 1 critical quality attributes. FDA suggests that one sample is randomly selected from each lot. The standard deviation of the reference product σ_R can be estimated based on the K test results. Let x_i, $i = 1,...,K$ be the test result of the ith lot. x_i, $i = 1,...,K$ are assumed to be independently and identically distributed with mean μ_R and variance σ_R^2. In other words, we assume that $\mu_{Ri} = \mu_{Rj} = \mu_R$ and $\sigma_{Ri}^2 = \sigma_{Rj}^2 = \sigma_R^2$ for $i \neq j$, $i, j = 1,...,K$. Thus, the expected value of $E(\bar{x}) = \mu_R$ and $Var(\bar{x}) = \sigma_R^2/K$. Under the assumption that $\mu_{Ri} \neq \mu_{Rj}$ and $\sigma_{Ri}^2 \neq \sigma_{Rj}^2$ for $i \neq j$, where μ_{Ri} and σ_{Ri}^2 be the mean and variance of the ith lot of the reference product. In this case, we have

$$\frac{\sigma_{(1)}^2}{K} \leq Var(\bar{x}) = \frac{\sigma_R^2}{K} \leq \frac{\sigma_{(K)}^2}{K},$$

where $\sigma_{(1)}^2$ and $\sigma_{(K)}^2$ are the smallest and largest within-lot variance among the K lots. Thus, it is recommended that the current approach of equivalence test for analytical similarity be modified as follows:

i. Randomly select at least two samples from each lot. The replicates will provide independent estimates of within-lot variability (σ_{WR}^2) and lot-to-lot variability (σ_{BR}^2). σ_R^2 is the sum of σ_{WR}^2 and σ_{RB}^2. In the interest of the same total number of tests, the sponsor can test on two samples from each lot among K/2 randomly selected lots.

ii. For the establishment of EAC, it is then suggested that $\sigma_{(K)}$ be used in order to take lot-to-lot and within-lot variabilities into consideration.

iii. In case only one sample from each lot is tested, it is suggested that the upper 95% confidence bound be used as σ_R for the establishment of EAC for equivalence testing of the identified CQAs in Tier 1. In other words, under the FDA's proposed approach, we will use the following to estimate σ_R:

$$\hat{\sigma}_R = \sqrt{\frac{n-1}{\chi_{\alpha/2,n-1}^2}}\,\hat{\sigma}_x,$$

where $\hat{\sigma}_x$ is the sample standard deviation obtained from the n reference lot test values and $\chi_{\alpha/2,n-1}^2$ is the $(\alpha/2)$th upper quantile of a chi-square distribution with n-1 degrees of freedom.

Alternative Approaches – Alternatively, we may consider a Bayesian approach with appropriate choices of priors for the mean and standard deviation of the reference product in order to take into consideration the heterogeneity in mean and variability. The Bayesian approach is to obtain a Bayesian creditable interval which will consider EAC for the assessment of analytical similarity.

6.6 Concluding Remarks

For Tier 1 equivalence test, different assumptions may lead to different conclusions due to the difference between mean responses of the various lots and the heterogeneity among lots. It should be noted that the difference between the mean responses of the lots may be offset by the heterogeneity across lots in the FDA's proposed equivalence test. Thus, one of the major criticisms of the FDA's proposed equivalence test procedure is the validity of the primary assumptions, especially the assumption that the difference in the mean responses between the reference product and the proposed biosimilar product is proportional to the variability of the reference product. In addition, for a given CQA, FDA only requires that a single sample obtained from a lot be tested. In this case, an independent estimate of the variability associated with the test result of the given lot is not available. Similar comments apply to the quality range approach for CQAs from Tier 2.

For the quality range approach for CQAs in Tier 2, FDA recommends to use x = 3 by default for 90% of values of test lots contained in the range. It allows approximately one standard deviation of reference for shifting, which may be adjusted based on biologist reviewers' recommendations. However, some sponsors propose using the concept of a tolerance interval in order to ensure that there is a high percentage of test values for the lots from the test product that fall within the quality range. It, however, should be noted that the percentage decreases when the difference in mean between the reference product and the proposed biosimilar product increases. This is also true when $\sigma_T \ll \sigma_R$. Even the tolerance interval is used as the quality range. This problem is commonly encountered mainly because the quality range approach does not take into consideration (i) the difference in means between the reference product and the proposed biosimilar product, and (ii) the heterogeneity among lots within and between products. In practice, it is very likely that a biosimilar product with small variability but a mean response which is away from the reference mean (e.g., within the acceptance range of $\sigma_R/8$ per FDA) will fall outside the quality range. In this case, a further evaluation of the data points that fall outside the quality range is necessary to rule out the possibility by chance alone.

FDA's current thinking for analytical similarity assessment using a three-tier analysis is encouraging. It provides a direction for statistical methodology development for a valid and reliable assessment toward providing the totality-of-the-evidence for demonstrating biosimilarity. The three-tier approach is currently under tremendous discussion within the pharmaceutical industry and academia. In addition to the challenging issues discussed above, there are some issues that remain unsolved and require further research. These issues include, but are not limited to, (i) the degree of similarity (i.e., how similar is considered highly similar?), (ii) multiplicity (i.e., is there a need to adjust α for controlling the overall type I error at a pre-specified level of significance?), (iii) acceptance criteria (e.g., about what percentage of CQAs in Tier 1 need to pass an equivalence test in order to pass the analytical similarity test for Tier 1?), (iv) multiple references (i.e., what if there are two reference products such as a US-licensed and an EU-approved reference product), and (v) credibility toward the totality-of-the-evidence.

7

Sample Size Requirement

7.1 Introduction

As discussed in the previous chapter, FDA recommends a stepwise approach for obtaining the totality-of-the-evidence for demonstrating biosimilarity between a proposed biosimilar (test) product and an innovative (reference) biological product (FDA, 2015a). The stepwise approach starts with analytical studies for functional and structural characterization of critical quality attributes (CQAs) that are relevant to clinical outcomes at various stages of the manufacturing process. FDA suggests that CQAs should be identified and classified into three tiers according to their criticality or risk ranking based on mechanism of action (MOA) or PK using appropriate statistical models or methods as discussed in Chapter 5. Identified CQAs that are most relevant to clinical outcomes are then classified to Tier 1 according to their criticality or risking ranking. FDA then proposes an equivalence test for CQAs in Tier 1. For a given biosimilar product development, there are usually a large number of critical quality attributes that may have an impact on clinical performance of the final product. In practice, it is often difficult, if not impossible, to conduct similarity assessment for each of these CQAs due to the fact that (i) analytical methods for some CQAs may not be fully developed, and/or (ii) there are only limited number of lots (test or reference or both) available. Thus, most sponsors focus on analytical similarity assessment on CQAs in Tier 1 as discussed in the previous chapter.

In practice, usually only a limited number of lots (test or reference or both) are available for an analytical similarity test. There are usually more reference lots than test lots. Regulatory experience indicates that not all lots were used, however, to measure each product quality attribute; the number of lots used to evaluate each quality attribute was determined by the sponsor and based on their assessment of the variability of the analytical method and availability of the United States (US)-licensed reference product and the European Union (EU)-approved reference product. This is because equivalence tests for CQAs in Tier 1 depend upon the equivalence acceptance criterion (EAC) which, in turn, depends upon the variability associated with the reference product. Thus, the investigator/sponsor is interested in the question of how to select test and reference lots for the equivalence test for Tier 1 CQAs. To assist the sponsors, several methods have been proposed by the FDA (e.g., Tsong, 2015) based on extensive simulation studies. In addition, Dong, Tsong, and Wang (2016) proposed a rule for selection of the number of reference lots for establishment of EAC and consequently the number of test lots for equivalence tests based on extensive simulation studies. This rule is easy to apply and yet lacks statistical justification (Chow et al., 2017).

In the next section, a traditional approach based on power analysis for sample size calculation is outlined. Section 7.3 summarizes FDA's current thinking on selection of test and reference lots for Tier 1 equivalence tests, as well as FDA's recommendation for sample size calculation. Following FDA's current thinking, a method for selection of test and reference lots proposed by Chow et al. (2017) is discussed in Section 7.4. Also included in this section is a recent development for sample size requirements for analytical similarity assessment. Section 7.5 provides some numerical studies for illustration of sample size calculation based on different criteria discussed in Section 7.4. Brief concluding remarks are given in the last section of this chapter. Also included in this chapter's appendix are SAS codes for power analysis for sample size calculation.

7.2 Traditional Approach

7.2.1 Power Calculation

For a given CQA in Tier 1, analytical similarity is concluded if the following null hypothesis of dis-similarity (or non-equivalence) is rejected:

$$H_0 : |\mu_T - \mu_R| \geq \delta \text{ vs. } H_a : |\mu_T - \mu_R| < \delta, \tag{7.1}$$

where $\delta > 0$ is the similarity margin or equivalence acceptance criterion (EAC), and μ_T and μ_R are the mean responses of the test (the proposed biosimilar) product and the reference (US-licensed) product, respectively. Hypothesis (7.1) can be re-written as

$$H_0 : \varepsilon \leq -\delta \text{ or } \varepsilon \geq \delta, \tag{7.2a}$$

where $\varepsilon = \mu_T - \mu_R$ is the difference between the mean responses of the proposed biosimilar product (μ_T) and the reference product lots (μ_R). Suppose that there are n_R reference lots and n_T test lots. A traditional procedure for sample size calculation involves the following steps. First, we need to derive an appropriate test statistic under the null hypothesis of dissimilarity. In practice, we intend to reject the null hypothesis in favor of the alternative hypothesis of similarity. Thus, the next step is to evaluate the performance (e.g., precision and/or power) of the test statistic under the alternative hypothesis. Then, based on certain criteria (e.g., achieving desired precision or power), select an appropriate sample size for meeting the selected criterion (e.g., achieving a desired power) at a pre-specified level of significance. In practice, one of the most frequently asked questions for analytical similarity assessment is probably how many reference lots are required for establishing an acceptable EAC in order for achieving a desired power.

For analytical similarity assessment of a given CQA in Tier 1, suppose that there are n_R reference lots and n_T test lots available and $n_R = k * n_T$, where k > 1. Equivalence (similarity) is concluded if the null hypothesis of in-equivalence (dis-similarity) is rejected at the α level of significance assuming that $\sigma_R \approx \sigma_T$. Thus, the null hypothesis would be rejected if and only if

$$\frac{\varepsilon + \delta}{\sigma_R \sqrt{\dfrac{1}{n_T} + \dfrac{1}{n_R}}} > z_\alpha \quad \text{or} \quad \frac{\varepsilon - \delta}{\sigma_R \sqrt{\dfrac{1}{n_T} + \dfrac{1}{n_R}}} < -z_\alpha.$$

Under the alternative hypothesis that $|\varepsilon| \le \delta$, the power of this test is

$$\Phi\left(\frac{\delta - \varepsilon}{\sigma_R \sqrt{\dfrac{1}{n_T} + \dfrac{1}{n_R}}} - z_\alpha\right) + \Phi\left(\frac{\delta + \varepsilon}{\sigma_R \sqrt{\dfrac{1}{n_T} + \dfrac{1}{n_R}}} - z_\alpha\right) - 1 \approx 2\Phi\left(\frac{\delta - |\varepsilon|}{\sigma_R \sqrt{\dfrac{1}{n_T} + \dfrac{1}{n_R}}} - z_\alpha\right) - 1$$

Then the required sample size to achieve power $1 - \beta$ can be obtained by solving the following equation

$$\frac{\delta - |\varepsilon|}{\sigma_R \sqrt{\dfrac{1}{n_T} + \dfrac{1}{n_R}}} - z_\alpha = z_{\beta/2}.$$

We then have (see also, Chow et al., 2008)

$$n_R = kn_T \text{ and } n_R = \frac{(z_\alpha + z_{\beta/2})^2 \sigma_R^2 \left(1 + \dfrac{1}{k}\right)}{(\delta - |\varepsilon|)^2} \tag{7.3a}$$

7.2.2 Alternative Criteria for Sample Size Calculation

Sample size required for achieving a desired power (7.3a) indicates that n_R is actually a function of α, β, ε, δ, σ_R, and k, i.e.,

$$n_R = f(\alpha, \beta, \varepsilon, \delta, \sigma_R, k) \tag{7.4a}$$

In other words, n_R (the number of reference lots) required is a function of overall type I error rate (α), type II error rate (β) or power ($1 - \beta$), clinically or scientifically meaningful difference ($\delta = \mu_T - \mu_R$), the variability associated with the reference product (σ_R) (assuming that $\sigma_T \approx \sigma_R$), and the allocation ratio between number of reference and testing product ($n_R = kn_T$). Note that ε is the true difference between μ_T and μ_R and δ is the clinically meaningful difference (or margin). In practice, ε and δ are often mixed used. In practice, it is not possible to select an appropriate sample size which controls all parameters. This leads to alternative criteria for sample size calculation. These criteria include, but are not limited to, (i) controlling σ_R, (ii) maintaining $\delta = \mu_T - \mu_R$, (iii) achieving a desired power (i.e., $1 - \beta$), and (iv) reaching desired probability of reproducibility (by controlling both σ_R and $\delta = \mu_T - \mu_R$).

For the criterion of controlling variability (σ_R), we simply choose an appropriate sample size n_R which controls the variability σ_R by fixing α, β, $\in$, δ, and k. In this case, (7.4a) reduces to $n_R = f(\sigma_R)$. Similarly, for the criterion of maintaining $\delta = \mu_T - \mu_R$, an appropriate sample size n_R will be selected to maintain clinically meaningful difference (or treatment effect) $\delta = \mu_T - \mu_R$ by fixing α, β, $\in$, σ_R, and k. In this case, (7.4a) becomes $n_R = f(\delta)$. In the interest of achieving a desired power, we typically fixed α, $\in$, δ, σ_R, and k and select an appropriate n_R for achieving a desired power (i.e., $1 - \beta$). The resultant required sample size is given in (7.3a). Thus, we have $n_R = f(\beta)$. It should be noted that criteria (i)–(iii) are considered single parameter problem because (7.4a) reduces to a function with a single parameter with other parameters fixed. On the other hand, the criterion of reaching desired probability of reproducibility (by controlling both σ_R and $\delta = \mu_T - \mu_R$) is in fact a two-parameter problem. More details can be found in Section 7.4.

7.3 FDA's Current Thinking and Recommendation

7.3.1 FDA's Current Thinking

For equivalence test for CQAs in Tier 1, under hypotheses (7.1), FDA recommended that the EAC be chosen as $1.5 * \sigma_R$, i.e., $\delta = \text{EAC} = 1.5 * \sigma_R$, where σ_R is the variability of the reference product. For sample size requirement, FDA suggested that EAC margin be established relying on the following three assumptions. First, it was assumed that the true difference in mean responses is proportional to the variability of reference product, i.e., $\mu_T - \mu_R \propto \sigma_R$. Second, FDA assumes that σ_R can be estimated from the sample standard deviation of test values from the selected reference lots. Third, FDA recommended that an appropriate sample size (number of reference product) should be selected by evaluating the power under the alternative hypothesis at $\mu_T - \mu_R = \dfrac{\sigma_R}{8}$ to achieve a desired power of the similarity test. $\dfrac{\sigma_R}{8}$ is considered maximum difference allowed for demonstrating biosimilarity. In other words, if $\varepsilon > \dfrac{\sigma_R}{8}$, we are unlikely to claim biosimilarity. However, these assumptions may not be verified in practice.

As a result, the estimate of σ_R will have great impact on the passage of the equivalence test, which also raise the following issues.

Raw data versus log-transformed data – One of the commonly asked questions regarding equivalence tests for CQAs in Tier 1 is whether the equivalence test should be applied to the raw data or log-transformed data. As indicated by the FDA, the current equivalence test is used to assess difference of two normal means. They are the means of lots instead of measurements within a lot. The common assumption of lognormal applies to measurements within a lot, not between lots. With a small sample size, FDA would not accept justification for log-transformation based on normality test. FDA, however, also indicates that some of the measurements are ratio of attribute of standard reference to the attribute of test (or reference). The data are considered lognormally distributed. In this case, the margin is determined by the standard deviation of reference after log transformation. In summary, FDA suggests that whether raw data or log-transformed data should be used depends upon the nature of the CQA rather than the data itself. FDA also indicates that the sponsor needs to obtain FDA's agreement on data transformation under such condition at early IND planning stage. Once agreed with FDA, it will not be changed even if the data are distributed like lognormal or chi-square.

Fixed versus random criteria – FDA's approach for establishment of EAC in the equivalence test is clearly a *fixed* approach. In other words, an estimate of σ_R (obtained based on the standard deviation of single test values from some selected reference lots) is treated as the true value of σ_R. In practice, however, EAC may vary depending upon the selection of reference lots. In this case, a test product may fail to pass equivalence test by chance (i.e., bad luck) depending upon the selected reference lots for establishment of EAC. For example, the test product is likely to fail the equivalence test if reference lots with smaller variabilities are selected for establishment of EAC. In this case, the resultant EAC will be narrower. To overcome this problem, Chow (2015) and Chow et al. (2016) suggest that a Bayesian approach assuming that the means and standard deviations of the selected reference lots follow some prior distributions be considered. The performances of the fixed approach and the random approach for establishment of EAC in equivalence test for CQAs in Tier 1 require further research.

Matching test and reference lots – For equivalence test for CQAs in Tier 1, the probability of passage depends upon the estimate of σ_R (or EAC margin, where EAC = $1.5 * \sigma_R$). Thus, the following questions are commonly asked: First, how many reference lots should be used to estimate σ_R? Second, how many test reference lots should be used for a valid and reliable equivalence test? Third, should the number of test lots be matched by reference lots for achieving optimal statistical property for the equivalence test? If there are a large number of reference lots and test lots available, then the above questions can be easily addressed. In practice, however, there may be more reference lots but only a limited number of test lots available. In this case, it is often a question to the investigator and/or sponsor how to perform a valid and reliable equivalence test for CQAs in Tier 1.

In the past few years, to address the above questions, several approaches have been proposed. Suppose there are N_R reference lots and N_T available, where $N_R > N_T$. The first approach is to randomly select N_T lots from the reference product to match the N_T test lots. The remaining $N_R - N_T$ reference lots are then used to estimate σ_R in order to establish the EAC margin for the equivalence test based on the N_T matched test and reference lots. This approach has been criticized for possible selection bias. Alternatively, it has been suggested to use all of the available reference and test lots: (i) all N_R reference lots are used to estimate σ_R and (ii) randomly select N_T reference lots to match the N_T lots for the equivalence test. This approach, however, is not scientifically justified either.

The selection of n_R and k = n_T/n_R – The selection of n_R for establishment of EAC and n_T for having a desired power of passing the equivalence test in probably one the most important questions to the sponsors of biosimilar products. To assist the sponsors, Dong, Tsong, and Wang (2016) proposes to select n_R as follows

$$n_R^* = min\{1.5 * N_T, N_R\}, \tag{7.2b}$$

where N_R and N_T are the available reference lots and the available test lots, respectively. The proposal of Dong, Tsong, and Wang seems to suggest that n_T be selected as $n_T^* = n_R^*/1.5$ if $1.5N_T < N_R$ based on some extensive clinical trial simulation. Their proposal does provide some guidance to the sponsors for selection of n_R for establishment of EAC. However, this proposal also raised the following questions; First, the selection of 1.5 (ratio between N_R and N_T) is not statistically justified. Second, how to select n_R^* from N_R if $1.5N_T > N_R$. Third, when $1.5N_T > N_R$, there is no rule for selecting n_T^*.

7.3.2 FDA's Recommendation

FDA recommended that an appropriate sample size (number of reference lots) should be selected for achieving 80% power for establishing equivalence (similarity) assuming that the difference between the proposed biosimilar product and the reference product is $\frac{\sigma_R}{8}$. Under this assumption, FDA intended to take variability into consideration by considering the effect size adjusted for variability:

$$\text{eff} = \frac{\mu_T - \mu_R}{\sigma_R} = \frac{\frac{1}{8}\sigma_R}{\sigma_R} = \frac{1}{8}.$$

In this case, parameters in (7.4a) have been reduced, and the number of reference lots becomes only a function of type I error (α) and power ($1 - \beta$) that is independent of σ_R. Through extensive clinical trial simulation, FDA further selected c, the constant term in the EAC margin $\delta = \pm c * \sigma_R$, to be 1.5 which gives 87% power when the difference $= \frac{\sigma_R}{8}$.

$n_T = n_R$	4	5	6	7	8	9	10	11	12	13	14
Power	67%	71%	74%	77%	79%	82%	84%	85%	87%	88%	89%
δ/σ_R	2.11	1.89	1.74	1.64	1.55	1.50	1.45	1.39	1.36	1.33	1.30
$n_T = n_R$	15	16	17	18	19	20	21	22	23	24	25
Power	90%	91%	91%	92%	93%	93%	93%	94%	94%	94%	95%
δ/σ_R	1.27	1.25	1.21	1.20	1.19	1.16	1.13	1.13	1.11	1.09	1.09

Source: Tsong et al., 'Equivalence Margin Determination for Analytical Biosimilar Assessment', Duke-Industry Statistics Symposium 2016.

However, in practice, the assumption that the clinical meaningful difference ($\mu_T - \mu_R$) is proportional to variability (σ_R) cannot be verified.

In practice, an appropriate samples size (number of reference lots) is often selected based on a desired power of $1 - \beta$ for detecting a clinically meaningful difference of $\mu_T - \mu_R$ at a pre-specified level of significance α assuming that the true variability is σ_R. If true variability (σ_R), clinically meaningful difference ($\mu_T - \mu_R$) and significance level (α) are fixed, the appropriate number of reference lots becomes a function of power. We can then select an appropriate sample size (number of reference lots) for achieving the desired power. FDA's recommendation attempts to select a sample size for achieving a desired power at a pre-specified level of significance by knowing that clinically meaningful difference and variability are varying. However, in practice, it's often difficult to control the above components at the same time.

Sample size is a critical component for Tier 1 equivalent test (FDA's current thinking). Chow, Song, and Bai (2016) proposed that under the assumption that $\delta = EAC = 1.5 * \sigma_R$ (FDA's recommendation), we can determine number of reference lots (i.e., n_R) for achieving a desired power (i.e., $1 - \beta$) at the α level of significance of various selection of k, where $k = \dfrac{n_R}{n_T}$.

7.4 Sample Size Requirement

7.4.1 Chow et al.'s Proposal for Test/Reference Lots Selection

One of the most commonly asked questions for analytical similarity assessment is probably how many reference lots are required for establishing an acceptable EAC for achieving a desired power. For a given EAC, formulas for sample size calculation under different study designs are available in Chow, Shao, and Wang (2008). In general, under a parallel-group design and the hypotheses (1), sample size (the number of reference lots, n_R, and the number of test lots, n_T) required is a function of (i) overall type I error rate (α), (ii) type II error rate (β) or power ($1 - \beta$), (iii) clinically or scientifically meaningful difference (i.e., $\mu_T - \mu_R$), and (iv) the variability associated with the reference product (i.e., σ_R) assuming that $\sigma_T = \sigma_R$. Thus, we have

$$
\begin{aligned}
n_T &= f(\alpha, \beta, \mu_T - \mu_R, k, \sigma_R) \\
&= \frac{(z_\alpha + z_{\beta/2})^2 \sigma_R^2 (1 + 1/k)}{\left(\delta - |\mu_T - \mu_R|\right)^2},
\end{aligned}
\tag{7.3b}
$$

where $k = n_T/n_R$. In practice, for a given n_R (e.g., available reference lots), we select an appropriate k (and consequently n_T) for achieving a desired power of $1 - \beta$ for detecting a clinically meaningful difference of $\mu_T - \mu_R$ at a pre-specified level of significance α assuming that the true variability is σ_R. FDA's recommendation attempts to control all parameters at the desired levels (e.g., $\alpha = 0.05$ and $1 - \beta = 0.8$) by *knowing* that $\mu_T - \mu_R$ and σ are varying. In practice, it is often difficult, if not impossible, to control (or find a balance point among) α (type I error rate), $1 - \beta$ (power), $\mu_T - \mu_R = \Delta$ (clinically meaningful difference), and σ_R (variability in observing the response) *at the same time*. For example, controlling α at a pre-specified level of significance may be at the risk of decreasing power with a selected sample size.

If α, $\mu_T - \mu_R$, and σ_R are fixed, the above equation becomes $n_T = f(\beta, k)$. We can then select an appropriate k for achieving the desired power. Under the assumptions that (i) $\mu_T - \mu_R$ is proportional to σ_R, i.e., $\mu_T - \mu_R = r\sigma_R$ and (ii) $\delta = EAC = 1.5 * \sigma_R$, Equation (7.3b) becomes

$$
n_T = \frac{A^2 \sigma_R^2 (1 + 1/k)}{(1.5\sigma_R - r\sigma_R)^2} = \frac{A^2 (1 + 1/k)}{(1.5 - r)^2},
\tag{7.4b}
$$

where $A = z_\alpha + z_{\beta/2}$. As indicated earlier, FDA allows a mean shift of $\frac{1}{8}\sigma_R = 0.125\sigma_R$. If we take this mean shift into consideration (i.e., $r = 0.125$) and assume that the desired power is 80% ($\beta = 0.2$) at the $\alpha = 5\%$ level of significance, i.e., $A = z_{0.05} + z_{0.1} = 1.645 + 0.84 = 2.285$, then the above sample size requirement (7.4b) becomes

$$n_T = \left(\frac{2.285}{1.5 - 0.125}\right)^2 \left(1 + \frac{1}{k}\right) = 2.762\left(1 + \frac{1}{k}\right),$$

where $k\frac{n_T}{n_R}$. If we choose $\frac{1}{k} = 1.5$, then $n_T = 6.905 \approx 7$. Thus, under the assumptions that (i) $\mu_T - \mu_R = r\sigma_R = \frac{1}{8}\sigma_R$ and (ii) $\delta = EAC = 1.5 * \sigma_R$, 7 test lots are required for achieving an 80% power at the 5% level of significance in an equivalence test for analytical similarity assessment for CQAs in Tier 1.

Based on the above discussion, we would like to propose the following strategy for selection of n_T, k, and then n_R assuming that N_R (available reference lots) is much larger than n_R (required reference lots for establishing EAC in equivalence test):

Step 1: Selection minimum number of test lots required, n_T

Under the assumptions that (i) $\mu_T - \mu_R = r\sigma_R = \frac{1}{8}\sigma_R$ and (ii) $\delta = EAC = 1.5 * \sigma_R$ (FDA's recommendation), we can use Equation (4) to determine n_T for achieving a desired power (i.e., $1 - \beta$) at the α level of significance for various selections of k.

Step 2: Determination of k

Depending upon the availability of the reference lots (N_R) and test lots (N_T), carefully evaluate the trade-off between controlling type I error rate and achieving desired power with various selection of different ks.

Step 3: Selection of n_R lots from N_R available reference lots

Once n_T and k have been determined, n_R can be obtained as $n_R = n_T/k$. The n_R lots, which will be randomly selected from the N_R available reference lots, will then be used for establishment of EAC for equivalence tests.

Note that the above strategy is developed under the assumptions that (i) $\mu_T - \mu_R = r\sigma_R = \frac{1}{8}\sigma_R$ and (ii) $\delta = EAC = 1.5 * \sigma_R$ (a fixed margin). In practice, however, assumption that $\mu_T - \mu_R$ is proportional to σ_R may not be met. As a result, the above power calculation for sample size may be biased. For the second assumption, since it is a fixed approach, it is very likely that different sponsors may come up with different EAC using different available reference lots. It should also be noted that the proposed strategy may result in the situation where $n_T \neq n_R$.

For the establishment of EAC, FDA made the following assumptions. First, FDA assumes that the true difference in means is proportional to σ_R, i.e., $\mu_T - \mu_R$ is proportional to σ_R. Second, σ_R is estimated by the sample standard deviation of test values from selected reference lots (one test value from each lot). Third, in the interest of achieving a desired power of the similarity test, FDA further recommends that an appropriate sample size be selected by evaluating the power under the alternative hypothesis at $\mu_T - \mu_R = r\sigma_R = \dfrac{1}{8}\sigma_R$. These assumptions, however, may not be true in practice. Thus, these debatable assumptions have generated tremendous discussion among FDA, biosimilar sponsors, and academia (see, e.g., Chow, Song, and He, 2016).

Evaluation of FDA's approach – Wang and Chow (2016) showed that FDA's approach (i.e., with single test value per lot) is an unbiased estimate of σ_R. However, when there are multiple test values per lot, the FDA's recommended approach is a biased estimate of σ_R. In the interest of having an unbiased estimate with multiple test samples per lot, Wang and Chow (2015) proposed an alternative approach to correct the biasedness. As indicated by Wang and Chow (2015), multiple test samples per lot provide valuable information regarding the heterogeneity across lots, which is useful especially when extreme lots (i.e., lots with extremely low or high variability) are selected for equivalence test.

7.4.2 Recent Development

Sample size is a critical component for Tier 1 equivalent test (FDA's current thinking). As indicated earlier, an appropriate number of reference and test lots can be determined by following equations:

$$n_R = kn_T \text{ and } n_R = \frac{(z_\alpha + z_{\beta/2})^2 \sigma_R^2 \left(1 + \dfrac{2}{k}\right)}{\left(\delta - |\epsilon|\right)^2}.$$

Similarly, we can control other single parameters or two parameters at the same time (see Table 7.1). Note that (i)–(iii) are considered one parameter problem and (iv) is a two-parameter problem (see Section 7.2.2).

TABLE 7.1

Sample Size Requirement for Controlling Specific Parameter(s)

x^{a}	$n_R = f(x)^{b}$	$x = g(n_R)^{c}$
EAC margin	$\dfrac{(z_\alpha + z_{\beta/2})^2 \sigma_R^2 \left(1 + \dfrac{1}{k}\right)}{(\delta - |\varepsilon|)^2}$	$\delta = \sqrt{\dfrac{(z_\alpha + z_{\beta/2})^2 \left(1 + \dfrac{1}{k}\right)\sigma_R^2}{n_R}} + |\varepsilon|$
σ_R	$\dfrac{(z_\alpha + z_{\beta/2})^2 \sigma_R^2 \left(1 + \dfrac{1}{k}\right)}{(\delta - |\varepsilon|)^2}$	$\sigma_R^2 = \dfrac{n_R * (\delta - |\varepsilon|)^2}{\left(1 + \dfrac{1}{k}\right)(z_\alpha + z_{\beta/2})^2}$
β	$\dfrac{(z_\alpha + z_{\beta/2})^2 \sigma_R^2 \left(1 + \dfrac{1}{k}\right)}{(\delta - |\varepsilon|)^2}$	$\beta = 2\Phi\left(\sqrt{\dfrac{n_R * (\delta - |\varepsilon|)^2}{\left(1 + \dfrac{1}{k}\right)\sigma_R^2}} - z_\alpha\right) = g(n_R)$
k	$\dfrac{(z_\alpha + z_{\beta/2})^2 \sigma_R^2 \left(1 + \dfrac{1}{k}\right)}{(\delta - |\varepsilon|)^2}$	$k = \dfrac{(z_\alpha + z_{\beta/2})^2 \sigma_R^2}{n_R (\delta - |\varepsilon|)^2 - (z_\alpha + z_{\beta/2})^2 \sigma_R^2}$
(δ, σ_R^2)	$\dfrac{(z_\alpha + z_{\beta/2})^2 \sigma_R^2 \left(1 + \dfrac{1}{k}\right)}{(\delta - |\varepsilon|)^2}$	$(\delta, \sigma_R^2)^T$

[a] Parameter(s) of Interest.
[b] Sample size formula.
[c] Inverse function of the parameter(s) of interest.

One-parameter problem

We first consider the situation that required sample size is a function of variability, EAC margin, or power.

Scenario 1: Controlling variability σ_R by fixing δ, α, k and β. Thus, Equation (7.4b) becomes

$$n_R = \frac{(z_\alpha + z_{\beta/2})^2 \sigma_R^2 \left(1 + \dfrac{1}{k}\right)}{(\delta - |\varepsilon|)^2} = f(\sigma_R^2).$$

In the interest of controlling σ_R, we can solve the following equation for n_R.

$$\sigma_R^2 = \frac{n_R {}^* \left(\delta - |\varepsilon|\right)^2}{\left(1+\dfrac{1}{k}\right)\left(z_\alpha + z_{\beta/2}\right)^2} = g(n_R)$$

Scenario 2: Maintaining EAC margin, fix σ_R^2, α, k, and β

$$n_R = \frac{\left(z_\alpha + z_{\beta/2}\right)^2 \sigma_R^2 \left(1+\dfrac{1}{k}\right)}{\left(\delta - |\varepsilon|\right)^2} = f(\delta)$$

$$\delta = \sqrt{\frac{\left(z_\alpha + z_{\beta/2}\right)^2 \left(1+\dfrac{1}{k}\right)\sigma_R^2}{n_R}} + |\varepsilon| = g(n_R)$$

Scenario 3: Achieving desired power, fix δ, α, k, and σ_R^2

$$n_R = \frac{\left(z_\alpha + z_{\beta/2}\right)^2 \sigma_R^2 \left(1+\dfrac{1}{k}\right)}{\left(\delta - |\varepsilon|\right)^2} = f(\beta)$$

$$\beta = 2\Phi\left(\sqrt{\frac{n_R {}^* \left(\delta - |\varepsilon|\right)^2}{\left(1+\dfrac{1}{k}\right)\sigma_R^2}} - z_\alpha\right) = g(n_R)$$

Scenario 4: Calculating allocation ratio, fix δ, α, β, and σ_R^2

$$n_R = \frac{\left(z_\alpha + z_{\beta/2}\right)^2 \sigma_R^2 \left(1+\dfrac{1}{k}\right)}{\left(\delta - |\varepsilon|\right)^2} = f(k)$$

$$k = \frac{\left(z_\alpha + z_{\beta/2}\right)^2 \sigma_R^2}{n_R \left(\delta - |\varepsilon|\right)^2 - \left(z_\alpha + z_{\beta/2}\right)^2 \sigma_R^2} = g(n_R)$$

Two-parameters problem

We then consider the situation that required sample size is a function both EAC margin and variability.

$$n_R = \frac{(z_\alpha + z_{\beta/2})^2 \sigma_R^2 \left(1 + \dfrac{1}{k}\right)}{(\delta - |\varepsilon|)^2} = f(\delta, \sigma_R^2)$$

$$(\delta, \sigma_R^2)^T = g(n_R)$$

An Example – Under scenario 1, controlling variability, δ, α, and β should be fixed, and estimated sample variance $\widehat{\sigma_R}^2$ is substituted for the nuisance parameter σ_R^2. Then the minimum required sample size can be re-written as a function of variance, i.e.,

$$\sigma_R^2 = \frac{n_R * (\delta - |\varepsilon|)^2}{\left(1 + \dfrac{1}{k}\right)(z_\alpha + z_{\beta/2})^2} = g(n_R)$$

In order to assess the performance of the selected sample size (i.e., n_R) based on the observed variance of reference lots (i.e., $\widehat{\sigma_R}$) for a given β and δ, the significance level (α) (or the critical value z_α) can be seen as a function of the nuisance parameter $\widehat{\sigma_R}^2$, i.e.,

$$z_\alpha = f(\widehat{\sigma_R}) = \frac{\sqrt{n_R} * \delta}{\sqrt{1 + \dfrac{1}{k} * \widehat{\sigma_R}}} - z_{\beta/2}.$$

We can evaluate whether α is controlled by looking at whether the corresponding critical value (i.e., z_α) is controlled.

By Taylor expansion, the function $z_\alpha = f(\widehat{\sigma_R})$ can be expanded as the following:

$$z_\alpha = f(\widehat{\sigma_R}) = \frac{\sqrt{n_R} * \delta}{\sqrt{1 + \dfrac{1}{k} * \widehat{\sigma_R}}} - z_{\frac{\beta}{2}} = f(\sigma_R) + \sum_{j=1}^{\infty} \frac{f^{(j)}(\sigma_R)}{j!}(\widehat{\sigma_R} - \sigma_R)^j$$

$$= f(\sigma_R) + \sum_{j=1}^{\infty} \frac{(-1)^j * \sqrt{n_R} * \delta}{\sqrt{1 + \dfrac{1}{k} * \sigma_R^{j+1} * j!}}(\widehat{\sigma_R} - \sigma_R)^j$$

Let $a_j = \left| \dfrac{(-1)^j * \sqrt{n_R} * \delta}{\sqrt{1 + \dfrac{1}{k} * \widehat{\sigma}_R^{\,j+1} * j!}} (\widehat{\sigma}_R - \sigma_R)^j \right|$. Then by ratio test,

$$\lim_{j \to \infty} \frac{|a_{j+1}|}{|a_j|} = \lim_{j \to \infty} \frac{\left| \dfrac{(-1)^{j+1} * \sqrt{n_R} * \delta}{\sqrt{1 + \dfrac{1}{k} * \sigma_R^{\,j+2} * (j+1)!}} (\widehat{\sigma}_R - \sigma_R)^{j+1} \right|}{\left| \dfrac{(-1)^j * \sqrt{n_R} * \delta}{\sqrt{1 + \dfrac{1}{k} * \sigma_R^{\,j+1} * j!}} (\widehat{\sigma}_R - \sigma_R)^j \right|}$$

$$= \lim_{j \to \infty} \frac{\left| \dfrac{(-1)^{j+1}}{\sigma_R^{\,j+2} * (j+1)!} (\widehat{\sigma}_R - \sigma_R)^{j+1} \right|}{\left| \dfrac{(-1)^j}{\sigma_R^{\,j+1} * j!} (\widehat{\sigma}_R - \sigma_R)^j \right|} = \lim_{j \to \infty} \frac{\widehat{\sigma}_R - \sigma_R}{\sigma_R * (j+1)} = 0,$$

while the remaining term converge. If we substitute the observed variance of reference lots (i.e., $\widehat{\sigma}_R$), we can obtain the estimated critical value $z_{\hat{\alpha}}$ (4). Therefore, the selected sample size is able to control α and achieve the desired power.

7.5 Numerical Studies

In this section, numerical studies are conducted to compare the required sample size for achieving desired power at the 5% level of significance under various combinations of study parameters.

For one parameter problem, Table 7.2 compares sample sizes obtained under various combination of study parameters: (i) $\varepsilon = 0$, 0.05, 0.1, and 0.15, (ii) $\sigma_R = 0.1$, 0.12, 0.40, (iii) $\beta = 0.05$, 0.1, 0.15, 0.20, and 0.25, and (iv) $k = 1, 2$, and 3. As it can be seen, sample size decrease as the true difference $\varepsilon = \mu_T - \mu_R$ decreases. For example, when $\sigma_R = 0.40$ and $\varepsilon = 0.05$, 10 lots per product are required for achieving and 80% power for demonstrating biosimilarity between a proposed biosimilar and a reference product.

On the other hand, for two-parameter problem, similarly, Table 7.3 summaries required sample sizes under various combinations of EAC margin (i.e., $\delta = 0.20$, 0.30, 0.40, and 0.50) and variability (i.e., $\sigma_R = 0.2$ and 0.4).

As it can be seen from Table 7.3, sample size increases are margin become narrower, while sample size decreases as σ_R increases.

TABLE 7.2

One-Parameter Problem: Required Sample Size Comparison under Various Combinations of Study Parameters

		k = 1				k = 2				k = 3		
ε =	0	0.05	0.1	0.15	0	0.05	0.1	0.15	0	0.05	0.1	0.15
Power = 75%												
σ_R 0.10	7	16	63	.	6	12	47	.	5	11	42	.
0.12	7	14	36	251	6	10	27	188	5	9	24	167
0.14	7	12	26	86	6	9	19	64	5	8	17	57
0.16	7	12	21	50	6	9	16	38	5	8	14	33
0.18	7	11	18	36	6	8	14	27	5	7	12	24
0.20	7	11	16	28	6	8	12	21	5	7	11	19
0.22	7	10	15	24	6	8	11	18	5	7	10	16
0.24	7	10	14	21	6	8	10	16	5	7	9	14
0.26	7	10	13	19	6	7	10	14	5	7	9	13
0.28	7	9	12	17	6	7	9	13	5	6	8	12
0.30	7	9	12	16	6	7	9	12	5	6	8	11
0.32	7	9	12	15	6	7	9	12	5	6	8	10
0.34	7	9	11	14	6	7	9	11	5	6	8	10
0.36	7	9	11	14	6	7	8	10	5	6	7	9
0.38	7	9	11	13	6	7	8	10	5	6	7	9
0.40	7	9	11	13	6	7	8	10	5	6	7	9
Power = 80%												
0.10	8	18	69	.	6	13	52	.	6	12	46	.
0.12	8	15	39	275	6	11	29	206	6	10	26	183
0.14	8	14	28	94	6	10	21	70	6	9	19	63
0.16	8	13	23	55	6	10	17	41	6	9	15	37
0.18	8	12	20	39	6	9	15	29	6	8	13	26
0.20	8	11	18	31	6	9	13	23	6	8	12	21
0.22	8	11	16	26	6	8	12	20	6	8	11	18
0.24	8	11	15	23	6	8	11	17	6	7	10	15
0.26	8	11	14	21	6	8	11	16	6	7	10	14
0.28	8	10	14	19	6	8	10	14	6	7	9	13
0.30	8	10	13	18	6	8	10	13	6	7	9	12
0.32	8	10	13	17	6	8	10	13	6	7	9	11
0.34	8	10	12	16	6	8	9	12	6	7	8	11
0.36	8	10	12	15	6	7	9	11	6	7	8	10
0.38	8	10	12	15	6	7	9	11	6	7	8	10
0.40	8	10	11	14	6	7	9	11	6	7	8	10
Power = 85%												
0.10	9	20	77	.	7	15	58	.	6	13	51	.
0.12	9	17	43	305	7	13	33	229	6	11	29	203

(Continued)

TABLE 7.2 (CONTINUED)

One-Parameter Problem: Required Sample Size Comparison under Various Combinations of Study Parameters

	$k = 1$				$k = 2$				$k = 3$			
$\varepsilon =$	0	0.05	0.1	0.15	0	0.05	0.1	0.15	0	0.05	0.1	0.15
0.14	9	15	31	104	7	11	24	78	6	10	21	70
0.16	9	14	25	61	7	11	19	46	6	9	17	41
0.18	9	13	22	43	7	10	16	33	6	9	15	29
0.20	9	13	20	34	7	10	15	26	6	9	13	23
0.22	9	12	18	29	7	9	14	22	6	8	12	19
0.24	9	12	17	25	7	9	13	19	6	8	11	17
0.26	9	12	16	23	7	9	12	17	6	8	11	15
0.28	9	11	15	21	7	9	11	16	6	8	10	14
0.30	9	11	14	20	7	9	11	15	6	8	10	13
0.32	9	11	14	18	7	8	11	14	6	8	9	12
0.34	9	11	14	17	7	8	10	13	6	7	9	12
0.36	9	11	13	17	7	8	10	13	6	7	9	11
0.38	9	11	13	16	7	8	10	12	6	7	9	11
0.40	9	11	13	16	7	8	10	12	6	7	9	11
Power = 90%												
0.10	10	22	87	.	8	17	65	.	7	15	58	.
0.12	10	19	49	347	8	14	37	260	7	13	33	231
0.14	10	17	36	118	8	13	27	89	7	12	24	79
0.16	10	16	29	69	8	12	22	52	7	11	19	46
0.18	10	15	25	49	8	11	19	37	7	10	17	33
0.20	10	14	22	39	8	11	17	29	7	10	15	26
0.22	10	14	20	33	8	11	15	25	7	9	14	22
0.24	10	13	19	29	8	10	14	22	7	9	13	19
0.26	10	13	18	26	8	10	14	20	7	9	12	17
0.28	10	13	17	24	8	10	13	18	7	9	12	16
0.30	10	13	16	22	8	10	12	17	7	9	11	15
0.32	10	12	16	21	8	9	12	16	7	8	11	14
0.34	10	12	15	20	8	9	12	15	7	8	10	13
0.36	10	12	15	19	8	9	11	14	7	8	10	13
0.38	10	12	15	18	8	9	11	14	7	8	10	12
0.40	10	12	14	18	8	9	11	13	7	8	10	12
Power = 95%												
0.10	12	26	104	.	9	20	78	.	8	18	70	.
0.12	12	23	59	416	9	17	44	312	8	15	39	278
0.14	12	20	43	142	9	15	32	107	8	14	29	95

(Continued)

TABLE 7.2 (CONTINUED)

One-Parameter Problem: Required Sample Size Comparison under Various Combinations of Study Parameters

	k = 1				k = 2				k = 3			
$\varepsilon =$	0	0.05	0.1	0.15	0	0.05	0.1	0.15	0	0.05	0.1	0.15
0.16	12	19	34	83	9	14	26	62	8	13	23	55
0.18	12	18	30	59	9	14	22	44	8	12	20	39
0.20	12	17	26	47	9	13	20	35	8	12	18	31
0.22	12	17	24	39	9	13	18	30	8	11	16	26
0.24	12	16	23	34	9	12	17	26	8	11	15	23
0.26	12	16	21	31	9	12	16	23	8	11	14	21
0.28	12	15	20	28	9	12	15	21	8	10	14	19
0.30	12	15	20	26	9	11	15	20	8	10	13	18
0.32	12	15	19	25	9	11	14	19	8	10	13	17
0.34	12	15	18	24	9	11	14	18	8	10	12	16
0.36	12	15	18	23	9	11	14	17	8	10	12	15
0.38	12	14	17	22	9	11	13	16	8	10	12	15
0.40	12	14	17	21	9	11	13	16	8	10	12	14

TABLE 7.3

Two-Parameter Problem: Required Sample Size Comparison under Various Combinations of EAC Margin and Variability

	$\sigma_R = 0.2$											
	k = 1				k = 2				k = 3			
$\varepsilon =$	0	0.05	0.1	0.15	0	0.05	0.1	0.15	0	0.05	0.1	0.15
Power = 90%												
$\delta =$ 0.20	22	39	87	347	17	29	65	260	15	26	58	231
0.30	10	14	22	39	8	11	17	29	7	10	15	26
0.40	6	8	10	14	5	6	8	11	4	5	7	10
0.50	4	5	6	8	3	4	5	6	3	3	4	5
Power = 80%												
0.20	18	31	69	275	13	23	52	206	12	21	46	183
0.30	8	11	18	31	6	9	13	23	6	8	12	21
0.40	5	6	8	11	4	5	6	9	3	4	6	8
0.50	3	4	5	6	3	3	4	5	2	3	3	4

(Continued)

TABLE 7.3 (CONTINUED)

Two-Parameter Problem: Required Sample Size Comparison under Various Combinations of EAC Margin and Variability

	$\sigma_R = 0.4$											
	k = 1				**k = 2**				**k = 3**			
$\varepsilon =$	0	0.05	0.1	0.15	0	0.05	0.1	0.15	0	0.05	0.1	0.15
Power = 90%												
$\delta=$ 0.20	87	154	347	1386	65	116	260	1039	58	103	231	924
0.30	39	56	87	154	29	42	65	116	26	37	58	103
0.40	22	29	39	56	17	22	29	42	15	19	26	37
0.50	14	18	22	29	11	13	17	22	10	12	15	19
Power = 80%												
0.20	69	122	275	1097	52	92	206	823	46	82	183	731
0.30	31	44	69	122	23	33	52	92	21	30	46	82
0.40	18	23	31	44	13	17	23	33	12	15	21	30
0.50	11	14	18	23	9	11	13	17	8	10	12	15

To provide a better understanding and illustration purpose, Figures 7.1 through 7.5 plot controlling factor against required sample size n_R for controlling variability, maintaining EAC margin, achieving desired power, optimizing allocation ratio, and reaching anticipated reproducibility, respectively.

7.6 Concluding Remarks

FDA indicated that the equivalence limit, δ, would be a function of the variability of the reference product (i.e., σ_R). That is, $\delta = EAC = 1.5 * \sigma_R$. In practice, however, it is not possible that we can select an appropriate sample size (number of reference product) for achieving a desired power at a pre-specified level of significance (e.g., $\alpha = 0.05$ and $1 - \beta = 0.8$) while other parameters (i.e., $\mu_T - \mu_R$ and σ_R) vary. Instead, we can determine n_R for achieving a desired power (i.e., $1 - \beta$) at the α level of significance for various selection of k (ratio between reference and test products) and variability under the assumption that $\delta = EAC = 1.5 * \sigma_R$.

For assessment of biosimilarity of biosimilar products, FDA recommends a stepwise approach for obtaining the totality-of-the-evidence in order for demonstration of biosimilarity between a proposed biosimilar product and

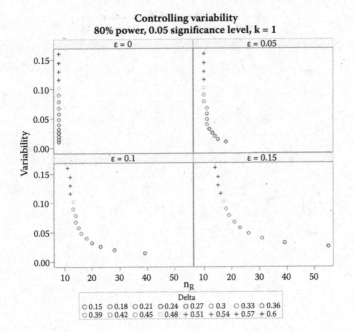

FIGURE 7.1
Sample size vs. variability plot.

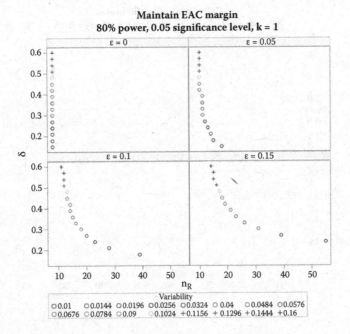

FIGURE 7.2
Sample size vs. EAC margin plot.

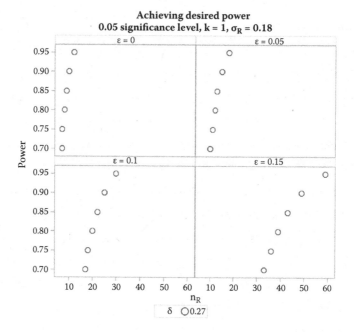

FIGURE 7.3
Sample size versus power plot.

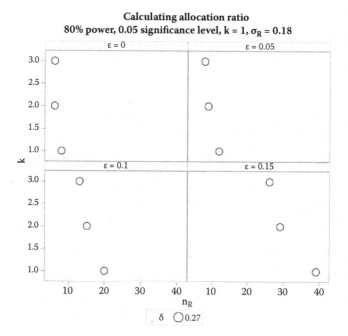

FIGURE 7.4
Sample size vs. allocation ratio plot.

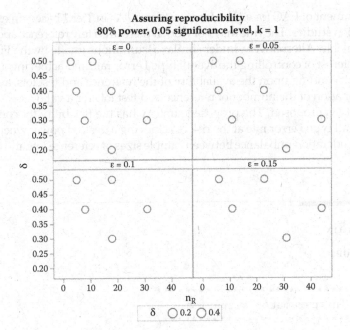

FIGURE 7.5
Sample size vs. EAC margin and variability plot.

an innovative biological product. The stepwise approach starts with structural and functional characterization of critical quality attributes that may be relevant to clinical outcomes. The totality-of-the-evidence consists of evidence from analytical studies for characterization of the molecule, animal studies for toxicity, pharmacokinetics and pharmacodynamics for pharmacological activities, clinical studies for safety/tolerability, immunogenicity, and efficacy.

For analytical similarity assessment, FDA recommends a tiered approach, that is, an equivalence test for CQAs in Tier 1 (which are considered most relevant to clinical outcomes), a quality range approach for CQAs in Tier 2 (which are considered mild-to-moderately relevant to clinical outcomes), and raw data or graphical presentations for CQAs in Tier 3 (which are considered least relevant to clinical outcomes). The FDA-recommended tiered approach has raised a number of scientific and/or controversial issues (Chow, Song, and He, 2016). These controversial issues are related to difference in population means and heterogeneity within and across lots within and between the test product and the reference product.

As equivalence test for similarity assessment for Tier 1 CQAs is currently required by the FDA for regulatory submission, how many reference and/ or test lots should be used for a valid assessment of similarity between a proposed biosimilar product and an innovative biologic product has become an interesting and important question to the sponsors. Dong, Tsong, and Wang (2016) proposed a rule for selection of the number of reference lots required for establishment of EAC for equivalence test for CQAs in Tier 1 based on extensive simulation studies. The rule may not work if there are a few reference and/or test lots available. Alternatively, under similar assumptions (made by the FDA) and in the interest of controlling the overall type I error rate and achieving a desired power, depending upon the availability of the reference and test lots, a strategy determination of the number of reference and test lots for a valid assessment of similarity is proposed. The proposed strategy has the flexibility for controlling the overall type I error rate at the risk of achieving a desired power when taking into consideration unbalance between sample sizes of reference lots and test lots.

Appendix

SAS Codes

```
/********************************
        One parameter situation
********************************/
data samplesize;
format power nR;
do beta=0.05 to 0.3 by 0.05;
        do k=1 to 3 by 1;
                do sigma=0.1 to 0.4 by 0.02;
                        do e=0 to 0.15 by 0.05;
                        power=1-beta;
                        delta=1.5*sigma;

        nR=ceil((((quantile('NORMAL',0.05)+quantile('NORMAL',
beta/2))**2)*(sigma**2)*(1+(1/k)))/((delta-abs(e))**2));
                                output;
                        end;
                end;
        end;
end;
drop beta;
run;

proc print data=samplesize;run;

%macro power(power);
title "Desired power=&power.%";
data power&power.;
set samplesize;where power=0.&power.;keep nR sigma k e;run;

proc sort data=power&power.;by sigma;run;
proc transpose data=power&power. out=power&power.(drop=_NAME_)
prefix=k_e_;
```

```
id k e;
var nR;
by sigma;
run;
%mend;

%power(75);
%power(80);
%power(85);
%power(90);
%power(95);

data final;set power75 power80 power85 power90 power95;run;
proc print data=final;run;

title;
ods rtf file='C:\Users\yizha\Documents\2016 Fall\biosimilar\
Master project\master project\power.rtf';
proc print data=final noobs;run;
options orientation=landscape;
ods rtf close;

proc sgplot data=samplesize;
series x=nR y=power/datalabel=k;
run;

proc sort data=samplesize;by k;run;
data test;set samplesize;where k=1;if nR~=.;run;
proc sgpanel data=test;
panelby e;
series x=nR y=power/group=sigma;
run;

/********************************
            Plotting
********************************/
title1 h=2.5 font=Verdana 'Controlling Variability';
title2 h=1.5 font=verdana '80% power, 0.05 significance level, k=1';
data p80;set samplesize;if power=0.8 & k=1& nR<=55;variability=
sigma**2;run;
proc sgpanel data=p80;
panelby e;
scatter x=nR y=variability/group=delta markerattrs=(size=10);
run;

title1 h=2.5 font=Verdana 'Maintaing EAC Margin';
title2 h=1.5 font=Verdana'80% power, 0.05 significance level, k=1';
```

```
data p80;set samplesize;if power=0.8 & k=1&nR<=55;variability=
sigma**2;run;
proc sgpanel data=p80;
panelby e;
scatter x=nR y=delta/group=variability markerattrs=(size=10);
run;

title1 h=2.5 font=verdana 'Achieving Desired Power';
title2 h=1.5 font=verdana '0.05 significance level,
k=1,sigma=0.18';
data p80;set samplesize;if k=1 & sigma=0.18 & nR<=60;variability=
sigma**2;run;
proc sgpanel data=p80;
panelby e;
scatter x=nR y=power/group=delta markerattrs=(size=15
color=blue);
run;

title1 h=2.5 font=verdana 'Calculating Allocation Ratio';
title2 h=1.5 font=verdana '80% power, 0.05 significance level,
k=1,sigma=0.18';
data p80;set samplesize;if power=0.8 & sigma=0.18 &
nR<=60;run;
proc sgpanel data=p80;
panelby e;
scatter x=nR y=k/group=delta markerattrs=(size=20 color=blue);
run;

/*********************************
      Two parameter situation
*********************************/

data samplesize2;
format power nR;
do sigma=0.2 to 0.4 by 0.2;
      do delta=0.2 to 0.5 by 0.1;
            do beta=0.1 to 0.2 by 0.1;
                  do k=1 to 3 by 1;
                        do e=0 to 0.15 by 0.05;
                        power=1-beta;
      nR=ceil(((((quantile('NORMAL',0.05)+quantile('NORMAL',be
ta/2))**2)*(sigma**2)*(1+(1/k)))/((delta-abs(e))**2)));
                              output;
                              end;
                        end;
                  end;
            end;
end;
```

```
drop beta;
run;
proc print data=samplesize2;run;
data test;set samplesize2;where sigma=0.2;run;

%macro sigma(sigma);
title "Sigma=0.&sigma.";
data sigma_&sigma.;
set samplesize2;where sigma=0.&sigma.;run;

proc sort data=sigma_&sigma.;by delta;run;
proc transpose data=sigma_&sigma. out=sigma_&sigma.(drop=_NAME_)
prefix=power_k_e_;
id power k e;
var nR;
by delta;
run;
%mend;

%sigma(2);
%sigma(4);

data final2;set sigma_2 sigma_4;run;
proc print data=final2;run;

title;
ods rtf file='C:\Users\yizha\Documents\2016 Fall\biosimilar\
Master project\master project\sigma2.rtf';
proc print data=sigma_2 noobs;run;
options orientation=landscape;
ods rtf close;

title;
ods rtf file='C:\Users\yizha\Documents\2016 Fall\biosimilar\
Master project\master project\sigma4.rtf';
proc print data=sigma_4 noobs;run;
options orientation=landscape;
ods rtf close;

/*Plotting*/
title1 h=2.5 font=verdana 'Assuring Reproducibility';
title2 h=1.5 font=verdana '80% power, 0.05 significance
level,k=1';
data s2;set samplesize2;if k=1 and power=0.8 and nR<=50;run;
proc sgpanel data=s2;
panelby e;
scatter x=nR y=delta/group=sigma markerattrs=(size=20);
run;
```

8

Analytical Studies with Multiple References

8.1 Background

As indicated in Chapter 1, when an innovative biological drug product is going off patent protection, biotechnology and/or pharmaceutical companies (sponsors) may seek regulatory approval for similar biological (biosimilar) products to the innovative product in European Union (through EMA) or the United States (through FDA). Thus, for assessment of biosimilarity between a proposed biosimilar product (test product) and an innovative biological product (reference product), there may be multiple references, e.g., a US-licensed reference product and an EU-approved reference product of the same product. When there are multiple references, the sponsors often obtain extensive analytical data intended not only to support a demonstration that the proposed biosimilar product and the US-licensed reference product are highly similar, but also to provide a justification of the relevance of the comparative data (e.g., pharmacokinetic and/or clinical data) generated using an EU-approved reference to support a demonstration of biosimilarity of the proposed biosimilar product to the US-licensed reference product.

In practice, however, the following questions are often encountered: Suppose there are two reference products: a US-licensed reference product and an EU-approved reference product. First, we may successfully demonstrate the proposed biosimilar product is highly similar to each of the two reference products, but fail to demonstrate that the two reference products are highly similar. Second, we are able to demonstrate that the proposed biosimilar product is highly similar to one of the two reference products but not the other. Third, it is an interesting question whether the two reference products should be combined (e.g., taking the average or adjust for their corresponding variabilities associated with the responses) for an overall biosimilarity assessment. To address the first two questions, the method of pairwise comparisons in conjunction with a head-to-head graphical comparison is often considered. For the third question, Kang and Chow (2013) proposed a three-arm study design for biosimilarity assessment under various scenarios of criteria related to multiple references.

At the 2017 July 13th Oncologic Drugs Advisory Committee (ODAC) meeting for review of biosimilar products of Avastin and Herceptin, the method of pairwise comparisons was criticized. For the method of pairwise comparisons, basically, there are three comparisons (i.e., a proposed biosimilar product versus a US-licensed reference product, the proposed biosimilar product versus an EU-approved reference product, and the US-licensed reference product versus the EU-approved reference product). The first criticism is related to the use of different EAC (equivalence acceptance criterion), which was developed based on data from test results from different reference products for the three comparisons. Different EACs may result in difference conclusions regarding the assessment of biosimilarity. The second criticism is related to the accuracy and reliability of each pairwise comparison because each comparison does not fully utilize all data collected from the three treatment groups. The third criticism is related to the justification of bridging PK and/or clinical data. In marginal cases, pairwise comparisons may not be sufficient evidence to scientifically/ statistically justify the validity of bridged PK and/or clinical data. As an alternative, the ODAC suggested the potential use of a simultaneous confidence approach, which has the advantages of utilizing all data collected from the study and using a single reference product. In other words, if submission occurs in the US, the US-licensed reference product will be selected as the single reference for the analytical similarity assessment with multiple reference products.

In the next section, the method of pairwise comparisons for analytical similarity assessment with multiple references is briefly outlined. The simultaneous confidence interval approach as suggested by the ODAC is described in Section 8.3. Also included in this section is a simulation study for evaluation of relative performances between the method of pairwise comparisons and the simultaneous confidence interval approach for various scenarios of different reference products. In Section 8.4, Kang and Chow's method for addressing the third question is discussed. Some concluding remarks are given in the last section of this chapter.

8.2 Method of Pairwise Comparisons

8.2.1 Equivalence Test for Tier 1 CQAs

For CQAs in Tier 1, FDA recommends that an equivalent test can be performed to assess analytical similarity. As indicated by the FDA, a potential approach could be a similar approach to the confidence interval method of bioequivalence testing for generic products under the raw data model.

In other words, for a given CQA, we may test for equivalence by the following interval (null) hypothesis:

$$H_0 : \mu_T - \mu_R \leq -\delta \ \text{ or } \ \mu_T - \mu_R \geq \delta, \tag{8.1}$$

where $\delta > 0$ is the equivalence limit (or similarity margin), and μ_T and μ_R are the mean responses of the test (the proposed biosimilar) product and the reference product lots, respectively. Analytical equivalence (similarity) is concluded if the null hypothesis of nonequivalence (dissimilarity) is rejected. Under the above null hypothesis, analytical similarity would be accepted for a given CQA if the $(1 - 2\alpha) \times 100\%$ two-sided confidence interval of the mean difference is within $(-\delta, \delta)$.

FDA further suggested that the equivalence acceptance criterion (EAC) as $\delta = 1.5 \, \sigma_R$, where σ_R is the population standard deviation associated with the reference product. In practice, σ_R can be estimated based on test values of some randomly sampled lots from a pool of reference lots. The suggested EAC margin is considered as fixed margin conditioned on the observed test values from different reference lots. In equivalence test for CQAs from Tier 1, it is very challenging for the sponsors and/or biostatisticians when there are only a limited number of lots available (for both reference product and test product). Thus, it is suggested that the sponsors provide a plan on how the reference standard deviation, σ_R, would be estimated with satisfactory scientific/statistical justification.

For a given CQA in Tier 1, denote $\Delta = \mu_T - \mu_R$ as the mean difference. Then null hypothesis (8.1) can be rewritten as:

$$H_0 : \Delta \leq -\delta \ \text{ or } \ \Delta \geq \delta. \tag{8.2}$$

Suppose there are n_R reference lots and n_T test lots for the equivalence test. Based on a two one-sided tests procedure, similarity is concluded if the null hypothesis of dissimilarity is rejected at the α level of significance, if

$$\frac{\hat{\Delta} + \delta}{\hat{\sigma}_R \sqrt{\dfrac{1}{n_T} + \dfrac{1}{n_R}}} > z_\alpha,$$

and

$$\frac{\hat{\Delta} + \delta}{\hat{\sigma}_R \sqrt{\dfrac{1}{n_T} + \dfrac{1}{n_R}}} < z_\alpha,$$

where $\hat{\Delta}$ is an estimator of Δ, z_α is the lower α quantile of the standard normal distribution, and $\hat{\sigma}_R$ is an estimator of σ_R. The statistical method is based on the assumption that $\sigma_R \approx \sigma_T$ where σ_T, is the population standard deviation associated with the test product. For estimating σ_R, FDA recommends testing one sample from each reference lot for obtaining an estimator of σ_R. This approach is an unbiased estimate of σ_R. $\hat{\Delta}$ is the difference of the arithmetic means between the test samples and reference samples.

Note that since a two one-sided tests procedure is operationally equivalence to a $(1 - 2\alpha) \times 100\%$ confidence interval approach in many cases, similarity is concluded if the $(1 - 2\alpha) \times 100\%$ confidence interval falls within the limits of $\left(\hat{\Delta} - 1.5\hat{\sigma}_R, \hat{\Delta} + 1.5\hat{\sigma}_R\right)$.

8.2.2 Pairwise Comparisons with Multiple References

Where there are multiple references, e.g., a US-licensed reference product and an EU-approved reference product of the same product, it is suggested pairwise comparisons be considered not only to (i) check whether the two reference products are highly similar, but also to (ii) compare the proposed biosimilar with each of the two references.

Denote T, R_1, and R_2 as the proposed biosimilar (test) product, the first reference product (e.g., a US-licensed reference product), and the second reference product (e.g., an EU-approved reference product), respectively. The pairwise comparisons deal with the following three sets of interval hypotheses:

$$H_{00} : \mu_{R_2} - \mu_{R_1} \leq -\delta \ \text{ or } \ \mu_{R_2} - \mu_{R_1} \geq \delta, \tag{8.3}$$

$$H_{01} : \mu_T - \mu_{R_1} \leq -\delta \ \text{ or } \ \mu_T - \mu_{R_1} \geq \delta, \tag{8.4}$$

$$H_{02} : \mu_T - \mu_{R_2} \leq -\delta \ \text{ or } \ \mu_T - \mu_{R_2} \geq \delta, \tag{8.5}$$

where the first two hypothesis use R_1 as the reference and the third uses R_2 as the reference. Each null hypothesis, i.e., (8.3)–(8.5) can be tested using the two one-sided tests procedure at the α level of significance described in the previous section. As indicated earlier, since the two one-sided tests procedure is operationally equivalence to a $(1 - 2\alpha) \times 100\%$ confidence interval approach in many cases, similarity is often concluded if the $(1 - 2\alpha) \times 100\%$ confidence interval falls within the equivalence limit. Intuitively, pairwise comparisons sound reasonable. However, as indicated by the ODAC panel at the 2017 July 13th ODAC meeting, pairwise comparisons may not be justifiable due to the following deficiencies:

First, the equivalence limits may be different from one comparison to another. As it can be seen from hypotheses (8.3)–(8.5), hypotheses (8.3) and (8.4) use R_1 as the reference product, which hypothesis uses R_2 as the reference product. As a result, pairwise comparisons may be biased because the equivalence limits are data-driven, which depends upon an estimated variability associated with the reference product. This may present a critical issue in assessing biosimilarity especially when the test product is highly similar to each of the reference products, but a notable difference is observed between the two reference products (i.e., the two reference products fail to pass the equivalence test).

The other criticism of pairwise comparisons is that each pairwise comparison does not fully utilize all data collected from the three treatment groups. That is, hypothesis (8.3) uses data obtained from both R_1 and R_2, hypothesis (8.4) is tested based on data from the test (T) product and the first reference product (R_1), while hypothesis (8.5) considers data obtained from the test (T) product and the second reference (R_2). This may present a critical issue in assessing biosimilarity when there is evidence of heterogeneity in mean and/or variance among the three groups with limited number of lots (both test and/or reference lots) available.

As a result, the feasibility and/or validity of pairwise comparisons have been challenged.

8.2.3 An Example

For illustration of the pairwise comparisons method for analytical similarity assessment of a given critical quality attribute which is considered most relevant to clinical outcomes when there are multiple reference products, consider an example concerning analytical similarity assessment of proliferation inhibition bioassay among a proposed biosimilar product (ABP215), US-licensed Avastin, and EU-approved bevacizumab in a recent biosimilar regulatory submission (BLA 761028) by Amgen.

The analytical similarity test for proliferation inhibition bioassay was performed on 13 lots of ABP 215, 24 lots of US-licensed Avastin, and 27 lots of EU-approved bevacizumab. The proliferation inhibition bioassay data distributions of ABP215, US-licensed Avastin, and EU-approved bevacizumab are displayed in Figure 8.1. Descriptive statistics for the proliferation inhibition bioassay are listed in Table 8.1. Table 8.2 summarizes the results of pairwise comparisons among ABP215, US-licensed Avastin, and EU-approved bevacizumab.

As it can be seen from Table 8.2, the 90% confidence interval for the mean difference in the proliferation inhibition bioassay between ABP215 and US-licensed Avastin is (0.55%, 5.29%), which falls entirely within the similarity margin of (−5.90%, 5.90%). Hence, the results of proliferation inhibition bioassay for ABP215 are equivalent to those for US-licensed Avastin. Also, the 90% confidence interval for the mean difference in proliferation inhibition bioassay between ABP215 and EU-approved bevacizumab is (0.98%, 5.64%), which falls within the similarity margin of (−5.64%, 5.64%). Therefore, the results of proliferation inhibition bioassay for ABP215 are equivalent to

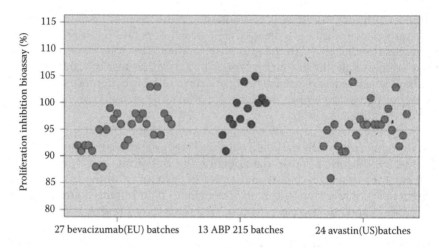

FIGURE 8.1

Scatter plots of proliferation inhibition bioassay for US-licensed Avastin, ABP215, and EU-approved bevacizumab. (Courtesy of US FDA regulatory submission BLA 761028.)

TABLE 8.1

Descriptive Statistics for the Proliferation Inhibition Bioassay Data

Product	Number of Batches	Sample Mean, %	Sample Standard Deviation, %	Min, %	Max, %
ABP215	13	98.5	3.86	91	105
US-licensed Avastin	24	95.5	3.93	86	104
EU-approved bevacizumab	27	98.1	3.76	88	103

Source: US FDA regulatory submission BLA 761028.

TABLE 8.2

Pairwise Comparison Equivalence Testing Results

Comparisont	Number of Batches	Mean Difference, %	90% Confidence Interval, %	Equivalence Margin, %	Equivalent
ABP215 vs US	(13, 24)	2.92	(0.55, 5.29)	(−5.90, +5.90)	Yes
ABP215 vs EU	(13, 24)	3.31	(0.98, 5.64)	(−5.64, +5.64)	Yes
EU vs US	(27, 24)	−0.39	(−2.21, 1.42)	(−5.90, +5.90)	Yes

Source: US FDA regulatory submission BLA 761028.

those for EU-approved bevacizumab. Similarly, the 90% confidence interval for the mean difference in the proliferation inhibition bioassay between EU-approved bevacizumab and US-licensed Avastin is (–2.21%, 1.42%), which falls entirely within the similarity margin of (–5.90%, 5.90%). Thus, the results of proliferation inhibition bioassay for EU-approved bevacizumab are equivalent to those for US-licensed Avastin.

8.3 Simultaneous Confidence Approach

As an alternative to the pairwise comparisons approach, the ODAC panel suggested the potential use of a simultaneous confidence approach, which allows us to fully utilize all data from the study with single reference product (e.g., the US-licensed product). In this section, we will describe the simultaneous confidence interval method proposed by Zheng et al. (2018) under a parallel-group design for analytical studies.

8.3.1 Assumptions and Statistical Framework

For illustration of the concept of simultaneous confidence interval and for simplicity, we will consider the case where there are one test product and two reference products, denoted by T, R_1, and R_2. Without loss of generality, let T, R_1, and R_2 be the test (proposed biosimilar) product, the US-licensed product, and the EU-approved product. We further assume that R_1 is the primary reference product and R_2 is the secondary reference product for regulatory submission.

For a given critical quality attribute (CQA), FDA recommends performing a single test on each lot. Let n_1 be the samples from the n_1 (primary) reference lots and let n_2 be the samples from the n_2 (secondary) reference lots. Test results from these samples are then used to obtain estimates of σ_{R_1} and σ_{R_2}, where σ_{R_1} and σ_{R_2} are the standard deviations associated with the primary reference product and secondary reference product, respectively. Furthermore, denote by σ_T the standard deviation associated with the test product. Now suppose there are N_T, N_{R_1}, and N_{R_2} lots for the test product, the primary reference product, and the secondary product, respectively. For a given test (primary reference, secondary reference) lot, assume that the test value follows a normal distribution with mean $\mu_{Ti}\left(\mu_{R_1i}, \mu_{R_2i}\right)$ and variance $\sigma_{Ti}^2\left(\sigma_{R_1i}^2, \sigma_{R_2i}^2\right)$. For equivalence test for CQAs in Tier 1, FDA's recommended approach assumes that

$$\mu_{Ti} = \mu_{Tj} \text{ and } \sigma_{Ti}^2 = \sigma_{Tj}^2 \text{ for } i \neq j, i, j = 1, \dots, N_T;$$

$$\mu_{R_1i} = \mu_{R_1j} \text{ and } \sigma_{R_1i}^2 = \sigma_{R_1j}^2 \text{ for } i \neq j, i, j = 1, \dots, N_{R_1};$$

and

$$\mu_{R_2i} = \mu_{R_2j} \text{ and } \sigma^2_{R_2i} = \sigma^2_{R_2j} \text{ for } i \neq j, i, j = 1, \ldots, N_{R_2}.$$

In other words,

$$\mu_T = \mu_{T1} = \cdots = \mu_{TN_T}, \sigma^2_T = \sigma^2_{T1} \cdots \sigma^2_{TN_T},$$

$$\mu_{R_1} = \mu_{R_11} = \cdots = \mu_{R_1N_{R_1}}, \sigma^2_{R_1} = \sigma^2_{R_11} \cdots \sigma^2_{R_1N_{R_1}},$$

and

$$\mu_{R_2} = \mu_{R_21} = \cdots = \mu_{R_2N_{R_2}}, \sigma^2_{R_2} = \sigma^2_{R_21} \cdots \sigma^2_{R_2N_{R_2}},$$

Thus,

$$n_T = N_T,$$

$$n_{R_1} = N_{R_1} - n_1$$

and

$$n_{R_2} = N_{R_2} - n_2$$

lots are used for testing hypotheses (8.3)–(8.5) with estimates (based on the test values) of σ_{R_1} and σ_{R_2}. These estimates are then considered as the true values for obtaining the EAC margins.

Following the sampling plan of one sample from each reference lot as recommended by the FDA, the empirical variance estimators of σ_{R_1} and σ_{R_2}, denoted as $\hat{\sigma}_{R_1}$ and $\hat{\sigma}_{R_2}$, respectively, follow the probability distributions below:

$$\frac{(n_1 - 1)\hat{\sigma}^2_{R_1}}{\sigma^2_{R_1}} \sim \chi^2(n_1 - 1),$$

$$\frac{(n_2 - 1)\hat{\sigma}^2_{R_2}}{\sigma^2_{R_2}} \sim \chi^2(n_2 - 1),$$

(8.6)

where $\chi^2(n_1 - 1)$ and $\chi^2(n_2 - 1)$ are chi-square distributions with the degree of freedom $n_1 - 1$ and $n_2 - 1$, respectively. For testing hypotheses with $\hat{\sigma}_{R_1}$ and $\hat{\sigma}_{R_2}$ obtained, denote $X_{iT}, i = 1, \ldots, n_T; X_{iR_1}, i = 1, \ldots, n_{n_{R_1}}$ and $X_{iR_2}, i = 1, \ldots, n_{n_{R_2}}$

as the observations (test results) of the CQA in Tier 1 of the test arm, the primary reference arm, and secondary reference arm, respectively.

To propose the simultaneous confidence interval methods under the framework described above, Zheng et al. (2018) considered the scenarios (i) under the assumption that $\sigma_T = \sigma_{R_1} = \sigma_{R_2}$ and (ii) without the assumption that $\sigma_T = \sigma_{R_1} = \sigma_{R_2}$, which are briefly described below.

8.3.2 Simultaneous Confidence Interval with the Assumption That $\sigma_T = \sigma_{R_1} = \sigma_{R_2}$

Assume $\sigma_T = \sigma_{R_1} = \sigma_{R_2}$ and samples for each arm are independent and identical distributed. Denote

$$\bar{X}_T = \sum_{i=1}^{n_T} X_{iT}/n_T,$$

$$\bar{X}_{R_1} = \frac{\sum_{i=1}^{n_{R_1}} X_{iR_1}}{n_{R_1}},$$

and

$$\bar{X}_{R_2} = \sum_{i=1}^{n_{R_2}} X_{iR_2}/n_{R_2}.$$

We have

$$\bar{X}_T \sim N\left(\mu_T, \sigma_{R_1}^2/n_T\right), \bar{X}_{R_1} \sim N\left(\mu_{R_1}, \sigma_{R_1}^2/n_{R_1}\right) \text{ and } \bar{X}_{R_2} \sim N\left(\mu_{R2}, \sigma_{R_1}^2/n_{R2}\right).$$

Follow similar idea of fiducial inference theory, the marginal fiducial distributions of the three location parameters can be obtained as follows:

$$\mu_T \sim N\left(\bar{X}_T, \sigma_{R_1}^2/n_T\right),$$

$$\mu_{R_1} \sim N\left(\bar{X}_{R_1}, \sigma_{R_1}^2/n_{R_1}\right),$$

$$\mu_{R_2} \sim N\left(\bar{X}_{R_2}, \sigma_{R_1}^2/n_{R_2}\right),$$

Denote $f_1(x), f_2(y)$, and $f_3(z)$ as the probability density functions of the above three normal distributions, respectively. Since the three groups of samples,

$$\left\{X_{iT}, i = 1, \ldots, n_T\right\},$$

$$\left\{X_{iR_1}, i = 1, \ldots, n_{n_{R_1}}\right\},$$

and

$$\left\{ X_{iR_2}, i = 1, \ldots, n_{n_{R_2}} \right\},$$

are statistically independent between each other, the joint fiducial probability density function of $\left(\mu_T, \mu_{R_1}, \mu_{R_2} \right)$ can be express as $f(x, y, z) = f_1(x)f_2(y)f_3(z)$. Now we define the first version of fiducial probability.

$$FP_1\left(\sigma_{R_1}\right)$$
$$= pr\left\{ \left|\mu_T - \mu_{R_1}\right| \le 1.5\sigma_{R_1}, \left|\mu_{R_1} - \mu_{R_2}\right| \le 1.5\sigma_{R_1}, \left|\mu_{R_1} - \mu_{R_2}\right| \le 1.5\sigma_{R_1} \middle| \sigma_{R_1}, f \right\} \quad (8.7)$$
$$= \iiint_{|x-y| \le 1.5\sigma_{R_1}, |x-z| \le 1.5\sigma_{R_1}, |y-z| \le 1.5\sigma_{R_1}} f(x, y, z)\, dx\, dy\, dz$$

If the above $FP_1\left(\sigma_{R_1}\right) \ge 1 - \alpha$, where $1 - \alpha$ is the pre-specified confidence level, the null hypothesis of (8.3) is rejected and analytical similarity between T and R_1 is concluded.

As indicated earlier, two one-sided tests procedure is operationally equivalent to the confidence interval approach in many cases. Under (8.7), Zheng et al. (2018) proposed the following two types of simultaneous confidence interval for $\mu_T - \mu_{R_1}$ namely type 1 restricted simultaneous confidence interval (RSCI I) and type 2 restricted simultaneous confidence interval (RSCI II), which are briefly outlined below.

Type 1 Restricted Simultaneous Confidence Interval (RSCI I)

For any $\Delta > 0$, we first calculate the following fiducial probability based on $f(x, y, z)$

$$pr\left\{ \left|\mu_T - \mu_{R_1}\right| \le \Delta, \left|\mu_T - \mu_{R_2}\right| \le 1.5\sigma_{R_1}, \left|\mu_{R_1} - \mu_{R_2}\right| \le 1.5\sigma_{R_1} \middle| \sigma_{R_1}, f \right\},$$

which is denoted as q_Δ. When $\Delta = 1.5\sigma_{R_1}$, q_Δ is equal to $FP_1\left(\sigma_{R_1}\right)$. For any $\Delta > 0$, we then find the minimal Δ that satisfies

$$pr\left\{ \left|\mu_T - \mu_{R_1}\right| \le \Delta, \left|\mu_T - \mu_{R_2}\right| \le 1.5\sigma_{R_1}, \left|\mu_{R_1} - \mu_{R_2}\right| \le 1.5\sigma_{R_1} \middle| \sigma_{R_1}, f \right\} \ge q.$$

Denote the minimal Δ by Δ_q^1 if it exists. Then the type 1 restricted simultaneous confidence interval (RSCI I) of $\mu_T - \mu_{R_1}$ can be obtained as $\left(-\Delta_q^1, \Delta_q^1\right)$, with the confidence level of q. If $\Delta_{1-\alpha}^1$ exists and $\Delta_{1-\alpha}^1 \le 1.5\sigma_{R_1}$, the analytical similarity between T and R_1 is concluded. In other words, in this case, we have

$$pr\left\{ \left|\mu_T - \mu_{R_1}\right| \le 1.5\sigma_{R_1}, \left|\mu_T - \mu_{R_2}\right| \le 1.5\sigma_{R_1}, \left|\mu_{R_1} - \mu_{R_2}\right| \le 1.5\sigma_{R_1} \middle| \sigma_{R_1}, f \right\} \ge 1 - \alpha.$$

Type 2 Restricted Simultaneous Confidence Interval (RSCI II)

For any $\Delta > 0$, the type 2 restricted simultaneous confidence interval (RSCI II) can be obtained similarly. We first calculate the follows fiducial probability based on $f(x, y, z)$

$$pr\left\{\left|\mu_T - \mu_{R_1}\right| \leq \Delta, \left|\mu_T - \mu_{R_2}\right| \leq \Delta, \left|\mu_{R_1} - \mu_{R_2}\right| \leq \Delta \mid \sigma_{R_1}, f\right\},$$

which is denoted as q_Δ. When $\Delta = 1.5\sigma_{R_1}$, q_Δ is equal to $FP_1\left(\sigma_{R_1}\right)$. For any $\Delta > 0$, find the minimal Δ satisfying

$$pr\left\{\left|\mu_T - \mu_{R_1}\right| \leq \Delta, \left|\mu_T - \mu_{R_2}\right| \leq \Delta, \left|\mu_{R_1} - \mu_{R_2}\right| \leq \Delta \mid \sigma_{R_1}, f\right\} \geq q.$$

Denote the minimal Δ by Δ_q^2 if it exists. The RSCI II confidence interval of $\mu_T - \mu_{R_1}$ can be obtained as $\left(-\Delta_q^2, \Delta_q^2\right)$, with the confidence level of q. If $\Delta_{1-\alpha}^2$ exists and $\Delta_{1-\alpha}^2 \leq 1.5\sigma_{R_1}$, the analytical similarity between T and R_1 is concluded. In this case, we have

$$pr\left\{\left|\mu_T - \mu_{R_1}\right| \leq 1.5\sigma_{R_1}, \left|\mu_T - \mu_{R_2}\right| \leq 1.5\sigma_{R_1}, \left|\mu_{R_1} - \mu_{R_2}\right| \leq 1.5\sigma_{R_1} \mid \sigma_{R_1}, f\right\} \geq 1 - \alpha.$$

Note: that in practice, the true value of σ_{R_1} is often unknown. In this case, we can simply replace σ_{R_1} by its estimate $\hat{\sigma}_{R_1}$ in all of the expressions above and obtained estimates for the fiducial probability in (8.7), i.e., $FP_1\left(\hat{\sigma}_{R_1}\right)$ and the two restricted simultaneous confidence intervals. (i.e., $\hat{\Delta}_{1-\alpha}^1, \hat{\Delta}_{1-\alpha}^2$). In practice, if $\hat{\sigma}_{R_1}$ is a good estimate of σ_{R_1}, it is expected that $\hat{\Delta}_{1-\alpha}^1$ and $\hat{\Delta}_{1-\alpha}^2$ would perform similarly as compared with the RSC I assuming that σ_{R_1} is known.

It can be easily verified that $\Delta_{1-\alpha}^2 \geq \Delta_{1-\alpha}^1$ and $\hat{\Delta}_{1-\alpha}^2 \geq \hat{\Delta}_{1-\alpha}^1$. Thus, the RSCI II confidence interval approach is more conservative than the RSCI I confidence interval approach. In other words, RSCI I confidence interval tends to, more favorably, conclude the rejection of all of the hypotheses as compared to that of RSCI II confidence interval.

Modified RSCI I and RSCI II Confidence Intervals

As discussed in the previous section, $\sigma_{R_1}\left(\hat{\sigma}_{R_1}\right)$ is considered know (its estimate is fixed as the true value). However, in real world, σ_{R_1} is often unknown and there exists variability associated with the estimate of σ_{R_1} (i.e., $\hat{\sigma}_{R_1}$). To take this variability into consideration, Zheng and Chow (2018) also proposed two modified simultaneous confidence intervals based on $\hat{\Delta}_{1-\alpha}^1, \hat{\Delta}_{1-\alpha}^2$ and $FP_1\left(\hat{\sigma}_{R_1}\right)$. One is referred to as the integrated version and the other is known as the least favorable version (Zheng et al. 2018). Both modified simultaneous confidence intervals proposed by Zheng et al. (2018) are derived based on the fiducial distribution of σ_{R_1} given in (8.6). As it can be seen from (8.6), the fiducial distribution of σ_{R_1} can be expressed as

$$\sqrt{\frac{(n_1-1)\hat{\sigma}_{R_1}^2}{\chi^2(n_1-1)}},$$

where $\hat{\sigma}_{R_1}^2$ is considered as fixed and $\chi^2(n_1-1)$ is chi-square distribution with degree of freedom n_1-1. Denote the probability density function of this fiducial distribution as e_1.

The integrated version – The integrated fiducial probability (IFP) can be expressed as

$$IFP_1 = \int FP_1(u)e_1(u)du. \tag{8.8}$$

Similarly, replace

$$pr\left\{\left|\mu_T-\mu_{R_1}\right|\le\Delta,\left|\mu_T-\mu_{R_2}\right|\le1.5\sigma_{R_1},\left|\mu_{R_1}-\mu_{R_2}\right|\le1.5\sigma_{R_1}\,|\,\sigma_{R_1},f\right\}$$

and

$$pr\left\{\left|\mu_T-\mu_{R_1}\right|\le\Delta,\left|\mu_T-\mu_{R_2}\right|\le\Delta,\left|\mu_{R_1}-\mu_{R_2}\right|\le\Delta\,|\,\sigma_{R_1},f\right\}$$

simply by their integrated versions

$$\int pr\left\{\left|\mu_T-\mu_{R_1}\right|\le\Delta,\left|\mu_T-\mu_{R_2}\right|\le1.5u,\left|\mu_{R_1}-\mu_{R_2}\right|\le1.5u\,|\,u,f\right\}e_1(u)du$$

and

$$\int pr\left\{\left|\mu_T-\mu_{R_1}\right|\le\Delta,\left|\mu_T-\mu_{R_2}\right|\le\Delta,\left|\mu_{R_1}-\mu_{R_2}\right|\le\Delta\,|\,u,f\right\}e_1(u)du$$

in the expressions above. Then with the same derivation, we have the type 1 integrated restricted simultaneous confidence interval

(IRSCI I) $\Delta^1_{I,1-\alpha}$ and the type 2 integrated restricted simultaneous confidence interval (IRSCI II) $\Delta^2_{I,1-\alpha}$.

> *The least favorable version* – it would be more conservative when the used value for σ_{R_1} is smaller, i.e., it's hard to reject all three hypotheses with smaller value of σ_{R_1}. Thus, we suggest another version using the $1 - \alpha$ lower fiducial confidence bound to estimate the least favorable values of σ_{R_1}, i.e.,

$$\hat{\sigma}'_{R_1} = \sqrt{\frac{(n_1 - 1)\hat{\sigma}^2_{R_1}}{\chi^2_{1-\alpha/2}(n_1 - 1)}},$$

where $\chi^2_{1-\alpha/2}(n_1 - 1)$ is the $1 - \alpha/2$ quantile of chi-square distribution with degree of freedom $n_1 - 1$. This leads to the least favorable fiducial probability (LFFP) $LFFP_1 = FP_1(\hat{\sigma}'_{R_1})$, the type 1 least favorable restricted simultaneous confidence interval (LFRSCI I) $\Delta^1_{LF,1-\alpha}$, and the type 2 least favorable restricted simultaneous confidence interval (LFRSCI II) $\Delta^2_{LF,1-\alpha}$.

8.3.3 Simultaneous Confidence Interval without the Assumption of $\sigma_T = \sigma_{R_1} = \sigma_{R_2}$

Now we do not assume $\sigma_T = \sigma_{R_1} = \sigma_{R_2}$ but still assume that samples for each arm are independent and identical distributed. Denote

$$\bar{X}_T = \sum_{i=1}^{n_T} X_{iT}/n_T,$$

$$\bar{X}_{R_1} = \sum_{i=1}^{n_{R_1}} X_{iR_1}/n_{R_1},$$

$$\bar{X}_{R_2} = \sum_{i=1}^{n_{R_2}} X_{iR_2}/n_{R_2},$$

$$\tilde{\sigma}_T = \sqrt{\sum_{i=1}^{n_T} (X_{iT} - \bar{X}_T)^2/(n_T - 1)},$$

$$\tilde{\sigma}_{R_1} = \sqrt{\sum_{i=1}^{n_{R_1}} (X_{iR_1} - \bar{X}_{R_1})^2/(n_{R_1} - 1)}.$$

and

$$\tilde{\sigma}_{R_2} = \sqrt{\sum_{i=1}^{n_{R_2}} (X_{iR_2} - \bar{X}_{R_2})^2/(n_{R_2} - 1)}.$$

We have

$$\sqrt{n_T}(\bar{X}_T - \mu_T)/\tilde{\sigma}_T \sim t(n_T - 1),$$
$$\sqrt{n_{R_1}}(\bar{X}_{R_1} - \mu_{R_1})/\tilde{\sigma}_{R_1} \sim t(n_{R_1} - 1),$$

and

$$\sqrt{n_{R_2}}(\bar{X}_{R_2} - \mu_{R_2})/\tilde{\sigma}_{R_2} \sim t(n_{R_2} - 1),$$

where $t(n - 1)$ is the t distribution with degree of freedom $n - 1$. Follow similar idea of fiducial inference theory, the marginal fiducial distributions of the three location parameters can be obtained as follows:

$$\mu_T \sim \bar{X}_T + \frac{\tilde{\sigma}_T t(n_T - 1)}{\sqrt{n_T}},$$

$$\mu_{R_1} \sim \bar{X}_{R_1} + \frac{\tilde{\sigma}_{R_1} t(n_{R_1} - 1)}{\sqrt{n_{R_1}}},$$

$$\mu_{R_2} \sim \bar{X}_{R_2} + \frac{\tilde{\sigma}_{R_2} t(n_{R_1} - 1)}{\sqrt{n_{R_2}}}.$$

Denote $g_1(x)$, $g_2(y)$ and $g_3(z)$ as the probability density functions of the above three fiducial distributions, respectively. Since the three groups of samples,

$$\{X_{iT}, i = 1,\ldots,n_T\},$$
$$\{X_{iR_1}, i = 1,\ldots,n_{n_{R_1}}\}$$

and

$$\{X_{iR_2}, i = 1,\ldots,n_{n_{R_2}}\},$$

are statistically independent between each other, the joint fiducial probability density function of $(\mu_T, \mu_{R_1}, \mu_{R_2})$ can be express as $g(x,y,z) = g_1(x)g_2(y)g_2(z)$. Now we define the second version of fiducial probability.

$$FP_2(\sigma_{R_1})$$

$$= pr\left\{\left|\mu_T - \mu_{R_1}\right| \leq 1.5\sigma_{R_1}, \left|\mu_{R_1} - \mu_{R_2}\right| \leq 1.5\sigma_{R_1}, \left|\mu_{R_1} - \mu_{R_2}\right| \leq 1.5\sigma_{R_1} \mid \sigma_{R_1}, g\right\}$$

$$= \iiint_{|x-y|\leq 1.5\sigma_{R_1},|x-z|\leq 1.5\sigma_{R_1},|y-z|\leq 1.5\sigma_{R_1}} (x,y,z)\,dx\,dy\,dz. \tag{8.9}$$

If $FP_2(\sigma_{R_1}) \geq 1-\alpha$, where $1-\alpha$ is the pre-specified confidence level, all hypotheses in (8.3) are rejected and analytical similarity between T and R_1 is concluded.

Based on (8.9), the following two types of simultaneous confidence interval of $\mu_T - \mu_{R_1}$ namely type 3 restricted simultaneous confidence interval (RSCI III) and type 4 restricted simultaneous confidence interval (RSCI IV) can be derived (Zheng et al. 2018).

Type 3 Restricted Simultaneous Confidence Interval (RSCI III)

Similarly, for any $\Delta > 0$, we calculate the following fiducial probability based on $g(x,y,z)$

$$pr\left\{\left|\mu_T - \mu_{R_1}\right| \leq \Delta, \left|\mu_T - \mu_{R_2}\right| \leq 1.5\sigma_{R_1}, \left|\mu_{R_1} - \mu_{R_2}\right| \leq 1.5\sigma_{R_1} \mid \sigma_{R_1}, g\right\},$$

which is denoted by q_Δ. Note that when $\Delta = 1.5\sigma_{R_1}$, q_Δ is equal to $FP_2(\sigma_{R_1})$. For any $\Delta > 0$, find the minimal Δ that satisfies

$$pr\left\{\left|\mu_T - \mu_{R_1}\right| \leq \Delta, \left|\mu_T - \mu_{R_2}\right| \leq 1.5\sigma_{R_1}, \left|\mu_{R_1} - \mu_{R_2}\right| \leq 1.5\sigma_{R_1} \mid \sigma_{R_1}, g\right\} \geq q.$$

Denote the minimal Δ by Δ_q^3 if it exists. Then the type 3 restricted simultaneous confidence interval (RSCI III) of $\mu_T - \mu_{R_1}$ denoted by $\left(-\Delta_q^3, \Delta_q^3\right)$ with the confidence level of q can be obtained. If $\Delta_{1-\alpha}^3$ exists and $\Delta_{1-\alpha}^3 \leq 1.5\sigma_{R_1}$, the analytical similarity between T and R_1 is concluded. In this case, we have

$$pr\left\{\left|\mu_T - \mu_{R_1}\right| \leq 1.5\sigma_{R_1}, \left|\mu_T - \mu_{R_2}\right| \leq 1.5\sigma_{R_1}, \left|\mu_{R_1} - \mu_{R_2}\right| \leq 1.5\sigma_{R_1} \mid \sigma_{R_1}, g\right\} \geq 1-\alpha.$$

Type 4 Restricted Simultaneous Confidence Interval (RSCI IV)

To obtain a type 4 restricted simultaneous confidence interval (RSCI IV), similarly, for any $\Delta > 0$, calculate the fiducial probability based on $g(x,y,z)$

$$pr\left\{\left|\mu_T - \mu_{R_1}\right| \leq \Delta, \left|\mu_T - \mu_{R_2}\right| \leq \Delta, \left|\mu_{R_1} - \mu_{R_2}\right| \leq \Delta \mid g\right\},$$

which is denoted as q_Δ. When $\Delta = 1.5\sigma_{R_1}$, q_Δ is equal to $FP_2\left(\sigma_{R_1}\right)$. For any $\Delta > 0$, then find the minimal Δ satisfying

$$pr\left\{\left|\mu_T - \mu_{R_1}\right| \leq \Delta, \left|\mu_T - \mu_{R_2}\right| \leq \Delta, \left|\mu_{R_1} - \mu_{R_2}\right| \leq \Delta \,|\, g\right\} \leq q.$$

Denote the minimal Δ by Δ_q^4 if it exists. Then we RSCI IV confidence interval of $\mu_T - \mu_{R_1}$ denoted by $\left(-\Delta_q^4, \Delta_q^4\right)$ with the confidence level of q can be obtained. Thus, if $\Delta_{1-\alpha}^4$ exists and $\Delta_{1-\alpha}^4 \leq 1.5\sigma_{R_1}$, the analytical similarity between T and R_1 is concluded. In other words, we have

$$pr\left\{\left|\mu_T - \mu_{R_1}\right| \leq 1.5\sigma_{R_1}, \left|\mu_T - \mu_{R_2}\right| \leq 1.5\sigma_{R_1}, \left|\mu_{R_1} - \mu_{R_2}\right| \leq 1.5\sigma_{R_1} \,|\, \sigma_{R_1}, g\right\} \geq 1 - \alpha.$$

Similarly, we can replace σ_{R_1} by its estimate $\hat{\sigma}_{R_1}$ in all expressions and obtain estimated versions of the fiducial probability and the two restricted simultaneous confidence intervals, which are denoted by $FP_2\left(\hat{\sigma}_{R_1}\right), \hat{\Delta}_{1-\alpha}^3$, and $\hat{\Delta}_{1-\alpha}^4$, respectively. Note that if $\hat{\sigma}_{R_1}$ is a good estimate of σ_{R_1}, it is expected that $\hat{\Delta}_{1-\alpha}^3$ and $\hat{\Delta}_{1-\alpha}^4$ would perform similarly as compared with the RSCI assuming that $\hat{\sigma}_{R_1}$ is known.

It can be easily verified that $\Delta_{1-\alpha}^4 \geq \Delta_{1-\alpha}^3$ and $\hat{\Delta}_{1-\alpha}^4 \geq \hat{\Delta}_{1-\alpha}^3$. Thus, RSCI IV confidence interval is considered more conservative than RSCI III confidence interval.

To take the variability associated with the estimate of σ_{R_1} into consideration, two modified versions for $FP_2\left(\hat{\sigma}_{R_1}\right), \hat{\Delta}_{1-\alpha}^3$, and $\hat{\Delta}_{1-\alpha}^4$ can be similarly derived. Both are based on the fiducial distribution of σ_{R_1} in (8.6). The fiducial distribution of σ_{R_1} can be expressed as

$$\sqrt{\frac{(n_1 - 1)\hat{\sigma}_{R_1}^2}{\chi^2(n_1 - 1)}},$$

where $\hat{\sigma}_{R_1}^2$ is considered as fixed and $\chi^2 (n_1 - 1)$ is chi-square distribution with degree of freedom $n_1 - 1$. Denote the probability density function of this fiducial distribution as e_1.

The integrated version – The integrated fiducial probability (IFP) can be expressed as

$$IFP_2 = \int FP_2(u)e_1(u)du. \tag{8.10}$$

Similarly, simply replace

$$pr\left\{\left|\mu_T-\mu_{R_1}\right|\leq\Delta,\left|\mu_T-\mu_{R_2}\right|\leq1.5\sigma_{R_1},\left|\mu_{R_1}-\mu_{R_2}\right|\leq1.5\sigma_{R_1}\Big|\sigma_{R_1},g\right\}$$

by its integrated version

$$\int pr\left\{\left|\mu_T-\mu_{R_1}\right|\leq\Delta,\left|\mu_T-\mu_{R_2}\right|\leq1.5u,\left|\mu_{R_1}-\mu_{R_2}\right|\leq1.5u\,\Big|\,u,g\right\}e_1(u)du$$

in the expressions above. Then with the same derivation, we have the type 3 integrated restricted simultaneous confidence interval (IRSCI III), i.e., $\Delta_{I,1-\alpha}^3$.

The least favorable version – It would be more conservative when the used value for σ_{R_1} is smaller, i.e., it's hard to reject all three hypotheses with smaller value of σ_{R_1}. Thus, we suggest another version using the $1-\alpha$ lower fiducial confidence bound to estimate the least favorable value of σ_{R_1}, i.e.,

$$\hat{\sigma}_{R_1}'=\sqrt{\frac{(n_1-1)\hat{\sigma}_{R_1}^2}{\chi_{1-\alpha/2}^2(n_1-1)}},$$

where $\chi_{1-\alpha/2}^2(n_1-1)$ is the $1-\alpha/2$ quantile of chi-square distribution with degree of freedom n_1-1. This leads to the least favorable fiducial probability (LFFP) $LFFP=FP_2(\hat{\sigma}_{R_1}')$, the type 3 least favorable restricted simultaneous confidence interval (LFRSCI III), i.e., $\Delta_{LF,1-\alpha}^3$.

The above methods all use single variance reference σ_{R_1} for EAC. For one of the three hypotheses, $H_{02}:\mu_T-\mu_{R_2}\leq-\delta$ *or* $\mu_T-\mu_{R_2}\geq\delta$, it may also be reasonable to use σ_{R_2} for EAC. To accommodate it, we propose another version of fiducial probabilities and the corresponding simultaneous confidence intervals.

$$FP_3\left(\sigma_{R_1},\sigma_{R_2}\right)$$
$$=pr\left\{\left|\mu_T-\mu_{R_1}\right|\leq1.5\sigma_{R_1},\left|\mu_T-\mu_{R_2}\right|\leq1.5\sigma_{R_2},\left|\mu_{R_1}-\mu_{R_2}\right|\leq1.5\sigma_{R_1}\Big|\sigma_{R_1},g\right\}$$
$$\iiint_{|x-y|\leq1.5\sigma_{R_1},|x-z|\leq1.5\sigma_{R_2},|y-z|\leq1.5\sigma_{R_1}}g(x,y,z)dxdydz. \qquad (8.11)$$

If the above $FP_3\left(\sigma_{R_1},\sigma_{R_2}\right)\geq1-\alpha$, where $1-\alpha$ is the pre-specified confidence level, all hypotheses in (8.3)–(8.5) are rejected and analytical similarity between T and R_1 is concluded.

In addition, we provide two types of simultaneous confidence interval of $\mu_T - \mu_{R_1}$ as follows.

Type 5 Restricted Simultaneous Confidence Interval (RSCI V)

For any $\Delta > 0$, calculate the follows fiducial probability based on $g(x,y,z)$

$$pr\left\{ \left| \mu_T - \mu_{R_1} \right| \leq \Delta, \left| \mu_T - \mu_{R_2} \right| \leq 1.5\sigma_{R_2}, \left| \mu_{R_1} - \mu_{R_2} \right| \leq 1.5\sigma_{R_1} \middle| \sigma_{R_1}, \sigma_{R_2}, g \right\},$$

denoted as q_Δ. When $\Delta = 1.5\sigma_{R_1}$, q_Δ is equal to $FP_3\left(\sigma_{R_1}, \sigma_{R_2}\right)$. For any $\Delta > 0$, look for the minimal Δ satisfying

$$pr\left\{ \left| \mu_T - \mu_{R_1} \right| \leq \Delta, \left| \mu_T - \mu_{R_2} \right| \leq 1.5\sigma_{R_2}, \left| \mu_{R_1} - \mu_{R_2} \right| \leq 1.5\sigma_{R_1} \middle| \sigma_{R_1}, \sigma_{R_2}, g \right\} \geq q.$$

Denote the minimal Δ by Δ_q^5 if it exists. Then we get the type 5 restricted simultaneous confidence interval of $\mu_T - \mu_{R_1}$ as $\left(-\Delta_q^5, \Delta_q^5\right)$, with the confidence level of q. If $\Delta_{1-\alpha}^5$ exists and $\Delta_{1-\alpha}^5 \leq 1.5\sigma_{R_1}$, the analytical similarity between T and R_1 is concluded. In other words, in this case, we have

$$pr\left\{ \left| \mu_T - \mu_{R_1} \right| \leq 1.5\sigma_{R_1}, \left| \mu_T - \mu_{R_2} \right| \leq 1.5\sigma_{R_2}, \left| \mu_{R_1} - \mu_{R_2} \right| \leq 1.5\sigma_{R_1} \middle| \sigma_{R_1}, \sigma_{R_2}, g \right\} \geq 1-\alpha.$$

Similarly, we can replace σ_{R_1} and σ_{R_2} by their estimator $\hat{\sigma}_{R_1}$ and $\hat{\sigma}_{R_2}$ in all expressions above and get estimated versions of the restricted simultaneous confidence interval, as well as the fiducial probability in (8.11). Denote them as $\hat{\Delta}_{1-\alpha}^5$ and $FP_3\left(\hat{\sigma}_{R_1}, \hat{\sigma}_{R_2}\right)$. If $\hat{\sigma}_{R_1}$ and $\hat{\sigma}_{R_2}$ are good estimators of σ_{R_1} and σ_{R_2}, the above results should perform similarly compared with the versions with the true value of σ_{R_1} and σ_{R_2}.

To taken this variability into consideration, two modified versions for above $\hat{\Delta}_{1-\alpha}^5$ and $FP_3\left(\hat{\sigma}_{R_1}, \hat{\sigma}_{R_2}\right)$ are also provided. One is the integrated version and the other is the least favorable version. Both are based on the fiducial distribution of σ_{R_1} and σ_{R_2} in (8.6). The fiducial distributions of σ_{R_1} and σ_{R_2} can be expressed as

$$\sqrt{\frac{(n_1 - 1)\hat{\sigma}_{R_1}^2}{\chi^2(n_1 - 1)}}$$

and

$$\sqrt{\frac{(n_2 - 1)\hat{\sigma}_{R_2}^2}{\chi^2(n_2 - 1)}},$$

respectively, where $\hat{\sigma}^2_{R_1}$ and $\hat{\sigma}^2_{R_2}$ are considered as fixed and $\chi^2(n-1)$ is chi-square distribution with degree of freedom $n-1$. Denote the probability density function of the two fiducial distributions as e_1 and e_2.

The integrated version – The integrated fiducial probability (IFP) can be expressed as

$$IFP_3 = \iint FP_3(u,w)e_1(u)e_2(w)\,du\,dw. \tag{8.12}$$

Similarly, simply replace

$$pr\left\{\left|\mu_T - \mu_{R_1}\right| \le \Delta, \left|\mu_T - \mu_{R_2}\right| \le 1.5\sigma_{R_2}, \left|\mu_{R_1} - \mu_{R_2}\right| \le 1.5\sigma_{R_1} \middle| \sigma_{R_1}, \sigma_{R_2}, g\right\}$$

with its integrated version

$$\iint pr\left\{\left|\mu_T - \mu_{R_1}\right| \le \Delta, \left|\mu_T - \mu_{R_2}\right| \le 1.5w, \left|\mu_{R_1} - \mu_{R_2}\right| \le 1.5u \middle| u, w, g\right\} e_1(u)e_2(w)\,du\,dw$$

in the expressions above. Then with the same derivation, we have the type 5 integrated restricted simultaneous confidence interval (IRSCI V) $\Delta^5_{I,1-\alpha}$.

The least favorable version – It would be more conservative when the used values for σ_{R_1} and/or σ_{R_2} are smaller, i.e., it's hard to reject all three hypotheses with smaller values of σ_{R_1} and/or σ_{R_2}. Thus, we suggest another version using the $\sqrt{1-\alpha}$ lower fiducial confidence bounds to estimate the least favorable values of σ_{R_1} and/or σ_{R_2}, i.e.,

$$\hat{\sigma}''_{R_1} = \sqrt{\frac{(n_1-1)\hat{\sigma}^2_{R_1}}{\chi^2_{1-(1-\sqrt{1-\alpha})/2}(n_1-1)}} \quad \text{and} \quad \hat{\sigma}''_{R_2} = \sqrt{\frac{(n_2-1)\hat{\sigma}^2_{R_2}}{\chi^2_{1-(1-\sqrt{1-\alpha})/2}(n_2-1)}},$$

where $\chi^2_{1-(1-\sqrt{1-\alpha})/2}(n-1)$ is the $1-(1-\sqrt{1-\alpha})/2$ quantile of chi-square distribution with degree of freedom $n-1$. This leads to the least favorable fiducial probability (LFFP)

$$LFFP_3 = FP_3\left(\hat{\sigma}''_{R_1}, \hat{\sigma}''_{R_2}\right),$$

the type 5 least favorable restricted simultaneous confidence interval (LFRSCI V) $\Delta^5_{LF,1-\alpha}$.

8.3.4 An Example

To illustrate the use of simultaneous confidence interval derived in this section, consider the following numerical example. Suppose the sponsor is interested in performing a Tier 1 equivalence test on a given critical quality attribute that is most relevant to clinical outcomes. The analytical study is only to demonstrate similarity between a proposed biosimilar product and US-licensed reference product, but also to establish similarity between the US-licensed reference product and an EU-approved reference product for bridging PK/clinical data. As suggested by the recent FDA draft guidance on analytical similarity assessment, 10 lots from each product are tested. The test results are summarized in Table 8.3.

A scatter plot for the distributions of the proposed biosimilar product, the US-licensed product, and the EU-approved product is illustrated in Figure 8.2.

Table 8.4 summarizes pairwise comparison equivalence test results. The results indicate that the proposed biosimilar product is highly similar to the US-licensed reference product and is highly similar to the EU-approved reference product. However, the EU-approved reference product fails to pass the Tier 1 equivalence test as compared to the US-licensed reference product.

On the other hand, the proposed simultaneous confidence interval pass all pairwise comparisons between the proposed biosimilar product, the US-licensed reference product, and the EU-reference product. The fiducial probability of biosimilarity between the proposed biosimilar product, the US-licensed reference product, and the EU-approved reference product is given by 92.65%, which is greater than the nominal level of 90%, with simultaneous confidence interval of (–10.79, 10.79).

TABLE 8.3

Analytical Test Results

Lot No.	Proposed Biosimilar	US-Licensed Product	EU-Approved Product
1	4.36	–1.75	12.53
2	4.17	–21.87	–1.99
3	14.92	–12.44	–3.09
4	18.64	1.35	1.64
5	–18.42	0.01	5.54
6	6.36	1.74	0.72
7	–2.95	–23.02	5.79
8	–7.51	2.49	–7.67
9	–2.78	5.26	–1.31
10	–4.81	0.95	–3.59

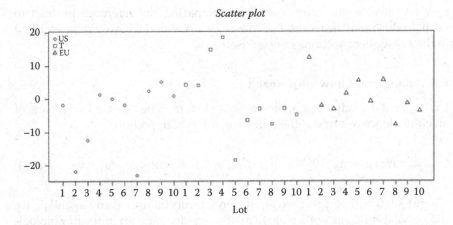

FIGURE 8.2
Scatter plots of US-licensed product, proposed biosimilar, and EU-approved reference product.

TABLE 8.4

Test Results of Pairwise Comparisons

Comparison	No. of Lots	Mean Difference	$1.5\sigma_R$	90% CI	Passage
				Equivalence Test	
T vs. US	10	5.00	11.28	(−0.53, 10.54)	Pass
T vs. EU	10	−0.79	9.18	(−5.29, 3.72)	Pass
EU vs. EU	10	5.79	11.28	(0.25, 11.33)	Fail

Abbreviations: T = Proposed biosimilar product; US = US-licensed product; EU = EU-approved reference product.

8.4 Reference Product Change

In analytical similarity assessment, it is not uncommon to observe change in reference products over time. The changes may be due to selection bias of available lots from the reference product. In other words, by chance, we may select several lots from the lower end of the reference products and several lots from the upper end of the reference product if the reference product is known to be highly variable. If a random sampling scheme is applied and the change (trend of change) in reference product persists, the changes may be due to, but are not limited to, changes in materials, assay methods, equipment, dates of manufacture, and modifications in manufacturing process. These changes and/or modifications may introduce bias and/or variability to the end-product of the reference products. Thus, the reference product may not pass a similarity test when comparing the reference product to itself. In this case, the method proposed by Kang and Chow (2013) described in the subsequent sections may be useful.

8.4.1 Kang and Chow's Approach

Under a valid study design, biosimilarity can then be assessed by means of an equivalence test under the following interval hypotheses:

$$H_0 : \mu_T - \mu_R \leq \theta_L \quad or \quad \mu_T - \mu_R \geq \theta_U \quad \text{vs.} \quad H_a : \theta_L < \mu_T - \mu_R < \theta_U, \quad (8.13)$$

where (θ_L, θ_U) are pre-specified equivalence limits (margins) and μ_T and μ_R are the population means of a biological (test) product and an innovator biological (reference) product, respectively. That is, biosimilarity is assessed in terms of the *absolute* difference between the two population means. Alternatively, biosimilarity can be assessed in terms of the *relative* difference (i.e., ratio) between the population means. Note that for the assessment of similarity between small-molecule drug products, average bioequivalence, population bioequivalence, and individual bioequivalence are defined in terms of ratios of appropriate parameters under a crossover design. In practice, since many biological products have long half-lives, a crossover design may not be appropriate for the assessment of biosimilarity. Instead, a parallel group design is considered more appropriate. In this section, statistical methods proposed by Kang and Chow (2013) under a newly proposed three-arm parallel design for investigation of biosimilarity of biosimilar products will be discussed. The statistical analysis methods proposed by Kang and Chow consider the relative distance based on the absolute mean differences. Under the three-arm design, patients who are randomly assigned to the first group receive a biosimilar (test) product while patients who are randomly assigned to the second and the third groups receive the innovator biological (reference) product from different batches. The distance between the test product and the reference product is defined by the absolute mean difference between two products. Similarly, the distance between the reference products from two different batches is defined. The relative distance is defined as the ratio of the two distances whose denominator is the distance between the two reference products from difference batches. Under the proposed design, Kang and Chow claim that the two products are biosimilar if the relative distance is less than a pre-specified margin.

Let T denote a proposed biosimilar product, and R_1 and R_2 denote the US-licensed reference product and the EU-approved reference product. Suppose that the N patients are randomized into the following three groups. The patients who are assigned into the first group receive the proposed biosimilar product T and the number of patients is denoted by n_1. The patients who are assigned into the second and the third groups receive the innovator biological products R_1 and R_2, respectively, and the number of patients in each group is denoted by n_2 for simplicity. The randomization ratio 2:1:1 is employed. So $n_1 = 2n_2$ and the total sample size is $N = n_1 + 2n_2$. Suppose that we have only one primary continuous response variable Y. If this trial is a pharmacokinetic study, Y can be either AUC or Cmax. If the trial is a pivotal trial, Y can be a clinical response.

8.4.2 Criteria for Biosimilarity

Let d(T,R) represent a distance between the biosimilar product T and the innovator biological product R. Similarly, $d(R_1,R_2)$ denote a distance between R_1 and R_2. There are many possible choices for specific forms of distance. For example,

$$d_1(T,R) = |\mu_T - \mu_R|$$
$$d_2(T,R) = |\mu_T/\mu_R|$$
$$d_3(T,R) = (\mu_T - \mu_R)^2$$
$$d_3(T,R) = E(\mu_T - \mu_R)^2$$

where μ_T is the population mean of Y in patients with the biosimilar product T, and μ_R is defined as $(\mu_{R_1} + \mu_{R_2})/2$ where μ_{R_1} and μ_{R_2} denote the population means of Y in patients who receive the reference product R_1 and R_2, respectively.

In this chapter, we consider the following relative distance to assess biosimilarity between the biosimilar product and the innovator biological product:

$$rd = \frac{d(T,R)}{d(R_1,R_2)}.$$

Since the distances take on only nonnegative values, the relative distances are also nonnegative numbers. If the relative distance *rd* is less than a pre-specified margin $\delta(\delta > 0)$ in the proposed design, we claim that the two products are biosimilar. Therefore, the hypotheses of interest are given by

$$H_0 : rd \geq \delta \quad \text{versus} \quad H_a : rd < \delta$$

Kang and Chow (2013) consider the absolute mean difference $d_1(T,R) = |\mu_T - \mu_R|$ to evaluate biosimilarity between the two products. Then, assuming that $\mu_{R_1} \neq \mu_{R_2}$, the relative distance is given by

$$rd = \left| \frac{d(T,R)}{d(R_1,R_2)} \right| = \left| \frac{\mu_T - \left(\mu_{R_1} + \mu_{R_2}\right)/2}{\mu_{R_1} - \mu_{R_2}} \right|.$$

Then the hypotheses in (8.13) can be rewritten as

$$H_0 : \theta \le -\delta \quad or \quad \theta \ge \delta \quad \text{vs.} \quad H_a : -\delta < \theta < \delta, \tag{8.14}$$

where

$$\theta = \frac{\mu_T - \left(\mu_{R_1} + \mu_{R_2}\right)/2}{\mu_{R_1} - \mu_{R_2}}. \tag{8.15}$$

It is well-known that the hypotheses in (8.14) can be decomposed into two one-sided hypotheses as follows.

$$H_{01} : \theta \le -\delta \quad \text{vs.} \quad H_{a1} : -\delta < \theta, \tag{8.16}$$

and

$$H_{02} : \theta \ge \delta \quad \text{vs.} \quad H_{a2} : \theta < \delta, \tag{8.17}$$

8.4.3 Statistical Tests for Biosimilarity

Let $Y_{T,i}$ $(i = 1, 2, \ldots, n_1)$ and $Y_{R_k,i}$ $(k = 1, 2, i = 1, 2, \ldots, n_2)$ denote the response variables from the biosimilar product in the first group and the innovator biological product in the second and the third group, respectively, with $n_1 = 2n_2$. First, it is assumed that $Y_{T,i}$ and $Y_{R_k,i}(k = 1, 2)$ follow independently the normal distribution with the mean μ_T and $\mu_{R_k}(k = 1, 2)$ and the common variance σ^2. When we need to derive an asymptotic distribution of a test statistic and the power functions, we assume that the sample size is large enough so that the central limit theorem can be employed. In the following two subsections, we propose two statistical tests in order to test the hypotheses in (8.16) and (8.17).

The Statistical Test based on the Ratio Estimator–A natural estimator of θ in (8.15) is to replace the population means with the corresponding sample means and is given by

$$\hat{\theta} = \frac{\bar{V}}{\bar{U}} \equiv \frac{\bar{Y} - \left(\bar{Y}_{R_1} + \bar{Y}_{R_2}\right)/2}{\bar{Y}_{R_1} - \bar{Y}_{R_2}}$$

where

$$\bar{Y}_T = \frac{1}{n_1} \sum_{i=1}^{n_1} Y_{T,i},$$

$$\bar{Y}_{R_k} = \frac{1}{n_2} \sum_{i=1}^{n_2} Y_{R_k,i} \text{ for k = 1,2.}$$

Since the exact distribution of $\hat{\theta}$ is very complicated, Kang and Chow (2013) obtained the asymptotic normality of $\sqrt{n_1}(\hat{\theta} - \theta)$ as follows. First, note that

$$\frac{\bar{V}}{\bar{U}} - \frac{v}{\mu} = \frac{\bar{V}_\mu - \bar{U}_v}{\bar{U}_\mu} = \frac{\mu(\bar{V} - v) - v(\bar{U} - \mu)}{\bar{U}_\mu} = \frac{(\bar{V} - v)}{\bar{U}} - \frac{v}{\mu} \frac{(\bar{U} - \mu)}{\bar{U}},$$

where

$$\mu = \mu_{R_1} - \mu_{R_2}, \quad \text{and} \quad v = \mu_T - (\mu_{R_1} + \mu_{R_2})/2.$$

Therefore, we have

$$\sqrt{n_1}(\hat{\theta} - \theta) = \sqrt{n_1}\left(\frac{\bar{V}}{\bar{U}} - \frac{v}{\mu}\right)$$

$$= \sqrt{n_1} \frac{(\bar{V} - v)}{\bar{U}} - \frac{\sqrt{n_1}}{\sqrt{n_2}} \frac{v}{\mu} \sqrt{n_2} \frac{(\bar{U} - \mu)}{\bar{U}}$$

Since $\bar{U} \xrightarrow{p} \mu$, we have

$$\sqrt{n_1}(\hat{\theta} - \theta) \sim \sqrt{n_1} \frac{(\bar{V} - v)}{\mu} - \sqrt{2} \frac{v}{\mu^2} \sqrt{n_2}(\bar{U} - \mu)$$

$$\xrightarrow{d} N\left(0, \frac{2\sigma^2}{\mu^2} + \frac{4v^2}{\mu^4}\sigma^2\right).$$

By using this asymptotic normality of $\sqrt{n_1}(\hat{\theta} - \theta)$, we can conduct hypothesis testing and establish an asymptotic confidence interval for θ. The null hypothesis H_{01} in (8.13) is rejected if $Z_1 > z_\alpha$ where

$$Z_1 = \frac{\hat{\theta} + \delta}{\frac{s}{\sqrt{n_1}}\sqrt{\frac{2}{(\bar{U})^2} + 4\frac{(\bar{V})^2}{(\bar{U})^4}}} = \frac{\frac{\bar{V}}{\bar{U}} + \delta}{\frac{s}{\sqrt{n_1}}\sqrt{\frac{2}{(\bar{U})^2} + 4\frac{(\bar{V})^2}{(\bar{U})^4}}}$$

and z_α is the upper α quantile of the standard normal distribution and

$$s^2 = \frac{1}{n_1 + 2n_2 - 3}\left[\sum_{i=1}^{n_1}(Y_{T,i} - \bar{Y}_T)^2 + \sum_{i=1}^{n_2}(Y_{R_1,i} - \bar{Y}_{R_1})^2 + \sum_{i=1}^{n_2}(Y_{R_2,i} - \bar{Y}_{R_2})^2\right].$$

Similarly, the null hypothesis H_{02} in (8.17) is rejected if $Z_2 < -z_\alpha$, where

$$Z_1 = \frac{\dfrac{\bar{V}}{\bar{\bar{U}}} - \delta}{\dfrac{s}{\sqrt{n_1}}\sqrt{\dfrac{2}{(\bar{U})^2} + 4\dfrac{(\bar{V})^2}{(\bar{U})^4}}}$$

If each null hypothesis in both (8.16) and (8.17) is rejected at the significance level α, we claim that the two products are biosimilar.

An alternative method to assess biosimilarity between the two products is to use a two-sided asymptotic confidence interval for θ. Since an $(1 - \alpha) \times 100\%$ asymptotic confidence interval for θ is given by

$$\left(\frac{\bar{V}}{\bar{\bar{U}}} \pm z_{\alpha/2}\frac{s}{\sqrt{n_1}}\sqrt{\frac{2}{(\bar{U})^2} + 4\frac{(\bar{V})^2}{(\bar{U})^4}}\right), \tag{8.18}$$

we claim that the two products are biosimilar if the $(1 - \alpha) \times 100\%$ asymptotic confidence interval for θ lies within the interval $(-\delta, \delta)$.

Although we have obtained the $(1 - \alpha) \times 100\%$ asymptotic confidence interval for θ, actually we can obtain an exact $(1 - \alpha) \times 100\%$ confidence interval for θ based on the Fieller's theorem (Fieller, 1940, 1944), because we assume that $Y_{T,i}$ and $Y_{R_k,i}$ $(k = 1, 2)$ follow the normal distribution. From the Fieller's theorem, since

$$Var(\bar{V}) = \left[\frac{1}{n_1} + \frac{1}{2n_2}\right]\sigma^2 \qquad Var(\bar{U}) = \frac{1}{n_2}\sigma^2 \qquad Cov(\bar{V},\bar{U}) = 0,$$

an exact $(1 - \alpha) \times 100\%$ confidence interval for θ is given by

$$\frac{1}{1-g}\left\{\frac{\bar{V}}{\bar{U}} \pm \frac{t_{\alpha,m}s}{\bar{U}}\left[\frac{1}{n_1} + \frac{1}{2n_2} + \left(\frac{\bar{V}}{\bar{U}}\right)^2\frac{2}{n_2} - g\left(\frac{1}{n_1} + \frac{1}{2n_2}\right)\right]^{1/2}\right\}, \tag{8.19}$$

where

$$g = \frac{t_{1-\alpha,m}^2 s^2}{(\bar{U})^2}\left(\frac{2}{n_2}\right), \quad m = n_1 + 2n_2 - 3$$

and $t_{\alpha,m}$ is the upper α quantile of the t distribution with m degrees of freedom.

The Linearization Method–When we conduct a hypothesis testing for a parameter which is the ratio of parameters, a popular method of constructing a hypothesis testing is the linearization method (Howe, 1974; Hyslop, Hsuan, and Holder, 2000; Pigeot et al., 2003). Both sides of the inequality in the hypothesis are multiplied by the parameter in the denominator of the ratio, and then the numerator of the ratio is moved to the opposite side of the inequality, so that the linearized parameter can be obtained.

The parameter of interest is θ and the denominator is $\mu_{R_1} - \mu_{R_2}$. First, we need to check the sign of the denominator, because the direction of inequalities in the hypotheses changes depending on the sign of the denominator. It is assumed that $\mu_{R_1} \neq \mu_{R_2}$. Otherwise, θ cannot be defined. In order to check the sign of the denominator $\mu_{R_1} - \mu_{R_2}$, we test the following preliminary hypotheses:

$$H_0 : \mu_{R_1} \langle \mu_{R_2} \quad \text{vs.} \quad H_a : \mu_{R_1} \rangle \mu_{R_2}. \tag{8.20}$$

The null hypothesis in (8.20) is rejected if

$$T_R = \frac{\bar{Y}_{R_1} - \bar{Y}_{R_2}}{s_R\sqrt{2/n_2}} > t_{\alpha,2n_2-2}$$

where

$$s_R^2 = \frac{1}{2n_2 - 2}\left[\sum_{i=1}^{n_2}\left(Y_{R_1,i} - \bar{Y}_{R_1}\right)^2 + \sum_{i=1}^{n_2}\left(Y_{R_2,i} - \bar{Y}_{R_2}\right)^2\right].$$

When the null hypothesis in (8.20) is rejected, the null hypothesis H_{01}: $\theta \leq -\delta$ can be expressed as

$$H_{01} : \mu_T - \frac{1}{2}\left(\mu_{R_1} + \mu_{R_2}\right) + \delta\left(\mu_{R_1} - \mu_{R_2}\right) \leq 0.$$

Similarly, the null hypothesis H_{02}: $\theta \geq \delta$ can be rewritten as

$$H_{02} : \mu_T - \frac{1}{2}\left(\mu_{R_1} + \mu_{R_2}\right) - \delta\left(\mu_{R_1} - \mu_{R_2}\right) \geq 0.$$

Therefore, the two products are claimed to be biosimilar if

$$T_1^L > t_{\alpha, n_1 + 2n_2 - 3} \quad \text{and} \quad T_2^L < -t_{\alpha, n_1 + 2n_2 - 3},$$

where

$$T_1^L \equiv \frac{\bar{Y}_T - \left(\frac{1}{2} - \delta\right)\bar{Y}_{R_1} - \left(\frac{1}{2} + \delta\right)\bar{Y}_{R_2}}{s\sqrt{\frac{1}{n_1} + \left(\frac{1}{2} - \delta\right)^2 \frac{1}{n_2} + \left(\frac{1}{2} + \delta\right)^2 \frac{1}{n_2}}},$$

$$T_2^L \equiv \frac{\bar{Y}_T - \left(\frac{1}{2} + \delta\right)\bar{Y}_{R_1} - \left(\frac{1}{2} - \delta\right)\bar{Y}_{R_2}}{s\sqrt{\frac{1}{n_1} + \left(\frac{1}{2} + \delta\right)^2 \frac{1}{n_2} + \left(\frac{1}{2} + \delta\right)^2 \frac{1}{n_2}}},$$

In a similar fashion, when the null hypothesis is accepted, the null hypothesis H_{01}: $\theta \leq -\delta$ can be expressed as

$$H_{01} : \mu_T - \frac{1}{2}\left(\mu_{R_1} + \mu_{R_2}\right) + \delta\left(\mu_{R_1} - \mu_{R_2}\right) \geq 0.$$

and the null hypothesis H_{02}: $\theta \geq \delta$ can be rewritten as

$$H_{02} : \mu_T - \frac{1}{2}\left(\mu_{R_1} + \mu_{R_2}\right) - \delta\left(\mu_{R_1} - \mu_{R_2}\right) \leq 0.$$

Hence, biosimilarity between the two products are claimed if

$$T_1^L \left\langle -t_{\alpha, n_1 + 2n_2 - 3} \quad \text{and} \quad T_2^L \right\rangle t_{\alpha, n_1 + 2n_2 - 3}.$$

8.4.4 Remarks

In this section, a lot of discussion was given to the design and analysis methods proposed by Kang and Chow (2013) for assessment of biosimilarity of biosimilar products. Kang and Chow's methods were derived based on relative distance of means observed from the test and reference products. The proposed design consists of three arms: one arm is for the test product and the other two arms are for the reference products from different batches. This design will allow us to assess relative distance in terms of the ratio of the difference between T and R and the difference between R_1 and R_2. Under the design, the performances of the derived statistical tests were evaluated both theoretically and through the simulation study. Since the statistical test based on the ratio estimator is more powerful than the linearization method, Kang and Chow recommend the statistical test based on the ratio estimator.

In practice, one of the most commonly used designs for assessing biosimilarity between a biosimilar product and an innovative biological product is probably a two-arm balanced design, and biosimilarity is assessed by using the difference between two population means. A disadvantage of this approach is that the variability of a reference product among different batches cannot be incorporated. Hence, when an equivalence margin is larger than the variability of a reference product, this approach may tend to conclude that two products are biosimilar, although there exists considerable difference between two products. This problem can be fixed if the equivalence margin can be determined by incorporating variability of a reference product prior to a clinical trial. However, in practice it may not be easy, because variability of a reference product may be unknown. This point is an advantage of the proposed design, because relative distance rather than absolute difference is employed to assess biosimilarity.

As mentioned earlier, there are many possible choices for specific forms of distance. For assessment of bioequivalence for small-molecule drug products, the average bioequivalence uses $d_2(T,R) = |\mu_T/\mu_R|$ and the population and the individual bioequivalence employ $d_4(T,R) = E(Y_T - Y_R)^2$. Since different distances may produce different conclusion for biosimilarity, it is of interest to develop statistical tests for several difference distances which have often been used in statistical literature. It seems that the absolute mean difference used as the distance in this chapter does not incorporate the variance. Another natural choice of the distance is the standardized absolute mean difference $d_1^s(T,R) = |\mu_T - \mu_R| / \sigma$. However, since the common variance is assumed, the common standard deviation σ is cancelled out in relative distance. Therefore, this chapter uses the standardized absolute mean difference as the distance. It is of interest to develop statistical tests when the common variance assumption does not hold.

In this chapter, the distance between a biosimilar product and an innovator biological product is defined as $d_1(t,R) = |\mu_T - (\mu_{R_1} + \mu_{R_2})/2|$. But, there are also other ways of defining the distance such as

$$d_1(T,R)' = \max\left(|\mu_T - \mu_{R_1}|, |\mu_T - \mu_{R_2}|\right)$$

and

$$d_1(T,R)'' = \min\left(\left|\mu_T - \mu_{R_1}\right|, \left|\mu_T - \mu_{R_2}\right|\right)$$

It is interesting to develop appropriate statistical tests for these new distances. Note that the randomization ratio 2:1 between n_1 and n_2 is employed in this chapter. However, a general randomization ratio k:1 can be used. Finding an optimal value of k might also be an interesting future study.

8.5 Concluding Remarks

In biosimilar product development, it is not uncommon to perform analytical similarity assessment with multiple references from different regions (e.g., US-licensed product and EU-approved product). The purpose is to be able to bridge PK/clinical data from one region (e.g., EU) to the other (e.g., US) if the US-licensed product is highly similar to the EU-approved product. In this case, pairwise comparisons (i.e., the proposed biosimilar product versus the US-licensed product, the proposed biosimilar product versus the EU-approved product, and the EU-approved product versus the US-licensed product) are often conducted.

The method of pairwise comparisons, however, suffers from a few drawbacks: (i) each comparison only utilizes data obtained from the corresponding two products, and (ii) the three pairwise comparisons use difference references for the equivalence test (i.e., the proposed biosimilar product versus the US-licensed product and the EU-approved product versus the US-licensed product use the US-licensed product as the reference product, while the comparison between the proposed biosimilar product and the EU-approved product uses the EU-approved product as the reference product.)

To overcome the drawbacks of the method of pairwise comparisons, Zheng et al. (2018) proposed the use of simultaneous confidence interval and assess similarity among the proposed biosimilar product, the US-licensed product, and the EU-approved product, based on the derived simultaneous confidence interval and the corresponding fiducial probability.

In practice, reference product change might be encountered due to changes in materials, assay methods, equipment, dates of manufacture, and modifications in manufacturing processes. These changes and/or modifications may introduce bias and/or variability to the end-product of the reference products. Thus, the reference product may not pass similarity tests when comparing the reference product to itself. In this case, Kang and Chow's method described in the previous section may be useful. From a regulatory perspective, it is necessary for guidance or policy for possible reference product change to be developed.

9

Extrapolation across Indications

9.1 Introduction

For development of a proposed biosimilar product, a stepwise approach consisting of analytical, PK/PD, and clinical similarity is often performed for obtaining totality-of-the-evidence for demonstration of highly similarity between the proposed biosimilar and its innovative biological product. Clinical studies are often conducted in patients with one or a couple of diseases (indications) to demonstrate that the proposed biosimilar product is highly similar to the innovative product and that there are no clinically meaningful differences between the proposed biosimilar product and the innovative product in terms of safety, purity, and potency. Although the intended clinical study is only conducted in patients with one or a couple of diseases (indications or conditions of use), the sponsor often seeks licensure for the indications for which the United States (US)-licensed product is licensed. In this case, FDA requires that sufficient scientific justification for extrapolation of data to support biosimilarity in each of the additional indications for which the sponsor is seeking licensure should be provided.

Under the assumption that these diseases have similar mechanism of action and/or PK/PD profiles, the concept of extrapolation across different indications can be assessed by evaluating a sensitivity index for generalizability. Generalizability is referred to as one of the following situations. First, it is referred to as whether the clinical results of the original target patient population under study (e.g., patients with one disease or indication) can be generalized to other similar but different patient populations (patients with another disease or indication). Second, it evaluates whether a newly developed or approved drug product in a region (e.g., EU) can be approved at another region (e.g., US), as it is a concern that differences in ethnic factors could alter the efficacy and safety of the drug product in the new region. Third, for a given disease or indication, it is often of interest to determine whether the clinical results of the original target patient population under study (e.g., adults) can be generalized to other similar but different patient populations (e.g., pediatrics or elderly). In practice, since it is of interest to determine whether the observed clinical results from the original target patient population (i.e., patients with the target disease or indication under study) can be generalized to a similar but different patient population (e.g., patients with another disease or indication), we will focus on the first scenario. Statistical methods for assessment of generalizability of clinical results, however, can also be applied to other scenarios.

For assessment of generalizability of clinical results from one population (e.g., adults) to another (e.g., pediatrics or elderly), Chow (2010) proposed to evaluate a so-called generalizability probability of a positive clinical result observed from the original patient population by studying the impact of shift in a target patient through a model that links the population means with some covariates (see, also Chow and Shao, 2005; Chow and Chang, 2006). However, in many cases, such covariates may not exist, or may exist but are not observable. In this case, it is suggested that the degree of shift in location and scale of patient population be studied based on a mixture distribution by assuming the location or scale parameter is random (Shao and Chow, 2002). The purpose of this chapter is to assess the generalizability of clinical results by evaluating the sensitivity index under different models that (i) shift in location parameter is random, (ii) shift in scale parameter is random, and (iii) shifts in both location and scale parameters are random.

The remaining of this chapter is organized as follows. In the next section, an example concerning a recent regulatory submission is given to illustrate the issue of extrapolation in biosimilar product development. The concept of a sensitivity index for measuring the degree of population shift is briefly introduced in Section 9.3. Section 9.4 outlines several scenarios for assessment of sensitivity index. Statistical inference for extrapolation based on mixture distribution for the cases (i) where shift in location parameter is random and change in scale parameter is fixed, (ii) where shift in location parameter is fixed and change in scale parameter is random, and (iii) where both shift in location parameter and change in scale parameter are random. Statistical inferences of the effect size for the three possible scenarios are given in Section 9.5. Also included in this section is an example concerning an asthma clinical trial. Brief concluding remarks are given in the last section of this chapter.

9.2 An Example

For illustration purpose, consider a recent biosimilar regulatory submission. This is a 351(k) BLA submitted by Amgen, Inc. for ABP215, a proposed biosimilar to US-licensed Avastin (bevacizumab). Although only a comparative clinical study between ABP215 and EU-approved bevacizumab in patients with advanced/metastatic non-small cell lung cancer (NSCLC) was conducted to support the demonstration of no clinically meaningful differences in terms of response, safety, purity, and potency between ABP215 and US-licensed Avastin, the sponsor seeks licensure for the following indications for which US-licensed Avastin is licensed:

1. Metastatic colorectal cancer (mCRC), in combination with intravenous (IV) 5-fluorouracil-(5-FU)-based chemotherapy for first- or second-line treatment;

2. MCRC, in combination with fluoropyrimidine-, irinotecan-, or fluoropyrimidine-oxaliplatin-based chemotherapy for second-line treatment in patients who have progressed on a first-line bevacizumab containing regimen;

3. Non-squamous non-small cell lung cancer (NSCLC), in combination with carboplatin and paclitaxel for first-line treatment of unresectable, locally advanced, recurrent, or metastatic disease;

4. Glioblastoma multiforme (GBM), as a single agent for adult patients with progressive disease following prior therapy;

5. Metastatic renal cell carcinoma (mRCC), in combination with interferon alfa;

6. Cervical cancer, in combination with paclitaxel and cisplatin or paclitaxel and topotecan in persistent, recurrent, or metastatic disease.

The ABP215 biosimilar program provided clinical data from a comparative PK study in healthy volunteers and a comparative clinical study in patients with NSCLC. Clinical data from patients with other diseases (indications) were not studied and collected. However, FDA has determined that it may be appropriate for a biosimilar product to be licensed for one or more conditions of use for which the reference product is licensed. FDA indicated that the extrapolation across indication may be determined based on data supporting a demonstration of biosimilarity, including data from clinical studies performed for another condition of use.

As described in the FDA 2015 guidance, if a biological product meets the statutory requirements for licensure as a biosimilar product under Section 351(k) of the PHS Act based on, among other things, data derived from a clinical study or studies sufficient to demonstrate safety, purity, and potency in an appropriate condition of use, the potential exists for that product to be licensed for one or more additional conditions of use for which the reference product is licensed (FDA, 2015b). However, the applicant is required to provide sufficient scientific rationale or justification for extrapolation, which should address the following issues for the tested and extrapolated conditions of use:

1. The mechanism(s) of action (MOA), if known or can reasonably be determined, in each condition of use for which licensure is sought,

2. The pharmacokinetics (PK) and bio-distribution of the product in different patient populations,

3. The immunogenicity of the product in different patient populations,
4. Differences in expected toxicities in each condition of use and patient population,
5. Any other factors that may affect the safety and efficacy of the product in each condition of use and patient population for which licensure is sought.

Following the above principles, the applicant provided the following scientific justifications to support extrapolation of the clinical data in NSCLC to other indications for which the applicant is seeking licensure (for details, see BLA 761028):

1. Bevacizumab binds circulating VEGF which prevents the interaction of VEGF to its receptors (Flt-1 [VEGFR-1] and KDR [VEGFR-2]) on the surface of endothelial cells. Neutralizing the biological activity of VEGF results in the regression of tumor vascularization, normalization of remaining tumor vasculature, and inhibition of the formation of new tumor vasculature, thereby inhibiting tumor growth (Avastin USPI). In each approved indication, the MOA of bevacizumab is to inhibit VEGF-induced angiogenesis and vascular permeability. The Applicant submitted an extensive analysis of the role of VEGF and VEGF inhibition in each one of the indications for which licensure is sought. FDA agrees that there is no evidence to support claims of a unique MOA in specific indications;

2. PK profiles of bevacizumab following IV infusions ranging from 0.1 mg/kg to 10 mg/kg have been evaluated in several dose escalation/dose finding studies in solid tumors (Gordon 2001; Margolis 2001, EMA 2006, Herbst 2008, Han 2016). The PK properties of bevacizumab across approved indications appear consistent.

3. Overall, FDA considers that Study 20110216 adequately demonstrated pharmacokinetic similarity among ABP215, US-licensed Avastin, and EU-approved bevacizumab. Since PK similarity was demonstrated between ABP215 and US-licensed Avastin, a similar PK profile would be expected for ABP215 in patients across the indications being sought for licensure.

4. As summarized in the labeling for US-licensed Avastin, 14 of 2233 evaluable subjects (0.63%) tested positive for treatment-emergent anti-bevacizumab antibodies as detected by an ECL-based assay. Further analysis of these 14 subjects using an ELISA assay concluded that 3 subjects were positive for neutralizing antibodies against bevacizumab. The clinical significance of these ADA responses to bevacizumab is unknown. The analysis of Studies 20110216 and 20120265 indicate that immunogenicity was low and that treatment of subjects with NSCLC with either ABP215, EU-approved bevacizumab,

or US-licensed Avastin resulted in similar rates of formation of bind-ing ADAs.

5. The expected toxicities of bevacizumab are well characterized and are summarized in the Avastin USPI, as well as multiple meta-analyses of earlier clinical trial data in various solid tumors. The MOA is common to all of the indications of use. While the incidence of specific toxicities may differ across indications (e.g., hyperten-sion is more frequent in patients with RCC while hemoptysis is more frequent in patients with NSCLC), due to the common MOA, the differing toxicities are predictable in each indication for which licensure is sought for ABP215 in this application. Data from Study 20120265 demonstrated that the type and incidence of treatment-emergent adverse events of special interest were similar for ABP215 and EU-approved bevacizumab and that there were no clinically meaningful differences between arms. No new safety signals were identified that would be indicative of new toxicities for the approved indications for US-licensed Avastin.

FDA then concluded that the applicant has provided justification for the proposed extrapolation of clinical data in NSCLC, as well as clinical phar-macology data from a healthy volunteer study, to each of the other indica-tions approved for US-licensed Avastin for which the Applicant is seeking licensure. In addition, FDA agrees with the applicant that the extensive ana-lytical characterization data support a demonstration that ABP215 is highly similar to US-licensed Avastin, and that clinical data support a demonstra-tion that there are no clinically meaningful differences between ABP215 and US-licensed Avastin based on similar clinical pharmacokinetics, anti-tumor activity, safety, and immunogenicity. Therefore, the evidence indicates that the extrapolation of biosimilarity to the indications for which the applicant is seeking licensure is scientifically justified. However, the ODAC panel still felt uncomfortable without seeing any clinical data collected from other indi-cations and suggested certain statistical assurance for the validity of extrapo-lation should be provided.

9.3 Development of Sensitivity Index

In clinical research, it is often of interest to generalize clinical results obtained from a given target patient population (or a medical center) to a similar but different patient population (or another medical center). Denote the origi-nal target patient population by (μ_0, σ_0), where μ_0 and σ_0 are the population mean and population standard deviation, respectively. Similarly, denote the similar but different patient population by (μ_1, σ_1). Since the two populations are similar but different, it is reasonable to assume that $\mu_1 = \mu_0 + \varepsilon$ and $\sigma_1 = C\sigma_0$ ($C > 0$), where ε is referred to as the shift in location parameter (population mean) and C is the inflation factor of the scale parameter (population standard deviation). Thus, the (treatment) effect size adjusted for standard deviation of population (μ_1, σ_1) can be expressed as follows:

$$E_1 = \left|\frac{\mu_1}{\sigma_1}\right| = \left|\frac{\mu_0 + \varepsilon}{C\sigma_0}\right| = |\Delta|\left|\frac{\mu_0}{\sigma_0}\right| = |\Delta|E_0, \qquad (9.1)$$

where $\Delta = (1 + \varepsilon/\mu_0)/C$ and E_0 and E_1 are the effect size (of clinically meaningful importance) of the original target patient population and the similar but different patient population, respectively. Δ is referred to as a sensitivity index measuring the change in effect size between patient populations (see, also Shao and Chow, 2002; Chow and Chang, 2006).

As it can be seen from (9.1), if $\varepsilon = 0$ and $C = 1$, $E_0 = E_1$. That is, the effect sizes of the two populations are identical. In this case, we claim that the results observed from the original target patient population (e.g., adults) can be generalized to the similar but different patient population (e.g., pediatrics or elderly). Applying the concept of bioequivalence assessment, we can claim that the effect sizes of the two patient populations are equivalent if the confidence interval of $|\Delta|$ is within (80%, 120%) of E_0. It should be noted that there is a masking effect between the location shift (ε) and scale change (C). In other words, shift in location parameter could be offset by the inflation or deflation of variability. As a result, the sensitivity index may remain unchanged while the target patient population has been shifted. Table 9.1 provides a summary of the impacts of various scenarios of location shift (i.e., change in ε) and scale change (i.e., change in C, either inflation or deflation of variability).

TABLE 9.1

Changes in Sensitivity Index

ε/μ(%)	Inflation of Variability		Deflation of Variability	
	C(%)	Δ	C(%)	Δ
−20	120	0.667	80	1.000
−10	120	0.750	80	1.125
−5	120	0.792	80	1.188
0	120	0.833	80	1.250
5	120	0.875	80	1.313
10	120	0.917	80	1.375
20	120	1.000	80	1.500

As indicated by Chow and Shao (2005), in many clinical trials, the effect sizes of the two populations could be linked by baseline demographics or patient characteristics if there is a relationship between the effect sizes and the baseline demographics and/or patient characteristics (e.g., a covariate vector). In practice, however, such covariates may not exist or may exist but are not observable. In this case, the sensitivity index may be assessed by simply replacing ε and C with their corresponding estimates (Chow and Shao, 2005). Intuitively, ε and C can be estimated by

$$\hat{\varepsilon} = \hat{\mu}_1 - \hat{\mu}_0 \quad \text{and} \quad \hat{C} = \hat{\sigma}_1/\hat{\sigma}_0,$$

where $(\hat{\mu}_0, \hat{\sigma}_0)$ and $(\hat{\mu}_1, \hat{\sigma}_1)$ are some estimates of (μ_0, σ_0) and (μ_1, σ_1), respectively. Thus, the sensitivity index can be estimated by

$$\hat{\Delta} = \frac{1 + \hat{\varepsilon}/\hat{\mu}_0}{\hat{C}}.$$

9.4 Assessment of Sensitivity Index

In practice, the shift in location parameter (ε) and/or the change in scale parameter (C) could be random. If both ε and C are fixed, the sensitivity index can be assessed based on the sample means and sample variances obtained from the two populations. In real world problems, however, ε and C could be either fixed or random variables. In other words, there are three possible scenarios: (i) the case where ε is random and C is fixed, (ii) the case where ε is fixed and C is random, and (iii) the case where both ε and C are random. These possible scenarios are discussed below.

9.4.1 The Case Where ε Is Random and C Is Fixed

Let $\{x_{0i}, i = 1, \dots, n_0\}$ be the responses observed from the original target patient population and $\{x_{1i}, i = 1, \dots, n_1\}$ be the responses observed from the similar but different population. If we split the data from the similar but different population into m segments of the same size n, then $\{x_{1i}, i = 1, \dots, n_1\}$ becomes $\{x_{ji}, j = 1, \dots, m; i = 1, \dots, n\}$. Combining $\{x_{0i}, i = 1, \dots, n_0\}$ and $\{x_{ji}, j = 1, \dots, m; i = 1, \dots, n\}$, we have

$$\left\{ x_{ji}, i = 1, \dots, n_0 \ (\text{if } j = 0) \quad \text{or} \quad n (\text{if } j > 0); j = 0, 1, \dots, m \right\}.$$

Note that $n_1 = \sum_{j=1}^{m} n = mn$. Under the normality assumption, estimates of μ_0 and σ_0^2 can be obtained. Based on $x_{0i}, i = 1, \ldots, n_0$, the maximum likelihood estimates of μ_0 and σ_0^2 can be obtained as follows

$$\hat{\mu}_0 = \frac{1}{n_0} \sum_{i=1}^{n_0} x_{0i}, \text{ and } \hat{\sigma}_0^2 = \frac{1}{n_0 - 1} \sum_{i=1}^{n_0} (x_{0i} - \hat{\mu}_0)^2. \tag{9.2}$$

To obtain estimates of μ_1 and σ_1^2, Chow, Chang, and Pong (2005) considered the case where $\mu_1 = \mu_0 + \varepsilon$ is random and $\sigma_1 = C\sigma_0$ is fixed. Assume that x conditional on $\mu = \mu_1$, i.e., $x|_{\mu=\mu_1}$ follows a normal distribution $N(\mu_1, \sigma_1^2)$. That is, $x|_{\mu=\mu_1} \sim N(\mu_1, \sigma_1^2)$ where μ_1 is distributed as $N(\mu_\mu, \sigma_\mu^2)$ and σ_1, μ_μ, and σ_μ are some unknown constants. Thus, the unconditional distribution of x is a mixed normal distribution given below

$$\int N(x; \mu_1, \sigma_1^2) N(\mu_1; \mu_\mu, \sigma_\mu^2) d\mu_1 = \frac{1}{\sqrt{2\pi\sigma_1^2}} \frac{1}{\sqrt{2\pi\sigma_\mu^2}} \int_{-\infty}^{\infty} e^{-\frac{(x-\mu_1)^2}{2\sigma_1^2} - \frac{(\mu_1-\mu_\mu)^2}{2\sigma_\mu^2}} d\mu_1,$$

where $x \in (-\infty, \infty)$, it can be verified that the above mixed normal distribution is a normal distribution with mean μ_μ and variance $\sigma_1^2 + \sigma_\mu^2$. In other words, x is distributed as $N(\mu_\mu, \sigma_1^2 + \sigma_\mu^2)$. For convenience's sake, we set $\sigma_1 = \sigma$.

Theorem 9.1. Suppose that $X|_{\mu=\mu_1} \sim N(\mu_1, \sigma^2)$ and $\mu_1 \sim N(\mu_\mu, \sigma_\mu^2)$, then we have

$$X \sim N(\mu_\mu, \sigma^2 + \sigma_\mu^2) \tag{9.3}$$

Proof: Consider the following characteristic function of a normal distribution $N(x, \mu, \sigma^2)$

$$\phi_0(w) = \frac{1}{\sqrt{2\pi\sigma^2}} \int_{-\infty}^{\infty} e^{iwt - \frac{1}{2\sigma^2}(t-\mu)^2} dt = e^{iw\mu - \frac{1}{2}\sigma^2 w^2}$$

For distribution $X|_{\mu=\mu_1} \sim N(\mu_1, \sigma^2)$ and $\mu_1 \sim N(\mu_\mu, \sigma_\mu^2)$ the characteristic function after exchange the order of the two integrations is given by

$$\phi(w) = \int_{-\infty}^{+\infty} e^{iw\mu - \frac{1}{2}\sigma^2 w^2} N(\mu, \mu_\mu, \sigma_\mu) d\mu$$

$$= \int_{-\infty}^{+\infty} e^{iw\mu - \frac{\mu - \mu_\mu}{2\sigma_\mu^2} - \frac{1}{2}\sigma^2 w^2} d\mu$$

Note that

$$\int_{-\infty}^{+\infty} e^{iw\mu - \frac{\mu - \mu_\mu}{2\sigma_\mu^2}} d\mu = e^{iw\mu - \frac{1}{2}\sigma_\mu^2 w^2}$$

is the characteristic function of the normal distribution. It follows that

$$\phi(w) = e^{iw\mu_\mu - \frac{1}{2}(\sigma^2 + \sigma_\mu^2)w^2},$$

which is the characteristic function of $N(\mu_\mu, \sigma^2 + \sigma_\mu^2)$. This completes the proof.

Based on the above theorem, the maximum likelihood estimates (MLEs) of σ^2, μ_μ, and σ_μ^2 can be obtained as follows:

$$\hat{\mu}_\mu = \frac{1}{m} \sum_{j=1}^{m} \hat{\mu}_j, \tag{9.4}$$

$$\hat{\sigma}_\mu^2 = \frac{1}{m} \sum_{j=1}^{m} (\hat{\mu}_j - \hat{\mu}_\mu)^2, \tag{9.5}$$

$$\hat{\sigma}^2 = \frac{1}{n_0} \sum_{j=1}^{m} \sum_{i=1}^{n} (x_{ji} - \hat{\mu}_j)^2, \tag{9.6}$$

where

$$\hat{\mu}_j = \frac{1}{n_j} \sum_{i=1}^{n_j} x_{ji}.$$

Based on these MLEs, estimates of the shift parameter (i.e., ε) and the scale parameter (i.e., C) can be obtained as follows: $\hat{\varepsilon} = \hat{\mu}_\mu - \hat{\mu}_0$ and $\hat{C} = (\hat{\sigma}_\mu^2 + \hat{\sigma}^2)/\hat{\sigma}_0^2$, respectively. Consequently, the sensitivity index can be estimated by simply replacing ε, μ and C with their corresponding estimates, i.e., $\hat{\varepsilon}$, $\hat{\mu}_0$, and $\hat{C}$. ∎

9.4.2 The Case Where ε Is Fixed and C Is Random

For the case where ε is fixed and C is random, similarly, set $\mu_1 \equiv \mu$ and $\sigma_1 \equiv \sigma_{new}$ and assume that $x|_{\sigma = \sigma_1}$ follows a normal distribution $N(\mu_1, \sigma_1^2)$, that is

$$x|_{\sigma_1 = \sigma_{new}} \sim N(\mu, \sigma_{new}^2),$$

where σ^2_{new} is distributed as $IG(\alpha, \beta)$, so μ_1, α, and β are unknown parameters. Thus, we have the following results.

Theorem 9.2. Suppose that $x \mid_\sigma \sim N(\mu, \sigma^2_{new})$ and $\sigma^2_{new} \sim IG(\alpha, \beta)$, then

$$x \sim f(x) = \frac{\Gamma\left(\alpha + \frac{1}{2}\right)}{\Gamma(\alpha)\sqrt{2\pi\beta}}\left[1 + \frac{(x-\mu)^2}{2\beta}\right]^{-\left(\alpha + \frac{1}{2}\right)} \tag{9.7}$$

That is, x is a non-central t-distribution, where $\mu \in R$ is location parameter, $\sqrt{\beta/\alpha}$ is scale parameter and 2α is degree of freedom.

Proof:

$$f(x) = \int_0^{+\infty} f(x \mid \sigma^2) f(\sigma^2) d\sigma^2$$

$$= \int_0^{+\infty} \frac{\beta^\alpha}{\sqrt{2\pi\sigma^2}\,\Gamma(\alpha)\sigma^{2(\alpha+1)}} \exp\left\{-\frac{(x-\mu)^2 + 2\beta}{2\sigma^2}\right\} d\sigma^2$$

$$= \frac{\beta^\alpha}{\sqrt{2\pi}\,\Gamma(\alpha)} \int_0^{+\infty} \left(\frac{1}{\sigma^2}\right)^{\alpha+\frac{3}{2}} \exp\left\{-\frac{(x-\mu)^2 + 2\beta}{2\sigma^2}\right\} d\sigma^2$$

$$= \frac{\beta^\alpha}{\sqrt{2\pi}\,\Gamma(\alpha)} \int_0^{+\infty} t^{\alpha-\frac{1}{2}} \exp\left\{-\frac{(x-\mu)^2 + 2\beta}{2}t\right\} dt$$

$$= \frac{\Gamma\left(\alpha + \frac{1}{2}\right)}{\Gamma(\alpha)\sqrt{2\pi\beta}}\left[1 + \frac{(x-\mu)^2}{2\beta}\right]^{-\left(\alpha + \frac{1}{2}\right)}$$

Thus, x follows a non-central t-distribution. Hence, we have $E(x) = \mu$ and $Var(x) = \beta/(\alpha - 1)$. This completes the proof.

Based the above theorem, the maximum likelihood estimates of the parameters μ, α, and β can be obtained as follows. Suppose that the observations satisfy the following conditions.

1. $\left(x_{ji} \mid \mu, \sigma_j^2\right) \sim N\left(\mu, \sigma_j^2\right)$, $j = 1, \ldots, m; i = 1, \ldots, n_j$ and given $\sigma_j^2, x_{j1}, \ldots, x_{jn_j}$ are independent, identically distributed (i.i.d).
2. $\{x_{ji}, i = 1, \ldots, n_j\}$ $j = 1, \ldots, m$ are independent.
3. $\sigma_j^2 \sim IG(\alpha, \beta), j = 1, \ldots, m$.

Thus, combining the response variable $x = (x_{11}, \ldots, x_{1n_1}, x_{21}, \ldots, x_{mn_m})$ and $\sigma^2 = \left(\sigma_1^2, \ldots, \sigma_m^2\right)$ as the complete data $z = \left(x_{11}, \ldots, x_{1n_1}, x_{21}, \ldots, x_{mn_m}, \sigma_1^2, \ldots, \sigma_m^2\right)$ to obtain the likelihood estimator of the parameters $\theta_1 = (\mu, \alpha, \beta)$ by expectation maximum algorithm (EM algorithm) (see, e.g., Wei, 1990; Dempster, 1977; Lange, 1989; Liu, 1995) as follows,

$$\log f(z|\theta_1) = \log f\left(x_{ji}\big|\sigma_j^2\right) + \log f\left(\sigma_j^2\big|\alpha,\beta\right).$$

Thus, we have

$$\log f(z|\theta_1) = \sum_{j=1}^{m}\left(\frac{n_j}{2}+\alpha+1\right)\log\frac{1}{\sigma_j^2}+m\alpha\log\beta - m\log\Gamma(\alpha)$$

$$-\sum_{j=1}^{m}\frac{2\beta+\sum_{i=1}^{n_j}(x_{ji}-\mu)^2}{2\sigma_j^2} \qquad (9.8)$$

Under (9.8) and by the EM algorithm, the E-step,

$$E\left(\frac{1}{\sigma_j^2}\bigg|x,\hat{\theta}_1^{(t)}\right) = \frac{\hat{\alpha}^{(t)}+n_j/2}{\hat{\beta}^{(t)}+\sum_{i=1}^{n_j}\left(x_{ji}-\hat{\mu}^{(t)}\right)^2/2}$$

$$E\left(\log\frac{1}{\sigma_j^2}\bigg|x,\hat{\theta}_1^{(t)}\right) = \phi\left(\hat{\alpha}^{(t)}+n_j/2\right)-\log\left(\hat{\beta}^{(t)}+\sum_{i=1}^{n_j}\left(x_{ji}-\hat{\mu}^{(t)}\right)^2/2\right)$$

where $\phi(*)$ is the digamma function, it is $\phi(*)=\dfrac{d}{dy}ln(\Gamma(*))$. Denote $E\left(\dfrac{1}{\sigma_j^2}\bigg|x,\hat{\theta}_1^{(t)}\right)\triangleq\hat{V}_{1j}^{(t)}$, $E\left(\log\dfrac{1}{\sigma_j^2}\bigg|x,\hat{\theta}_1^{(t)}\right)\triangleq\hat{V}_{2j}^{(t)}$, and the expectation of the log-likelihood function $E\left(logf\left(z\big|\hat{\theta}_1^{(t)}\right)\right)\triangleq Q\left(z\big|x,\hat{\theta}_1^{(t)}\right)$. By the Newton algorithm, the estimates of the parameters after the $(t + 1)^{th}$ iteration, denoted by $\hat{\theta}=\left(\hat{\mu}^{(t+1)},\hat{\alpha}^{(t+1)},\hat{\beta}^{(t+1)}\right)$ can be obtained as follows

$$\hat{\mu}^{(t+1)} = \sum_{j=1}^{m}\sum_{i=1}^{n_j}x_{ji}\hat{V}_{1j}^{(t)}\bigg/\sum_{j=1}^{m}n_j\hat{V}_{1j}^{(t)} \qquad (9.9)$$

$$\hat{\alpha}^{(t+1)} = \hat{\alpha}^{(t+1)} + \frac{\sum_{j=1}^{m} \hat{V}_{2j}^{(t)} - m\phi(\hat{\alpha}^{(t)}) + m\log\hat{\beta}^{(t)}}{m\phi'(\hat{\alpha}^{(t)})} \tag{9.10}$$

$$\hat{\beta}^{(t+1)} = m\hat{\alpha}^{(t+1)} \Big/ \sum_{j=1}^{m} \hat{V}_{1j}^{(t)} \tag{9.11}$$

The maximum likelihood estimates of μ, α and β can be obtained by (9.9)–(9.11).

Thus, the sensitivity index between the two regions could be estimated as follows

$$\hat{\Delta} = \hat{\sigma}_0^2 [1 + (\hat{\mu} - \hat{\mu}_0) / \hat{\mu}_0] / \sigma^2 \tag{9.12}$$

where $\hat{\sigma}^2 = \hat{\beta} / (\hat{\alpha} - 1)$. ∎

9.4.3 The Case Where Both ε and C Are Random

Now, consider the case when both ε and C are random. The following theorem is useful, it is possible there are differences in population mean and variance, assume that the clinical data based on the original region is distributed as (μ_1, σ_1^2) in new region the population mean μ_1 and variance are random, follows a normal-scaled inverse gamma distribution $(\mu_1, \sigma_1^2) \sim N - \Gamma^{-1}(\mu_{new}, v, a, \beta)$, that is, both ε and C are random.

Theorem 9.3. Suppose that $x\big|_{\mu_1, \sigma_1^2} \sim N(\mu_1, \sigma_1^2)$ and $(\mu_1, \sigma_1^2) \sim N - \Gamma^{-1}(\mu_{new}, v, \alpha, \beta)$, then x is distributed as

$$f(x) = \frac{\Gamma(\alpha + 1/2)}{\Gamma(\alpha)\sqrt{2\pi\beta(v+1)/v}} \left[1 + \frac{v(x - \mu_{new})^2}{2\beta(v+1)} \right]^{-\left(\alpha + \frac{1}{2}\right)}$$

It is a non-central t-distribution with a location parameter μ_{new}, scale parameter $\beta(v+1)/v$ and the degree of freedom 2α, and the $Var(x) = \beta(v+1)/(\alpha v)$. For simplicity, μ_1 and σ_1^2 are denoted by μ and σ^2, respectively.

Proof:

$$f(x,\mu,\sigma^2) = \int_0^\infty \int_{-\infty}^{+\infty} f\left(x|\mu,\sigma^2\right) f(\mu,\sigma^2) d\mu d\sigma^2$$

$$= \int_0^\infty \int_{-\infty}^\infty \frac{1}{\sqrt{2\pi\sigma}} \exp\left\{-\frac{(x-\mu)^2}{2\sigma^2}\right\} \frac{\sqrt{\upsilon}}{\sigma\sqrt{2\pi}} \frac{\beta^\alpha}{\Gamma(\alpha)} \left(\frac{1}{\sigma^2}\right)^{\alpha+1}$$

$$\exp\left\{-\frac{2\beta+\upsilon(\mu-\mu_{new})^2}{2\sigma^2}\right\} d\mu d\sigma^2$$

$$= \frac{\sqrt{\upsilon}}{2\pi} \frac{\beta^\alpha}{\Gamma(\alpha)} \int_0^\infty \int_{-\infty}^\infty \exp\left\{-\left[\frac{(x-\mu)^2}{2\sigma^2} + \frac{2\beta+\upsilon(\mu-\mu_{new})^2}{2\sigma^2}\right]\right\} d\mu d\sigma^2$$

$$= \frac{\sqrt{\upsilon}}{2\pi} \frac{\beta^\alpha}{\Gamma(\alpha)} \int_0^\infty \left(\frac{1}{\sigma^2}\right)^{\alpha+2} \exp\left\{-\frac{\upsilon[x-\mu_{new}]^2}{2(\upsilon+1)\sigma^2}\right\}$$

$$\int_{-\infty}^\infty \exp\left\{-\frac{\left[\sqrt{1+\upsilon}\mu - \frac{x\upsilon\mu_{new}}{\sqrt{1+\upsilon}}\right]^2}{2\sigma^2}\right\} d\mu d\sigma^2$$

$$= \frac{\sqrt{\upsilon}}{\sqrt{2\pi(1+\upsilon)}} \frac{\beta^\alpha}{\Gamma(\alpha)} \int_0^\infty \left(\frac{1}{\sigma^2}\right)^{\alpha+3/2} \exp\left\{-\frac{2\beta+\upsilon\left[x-\mu_{new}\right]^2/(\upsilon+1)}{2\sigma^2}\right\} d\sigma^2$$

$$= \frac{\sqrt{\upsilon}}{\sqrt{2\pi\beta(1+\upsilon)}} \frac{\Gamma\left(\alpha+\frac{1}{2}\right)}{\Gamma(\alpha)} \left[1+\frac{\upsilon[x-\mu_{new}]^2}{2\beta(\upsilon+1)}\right]^{-\left(\alpha+\frac{1}{2}\right)}.$$

Thus, the distribution of response variable x follows a non-central t-distribution. This completes the proof.

The observed response variable $x = (x_{ji})$, $j = 1, \ldots, m$; $i = 1, \ldots, n_j$ from N_2 patients in the new region, where $\sum_{j=1}^m n_j = N_2$, and the latent vector $(\mu,\sigma^2) = \left\{\left(\mu_j,\sigma_j^2\right), j = 1,\ldots,m\right\}$, the response variable x_{ji} satisfies the following conditions:

1. $\left(x_{ji}|\mu_j,\sigma_j^2\right) \sim N\left(\mu_j,\sigma_j^2\right), j = 1,\ldots,m; i = 1,\ldots,n_j$, and given $\left(\mu_j;\sigma_j^2\right)$, $x_{j1},\ldots,$ x_{jn_j} are i.i.d.;
2. $\{x_{ji}, i = 1,\ldots,n\}$, $j = 1,\ldots,m$ are independent;
3. $\left(\mu_j,\sigma_j^2\right) \sim N-\Gamma^{-1}(\mu_{new},\upsilon,a,\beta), j = 1,\ldots,m$.

Let $Z = (x, \mu, \sigma^2)$ be the complete data. The maximum likelihood estimates of $\theta_2 = (\mu_{new}, v, a, \beta)$ can be derived by expectation maximum (EM) algorithm, the log-likelihood function as follows

$$\log f(z \mid \theta_2) = \log f(x \mid \mu, \sigma^2) + \log(\mu, \sigma^2 \mid \theta_2)$$

$$= \sum_{j=1}^{m}\left(\frac{n_j + 3}{2} + a\right)\log\frac{1}{\sigma_j^2} - \sum_{j=1}^{m}\frac{2\beta + v(\mu_j - \mu_{new})^2}{2\sigma_j^2} - \sum_{j=1}^{m}\sum_{i=1}^{n_j}\frac{(x_{ji} - \mu_j)^2}{2\sigma_j^2}$$

$$- m\log\Gamma(a) + \frac{m}{2}\log v + ma\log\beta$$

conditional on the observed data x and the tth iterative estimates $\hat\theta_2^{(t)} = \left(\mu_{new}^{(t)}, v^{(t)}, a^{(t)}, \beta^{(t)}\right)$, the expectation with regard to latent data μ and σ^2 is given by

$$E\left\{\frac{1}{\sigma_j^2}\,\middle|\, x, \theta_2^{(t)}\right\} = \frac{\Gamma\left(\dfrac{n_j}{2} + a^{(t)} + 1\right)}{\left(\hat\beta^{(t)} + \dfrac{1}{2}\sum_{i=1}^{n_j}(x_{ji} - \bar x_{j\cdot})^2 + \dfrac{\hat v^{(t)} n_j}{2(n_j + \hat v^{(t)})}\left(\bar x_{j\cdot} - \hat\mu_{new}^{(t)}\right)\Gamma\left(\dfrac{n_j}{2} + a^{(t)}\right)\right)}$$

where $\bar x_{j\cdot} = \dfrac{1}{n_j}\sum_{i=1}^{n} x_{ji}$. Denote $E\left\{\dfrac{1}{\sigma_j^2}\,\middle|\, x, \hat\theta_2^{(t)}\right\} \triangleq \widehat{W}_{1j}^{(t)}$,

$$E\left\{\frac{(\mu_j - c)^2}{\sigma_j^2}\,\middle|\, x, \hat\theta_2^{(t)}\right\} = \frac{1}{n_j + \hat v^{(t)}} + \left(\frac{\sum_{i=1}^{n} x_{ji} + v^{(t)}\hat\mu_{new}^{(t)}}{n_j + \hat v^{(t)}} - c\right)^2 \hat W_{1j}^{(t)},$$

$$E\left\{\log\frac{1}{\sigma_j}\,\middle|\, x, \hat\theta_2^{(t)}\right\} = \phi\left(\frac{n}{2} + \hat\alpha^{(t)}\right) - \ln\left(\hat\beta^{(t)} + \frac{1}{2}\sum_{i=1}^{n}(x_{ji} - \bar x_{j\cdot})^2 + \frac{\hat v^{(t)} n_j}{2(n_j + \hat v^{(t)})}\left(\bar x_{j\cdot} - \hat\mu_{new}^{(t)}\right)\right).$$

Also, denote

$$E\left\{\log\frac{1}{\sigma_j^2}\,\middle|\, x, \hat\theta_2^{(t)}\right\} \triangleq \hat W_{2j}^{(t)} \quad\text{and}\quad E\left\{\frac{(\mu_j - c)^2}{\sigma_j^2}\,\middle|\, x, \hat\theta_2^{(t)}\right\} \triangleq \hat W_{3j}^{(t)}.$$

The maximum likelihood estimates of the vector θ_2:

$$\hat{\mu}_{new}^{(t+1)} = \sum_{j=1}^{m} \frac{\sum_{i=1}^{n_j} x_{ji} + \hat{v}^{(t)}\hat{\mu}_{new}^{(t)}}{n_j + \hat{v}^{(t)}} \bigg/ \sum_{j=1}^{m} \hat{W}_{1j}^{(t)}, \qquad (9.13)$$

$$\hat{v}^{(t+1)} = m \bigg/ \sum_{j=1}^{m} \widehat{W}_{3j}^{(t)}, \qquad (9.14)$$

$$\hat{\alpha}^{(t+1)} = \hat{\alpha}^{(t)} + \frac{\sum_{j=1}^{m} \hat{W}_{2j}^{(t)} - m\phi(\hat{\alpha}^{(t)}) + m\log\hat{\beta}^{(t)}}{m\phi'(\hat{\alpha}^{(t)})}, \qquad (9.15)$$

$$\hat{\beta}^{(t+1)} = m\hat{\alpha}^{(t+1)} \bigg/ \sum_{j=1}^{m} \hat{W}_{1j}^{(t)}. \qquad (9.16)$$

Without loss generality, assuming the $(t + 1)^{th}$ iteration is convergent, the maximum likelihood estimates of the unknown constant vector θ_2 is given by $\left(\hat{\mu}_{new}^{(t+1)}, \widehat{v}^{(t+1)}, \hat{\alpha}^{(t+1)}, \widehat{\beta}^{(t+1)}\right)$.

As a result, when both ε and C are random, their estimates can be obtained

$$\hat{\varepsilon} = \hat{\mu}_{new}^{(t+1)} - \hat{\mu}_0 \quad \text{and} \quad \hat{C} = \hat{\beta}(\hat{v} + 1) / (\hat{\alpha}\widehat{v\sigma_0^2}). \qquad (9.17)$$

and the estimate of the sensitivity index obtained by Eq. (9.16) between the original and the new region, the effect of ethnic factors resorting to sensitivity index will be measured, if it falls a pre-specified interval, the result of similarity between the two regions can be confirmed, otherwise, similarity is denied. ∎

9.5 Statistical Inference of Extrapolation

For the generalization of the clinical trial data from the original target patient population to the similar patient population, based on the sensitivity index Δ, we have the following hypothesis,

$$H_0 : \Delta \leq 1 - \delta \text{ or } \Delta \geq 1 + \delta \quad \text{vs.} \quad H_1 : 1 - \delta < \Delta < 1 + \delta \qquad (9.18)$$

and δ is the generalization margin, if it is applied to the concept of bioequivalence, usually $\delta = 20\%$. We can claim that the effect sizes of the two patient populations are equivalent if the null hypothesis is rejected.

According to the expression of Δ in Section 9.3, under the maximum likelihood estimates $\hat{\mu}_0, \hat{\sigma}_0^2$ and $\hat{\mu}_1, \hat{\sigma}_1^2$, respectively, of the parameters in the original target patient population and in the similar patient population, we have the estimate of the sensitivity index,

$$\hat{\Delta} = \frac{\hat{\mu}_1}{\hat{\sigma}_1} / \frac{\hat{\mu}_0}{\hat{\sigma}_0}$$

Using the nature log-transformation of $\hat{\Delta}$, the hypothesis becomes,

$$H_{01}: \ \log \Delta \le \log(1-\delta) or \ log\Delta \ge \log(1+\delta) \quad v.s. \quad H_{11}: \log(1-\delta) < log\Delta < \log(1+\delta)$$

$$(9.19)$$

due to the property of the maximum likelihood estimates, the estimate of the log-transformation $log\Delta$

$$log\,\hat{\Delta} = log\hat{\mu}_1 - log\hat{\sigma}_1^2 - log\hat{\mu}_0 + log\hat{\sigma}_0^2$$

and $E(log\Delta) = log\Delta$. We will discuss the test of sensitivity in the three cases as follows.

9.5.1 The Case Where ε Is Random and C Is Fixed

When the case where ε is random and C is fixed, the estimate of the log-transformation $log\Delta$ is,

$$log\,\hat{\Delta} = log\hat{\mu}_1 - \frac{1}{2}log\hat{\sigma}_1^2 - log\hat{\mu}_0 + log\hat{\sigma}_0^2$$

$$= log\hat{\mu}_1 - \frac{1}{2}log\left(\hat{\sigma}_\mu^2 + \hat{\sigma}^2\right) - log\hat{\mu}_0 + log\hat{\sigma}_0^2$$

Assuming $N = \dfrac{n_0}{\sum_{j=1}^{m} n_j} \to \gamma$, $n_0 \to \infty$ *and* $n_1 \to \infty$, where $0 < \gamma < \infty$, by application of the multivariate central limit theorem,

$$\sqrt{N}\left[\begin{pmatrix} \hat{\mu}_1 \\ \hat{\sigma} \\ \hat{\sigma}_\mu^2 \\ \hat{\mu}_0 \\ \hat{\sigma}_0 \end{pmatrix} - \begin{pmatrix} \mu_1 \\ \sigma \\ \sigma_\mu^2 \\ \mu_0 \\ \sigma_0^2 \end{pmatrix}\right] \xrightarrow{d} \left(\begin{pmatrix} 0 \\ 0 \\ 0 \\ 0 \\ 0 \end{pmatrix}, \; \Sigma_\varepsilon\right)$$

where Σ_ε is the covariance matrix of the parameter vector, and the diagonal matrix as follows,

$$\Sigma_\varepsilon = \begin{pmatrix} I_x^{-1}(\theta) & 0 \\ 0 & I_0(\mu_0, \sigma_0) \end{pmatrix}$$

$I_x^{-1}(\theta)$ and $I_0(\mu_0, \sigma_0)$ is respectively the covariance matrix of $\theta = (\mu_1, \sigma, \sigma_\mu)$ and (μ_0, σ_0).

From the multivariate central limit theorem and the multivariate delta method, it follows that, asymptotically,

$$\sqrt{N}(\log \hat{\Delta} - \log \Delta) \xrightarrow{d} N\left(0, \; \sigma_\varepsilon^2\right), \tag{9.20}$$

where

$$\sigma_\varepsilon^2 = B_\varepsilon \Sigma_\varepsilon B_\varepsilon^T$$

$$B_\varepsilon = \left(\frac{\partial \log \Delta}{\partial \mu_1}, \frac{\partial \log \Delta}{\partial \sigma}, \frac{\partial \log \Delta}{\partial \sigma_\mu}, \frac{\partial \log \Delta}{\partial \mu_0}, \frac{\partial \log \Delta}{\partial \sigma_0}\right) = \left(\frac{1}{\mu_1}, \frac{\sigma_\mu}{\sigma_\mu^2 + \sigma^2}, \frac{\sigma}{\sigma_\mu^2 + \sigma^2}, \frac{1}{\mu_0}, \frac{1}{\sigma_0}\right)$$

where the estimates of parameters $\mu_1, \sigma_\mu, \sigma, \mu_0, \sigma_0$ are given in Eq (9.4–9.6).

For the hypothesis, the statistics $z_\varepsilon = \dfrac{\sqrt{N}\left(log\hat{\Delta} - log\Delta\right)}{\sigma_\varepsilon}$, if $-z_{a/2} < z_\varepsilon < z_{a/2}$, the null hypothesis H_{01} is rejected, and the generalization is accepted.

9.5.2 The Case Where ε Is Fixed and C Is Random

When the case where ε is random and C is fixed, the estimate of the log-transformation $log\Delta$ is,

$$log\,\hat{\Delta} = log\hat{\mu}_1 - \frac{1}{2}log\hat{\sigma}_1^2 - log\hat{\mu}_0 + \frac{1}{2}log\hat{\sigma}_0^2$$

$$= log\hat{\mu}_1 - \frac{1}{2}log\left(\hat{\beta}\right) + \frac{1}{2}log\left(\hat{\alpha} - 1\right) - log\hat{\mu}_0 + \frac{1}{2}log\hat{\sigma}_0^2$$

Assuming $N = \dfrac{n_0}{\sum\limits_{j=1}^{m} n_j} \to \gamma$, $n_0 \to \infty$ and $n_1 \to \infty$, where $0 < \gamma < \infty$, by application of the multivariate central limit theorem,

$$\sqrt{N}\left[\begin{pmatrix}\hat{\mu}_1 \\ \hat{\beta} \\ \hat{\alpha} \\ \hat{\mu}_0 \\ \hat{\sigma}_0\end{pmatrix} - \begin{pmatrix}\hat{\mu}_1 \\ \hat{\beta} \\ \hat{\alpha} \\ \hat{\mu}_0 \\ \hat{\sigma}_0\end{pmatrix}\right] \xrightarrow{d} \left(\begin{pmatrix}0 \\ 0 \\ 0 \\ 0 \\ 0\end{pmatrix}, \Sigma_C\right)$$

For the estimate of the covariance matrix $\pounds_C$, the procedure is given in the appendix. We have

$$\Sigma_C = \begin{pmatrix} I_x^{-1}(\theta_1) & 0 \\ 0 & I_0(\mu_0, \sigma_0) \end{pmatrix}$$

where $I_x^{-1}(\theta_1)$ is given in the appendix (Thomas, 1982).

From the multivariate central limit theorem and the multivariate delta method, it follows that, asymptotically,

$$\sqrt{N}\left(log\hat{\Delta} - log\Delta\right) \xrightarrow{d} N\left(0,\ \sigma_C^2\right) \qquad (9.21)$$

where $\sigma_C^2 = B_C \Sigma_C B_C^T$ and

$$B_C = \left(\frac{\partial log\Delta}{\partial \mu_1}, \frac{\partial log\Delta}{\partial \beta}, \frac{\partial log\Delta}{\partial \alpha}, \frac{\partial log\Delta}{\partial \mu_0}, \frac{\partial log\Delta}{\partial \sigma_0} \right) = \left(\frac{1}{\mu_1}, \frac{1}{2\beta}, \frac{1}{2(\alpha-1)}, \frac{1}{\mu_0}, \frac{1}{\sigma_0} \right)$$

where the estimates of parameters μ_1, σ_μ, σ, μ_0, σ_0 is given by (9.8)–(9.10) and (9.2).

For the hypothesis, the statistics

$$z_C = \frac{\sqrt{N}\left(log\Delta - log\Delta\right)}{\sigma_\varepsilon}, \text{ if } -z_{\alpha/2} < z_C < z_{\alpha/2},$$

the null hypothesis H_{01} is rejected, and the generalization is accepted.

9.5.3 The Case Where ε and C Are Random

When the case where ε is random and C is fixed, the estimate of the log-transformation $log\Delta$ is,

$$log\hat{\Delta} = log\hat{\mu}_1 - \frac{1}{2}log\hat{\sigma}_1^2 \beta(v+1)/(\alpha v) - log\hat{\mu}_0 + \frac{1}{2}log\hat{\sigma}_0^2$$

$$= log\hat{\mu}_1 - \frac{1}{2}log\left(\hat{\beta}\right) - \frac{1}{2}log(\hat{v}+1) + \frac{1}{2}log(\hat{\alpha}) + \frac{1}{2}log(\hat{v}) - log\hat{\mu}_0 + \frac{1}{2}log\hat{\sigma}_0^2$$

Assuming $N = \dfrac{n_0}{\sum_{j=1}^{m} n_j} \to \gamma$, $n_0 \to \infty$ and $n_1 \to \infty$, where $0 < \gamma < \infty$, by application of the multivariate central limit theorem,

$$\sqrt{N}\left[\begin{pmatrix} \hat{\mu}_1 \\ \hat{v} \\ \hat{\alpha} \\ \hat{\beta} \\ \hat{\mu}_0 \\ \hat{\sigma}_0 \end{pmatrix} - \begin{pmatrix} \mu_1 \\ v \\ \alpha \\ \beta \\ \mu_0 \\ \sigma_0 \end{pmatrix} \right] \xrightarrow{d} \left(\begin{pmatrix} 0 \\ 0 \\ 0 \\ 0 \\ 0 \\ 0 \end{pmatrix}, \Sigma_{\varepsilon,C} \right),$$

For the estimate of the covariance matrix Σ_C, it is a bit complicated, and the procedure is given in the appendix. We have

$$\Sigma_{\varepsilon,C} = \begin{pmatrix} I_x^{-1}(\theta_2) & 0 \\ 0 & I_0(\mu_0, \sigma_0) \end{pmatrix},$$

where $I_x^{-1}(\theta_2)$ and $I_0(\mu_0, \sigma_0)$ is given in the appendix.

From the multivariate central limit theorem and the multivariate delta method, it follows that, asymptotically,

$$\sqrt{N}\left(\log\hat{\Delta} - \log\Delta\right) \xrightarrow{d} N\left(0, \ \sigma_{\varepsilon,C}^2\right), \tag{9.22}$$

where $\sigma_{\varepsilon,C}^2 = B_{\varepsilon,C} \Sigma_{\varepsilon,C} B_{\varepsilon,C}^T$ and

$$B_{\varepsilon,C} = \left(\frac{\partial \log\Delta}{\partial \mu_1}, \quad \frac{\partial \log\Delta}{\partial v}, \quad \frac{\partial \log\Delta}{\partial \alpha}, \quad \frac{\partial \log\Delta}{\partial \beta}, \quad \frac{\partial \log\Delta}{\partial \mu_0}, \quad \frac{\partial \log\Delta}{\partial \sigma_0} \right)$$

$$= \left(\frac{1}{\mu_1}, \quad \frac{1}{2v} - \frac{1}{2(v+1)}, \quad \frac{1}{2\alpha}, \quad -\frac{1}{2\beta}, \quad -\frac{1}{\mu_0}, \quad \frac{1}{\sigma_0} \right).$$

Note that estimates of parameters μ_1, σ_μ, σ, μ_0, σ_0 are given by (9.13)–(9.16) and (9.2).

For the hypothesis, the statistics

$$z_C = \frac{\sqrt{N}\left(\log\hat{\Delta} - \log\Delta\right)}{\sigma_{\varepsilon,C}}, \text{ if } -z_{\alpha/2} < z_{\varepsilon,C} < z_{\alpha/2},$$

the null hypothesis H_{01} is rejected, and the generalization is accepted.

9.5.4 The Confidence Interval of the Effect Size in Original Population

On the other hand, considering the confidence interval of the effect size $\frac{\mu_0}{\sigma_0}$ in the original population based on the given accepted ranges for sensitivity index, relative to the effect size $\frac{\mu_0}{\sigma_0}$ in original population while the effect size $\frac{\mu_1}{\sigma_1}$ in the new population is dropped into the confident interval, it is thought generalizable for the data in the new population. Given a $1 - \alpha$

confidence interval, we have the confidence interval of the log-transformation of the effect size $\log(\hat{\mu}_1/\hat{\sigma}_1)$ as follows,

$$\left(\log(1+\delta) + \log\left(\frac{\widehat{\mu_0}}{\sigma_0}\right) - \frac{\sigma_*}{\sqrt{N}} z_{\alpha/2}, \quad \log(1+\delta) + \log\left(\frac{\widehat{\mu_0}}{\sigma_0}\right) + \frac{\sigma_*}{\sqrt{N}} z_{\alpha/2} \right), \quad (9.23)$$

where δ is the sensitivity index, $\delta \in (-20\%, 20\%)$, and σ_* is respectively the σ_ε, σ_C and $\sigma_{\varepsilon,C}$ in (9.20)–(9.22).

9.5.5 An Example

To illustrate the proposed methods for analysis of the sensitivity index for addressing the generalizability of the observed clinical data from an original target patient population to a similar but different patient population, we consider the following example concerning an asthma clinical trial as described in Chow and Chang (2006).

A placebo-control clinical trial was conducted to evaluate the efficacy of an investigational drug product for treatment of patients with asthma. The primary study endpoint is the change in FEV1 (forced volume per second), which is defined to be the difference between the FEV1 after treatment and the baseline FEV1. Since the raw data are not available, without loss of generality and for illustration purpose, we simulate the asthma data according to summary statistics given in Table 9.2 (see also Chow and Chang, 2006).

Case 1: ε is random and C is fixed

The data of original target patients are generated from the population $N(0.34, 0.15^2)$, there are 40 adult patients data, and the data of similar but different target patient from the new patients $N(\mu_j, 0.143)$ and $\mu_j \sim N(0.32, 0.04)$, $j = 1,\dots,5$; we generate 8 data while $\mu_j = 0.3996, 0.3526, 0.3125, 0.3102, 0.3488$ respectively.

TABLE 9.2

Summary Statistics of an Asthma Trial

	Baseline FEV1 Range	Number of Patients	Baseline FEV1 Mean	FEV1 Change Mean	FEV1 Change SD
Test drug	1.5 – 2.0	9	1.86	0.31	0.14
	1.5 – 2.5	15	2.30	0.42	0.14
	1.5 – 3.0	16	2.79	0.54	0.16
Placebo	1.5 – 2.0	8	1.82	0.16	0.15
	1.5 – 2.5	16	2.29	0.19	0.13
	1.5 – 3.0	16	2.84	0.20	0.14

Source: Chow, S.C., Chang, M. (2006) *Adaptive Design Methodss in Clinical Trials*. Taylor & Francis, New York., Table 2.3.

The estimate value $\hat{\mu}_1 = 0.3107$ and $\hat{\sigma}_1 = 0.1370$ in the similar but different target patient, and $\hat{\mu}_0 = 0.3471$ and $\hat{\sigma}_0 = 0.1652$ in the adult patients population, so the estimate of the sensitivity index $\Delta = 1.0791$. The result from clinical data in the original target population could be generalized to the new target population.

Case 2: ε is fixed and C is random

The data of original target patient are generated from the population N(0.34, 0.15^2), there are 40 adult patients data, and the data of similar but different target patient from the new population $N\left(0.32,\ \sigma_j^2\right)$ and $\sigma_j^2 \sim IG(8,1)$, $j = 1,\ldots,5$; we generate 8 data while σ_j = 0.1490, 0.1005, 0.1655, 0.1234, 0.0890 from $\sigma_j^2 \sim IG(8,1)$, $j = 1,\ldots,5$, respectively. The estimated value $\hat{\mu}_1 = 0.3136$ and $\hat{\sigma}_1 = 0.1546$ in the similar but different target patient, and $\hat{\mu}_0 = 0.3280$ and $\hat{\sigma}_0 = 0.1546$ from the original adult patients, and the estimate of the sensitivity index $\Delta = 1.0566$ between the two populations. The result from clinical data in the original target population could be generalized to the new target population.

Case 3: ε and C are both random

The data of original target patient are generated from the population N(0.34, 0.15^2), there are 40 adult patients data, and the data of a similar but different target patient from the new population $N\left(\mu_j,\ \sigma_j^2\right)$ and $\sigma_j^2 \sim IG(8,1)$, $j = 1,\ldots,5$; we generate 8 data while σj = 0.1057, 0.1590, 0.0900, 0.1085, from $\sigma_j^2 \sim IG(8, 1)$, $j = 1,\ldots,5$ respectively, and the corresponding μ_j = 0.3417, −0.2132, 0.5486, 0.3410, 0.1179. The estimate value $\hat{\mu}_1 = 0.3173$ and $\hat{\sigma}_1 = 0.3024$ in the similar but different target patient, and $\hat{\mu}_0 = 0.3533$ and $\hat{\sigma}_0 = 0.1591$ from the original adult patients, and the estimate of the sensitivity index $\Delta = 0.4726$ between the two population. The result from clinical data in the original target population can't be generalized to the new target population.

On the other side, we give the simulation on the confidence interval of the effect size from the similar but different target patient, under the same parameter conditions as the above mentioned, and we have the results in Table 9.3.

TABLE 9.3

The Confidence Interval and the Estimate Value of the Log-Effect Size in the Similar but Different Target Patient Population

	δ = 0.1		δ = 0.2	
	CI	Estimate value	CI	Estimate Value
Case 1	(0.7989, 1.1709)	1.0075	(0.7037, 1.0756)	0.9233
Case 2	(0.6762, 1.4081)	0.7238	(0.7617, 1.1335)	0.9976
Case 3	(0.7474, 1.1193)	0.5907	(0.6870, 1.0588)	0.5986

Note: CI is the confidence interval of the log-effect size.

9.6 Concluding Remarks

In clinical research, it is often of interest to determine whether the observed clinical results can be generalized from the original target patient population (e.g., adults) to a similar but different patient population (e.g., pediatrics or elderly). Following a similar idea of population shift due to protocol amendments, analysis with covariate adjustment and the assessment of sensitivity index are the two commonly considered approaches for assessment of generalizability of clinical results from the target patient population to a similar but different patient population. For the method of analysis with covariate adjustment, since such covariates may not exist or exist but are not observable, we focus on the assessment of generalizability through the evaluation of the sensitivity index. For the assessment of sensitivity index, in addition to the cases where (i) ε is random and C is fixed, (ii) ε is fixed and C is random, and (iii) both ε and C are random, there are other cases such as (i) random split of the similar and different patient population (for assessment of shift in location parameter and change in scale parameter) and (ii) the number of segments in the split samples is also a random variable which remains unsolved.

In addition, statistically it is also a challenge to clinical researchers when there are differences in demographics and/or patient characteristics. The imbalances between the original target patient population and the similar but different patient population could have a negative impact on the generalizability of the clinical results from the original target patient population to the similar but different patient population (Shao and Chow, 2002). In this case, these imbalances must be properly adjusted in order to provide an unbiased and reliable assessment and interpretation of the treatment effect of the similar but different patient population.

Appendix

Consider the covariance-variance matrix of the parameter vector $\theta_1 = (\mu, \alpha, \beta)$. Louis (1982) derived the observer information matrix when the EM algorithm is used to find maximum likelihood estimates in incomplete data problems. According to Louis's method, the information matrix of vector $\theta_1 = (\mu, \alpha, \beta)$ could be given

$$I_x(\theta_1) = E_{\theta_1}\left\{B(z, \theta_1)|z \in R\right\} - E_{\theta_1}\left\{S(z, \theta_1)S^T(z, \theta_1)|z \in R\right\} + S^*(x, \theta_1)S^{*T}(x, \theta_1),$$

$$(9.A1)$$

where $S(z, \theta_1)$ are the gradient vectors of log-likelihood function $\log f(z|\theta_1)$, $B(z, \theta_1)$ are the negatives of the associated second derivative matrices. Of course, they need be evaluated only on the last iteration of the EM procedure, where $S^*(x, \theta_1) = E\{S(z, \theta_1)|z \in R\}$ is *zero. Here

$$S^T(z, \theta_1) = \left(\sum_{j=1}^{m}\sum_{i=1}^{n_j}(x_{ji} - \mu)/\sigma_j^2, \sum_{j=1}^{m}\log\frac{1}{j} + m\log\beta - m\phi(\alpha), \frac{m\alpha}{\beta} - \sum_{j=1}^{m}\frac{1}{\sigma_j^2} \right)$$

(9.A2)

$$B(z, \theta_1) = \begin{pmatrix} \sum_{j=1}^{m}\dfrac{n}{\sigma_j^2} & 0 & 0 \\[2ex] 0 & m\phi'(\alpha) & -\dfrac{m}{\beta} \\[2ex] 0 & -\dfrac{m}{\beta} & \dfrac{m\alpha}{\beta^2} \end{pmatrix}$$

(9.A3)

and

$$S^{*T}(z, \theta_1) = \left(\sum_{j=1}^{m}\sum_{i=1}^{n_j}\hat{V}_{1j}^{(t)}(x_{ji} - \mu), \sum_{j=1}^{m}\hat{V}_{2j}^{(t)} + m\log\beta - m\phi(\alpha), \frac{m\alpha}{\beta} - \sum_{j=1}^{m}\hat{V}_{1j}^{(t)} \right)$$ (9.A4)

substituting equations (9.A2)–(9.A4) for equation (9.A1), the information matrix $I_Y(\theta_1)$ of vector $\theta_1 = (\mu, \alpha, \beta)$ can be obtained. $I_x(\theta_1)$ can be inverted to find the covariance matrix of $\hat{\theta}$, that is,

$$Cov\left(\widehat{\theta_1}\right) = I_x^{-1}\left(\widehat{\theta_1}\right).$$

For the case 3 when the ε and C are random, the information matrix of the parameters vector θ_2 will be obtained by Louis (1982) method, where

$$S^T(z, \theta_2) =$$

$$\left(\sum_{j=1}^{m}\frac{v(\mu_j - \mu_{new})}{\sigma_j^2}, -\sum_{j=1}^{m}\frac{(\mu_j - \mu_{new})^2}{\sigma_j^2} + \frac{m}{2v}\sum_{j=1}^{m}\log\frac{1}{\sigma_j^2} - m\phi(\alpha) + m\log\beta, -\sum_{j=1}^{m}\frac{1}{\sigma_j^2} + \frac{m\alpha}{\beta} \right),$$

(9.A5)

$$B(z, \theta_2) = \begin{pmatrix} \displaystyle\sum_{j=1}^{m} \frac{v}{\sigma_j^2} & -\displaystyle\sum_{j=1}^{m} \frac{(\mu_j - \mu_{new})}{\sigma_j^2} & 0 & 0 \\[2ex] -\displaystyle\sum_{j=1}^{m} \frac{(\mu_j - \mu_{new})}{\sigma_j^2} & \dfrac{m}{2v^2} & 0 & 0 \\[2ex] 0 & 0 & m\phi'(\alpha) & -\dfrac{m}{\beta} \\[2ex] 0 & 0 & -\dfrac{m}{\beta} & \dfrac{m\alpha}{\beta^2} \end{pmatrix} \quad (9.A6)$$

and

$$S^T(x, \theta_2) = \left(\Sigma_{j=1}^{m} \left(\frac{\Sigma_{i=1}^{n} x_{ji} + v^{(t)} \mu_{new}}{n_j + v^{(t)}} - \mu_{new} \right) v \widehat{W}_{1i}^{(t)}, -\Sigma_{j=1}^{m} \frac{\widehat{W}_{2j}^{(t)}}{2} + \frac{m}{2v}, \right.$$
$$\left. \Sigma_{j=1}^{m} \widehat{W}_{2j}^{(t)} - m\phi(a) + m\log\beta, -\Sigma_{j=1}^{m} \widehat{W}_{j}^{(t)} + \frac{ma}{\beta} \right)$$

$$(9.A7)$$

substituting (9.A5)–(9.A7) to (9.A1), the covariance-variance matrix of vector $\theta_2 = (\mu_{new}, v, \alpha, \beta)$ can be obtained by the inverse matrix $I_{11}^{-1}(\widehat{\theta_2})$.

10

Case Studies – Recent FDA Biosimilar Submissions

10.1 FDA Abbreviated Licensure Pathway

Since the passage of the Biologics Price Competition and Innovation (BPCI) Act of 2009, the United States Food and Drug Administration (FDA) have created an abbreviated licensure pathway for biological products shown to be biosimilar to a US-licensed biological product (the reference product). This abbreviated licensure pathway was developed under section 351(k) of the PHS Act which permits reliance on certain existing scientific knowledge about the safety, purity, and potency of the reference product. Most importantly, this abbreviated licensure pathway also enables a biosimilar biological product to be licensed based on less than a full complement of product specific nonclinical and clinical data.

As indicated in previous Chapters, a biosimilar product is defined as a biological product that is *highly similar* to the reference product notwithstanding minor differences in clinically inactive components and that there are no clinically meaningful differences between the proposed biosimilar product and the reference product in terms of the safety, purity, and potency of the product. As a result, a 351(k) application must contain sufficient information in order to demonstrate that the proposed biosimilar product is highly similar to a reference product based upon data obtained from analytical studies, animal studies, and a clinical study or studies, unless the regulatory agency determines that certain studies are unnecessary in a 351(k) application (see section 351(k)(2) of the PHS Act). It should be noted that development of a biosimilar product differs from development of a biological product intended for submission under section 351(a) of the PHS Act (i.e., a *stand-alone* marketing application). The purpose of a stand-alone development program of a biological product is to demonstrate safety, purity, and potency of the proposed product based on data derived from a *full* complement of clinical and nonclinical studies. On the other hand, the goal of a biosimilar development program is to demonstrate that the proposed biosimilar product is highly similar to the reference product in terms of safety, purity, and potency. Both stand-alone biological product and proposed biosimilar product development programs are required to generate analytical, nonclinical, and clinical data from a number of different types of studies conducted based on differing goals and the different statutory standards for licensure.

In her presentation at an Oncologic Drugs Advisory Committee (ODAC) meeting for review of regulatory submissions of Avastin (sponsored by Amgen) and Herceptin (sponsored by Mylan) held on July 13, 2017, Lim (2017), an FDA representative, indicated that FDA recommends the use of a stepwise approach for obtaining the totality-of-the-evidence to support a demonstration of biosimilarity between a proposed biosimilar product and a reference product.

The totality-of-the-evidence generally includes structural and functional characterization, animal study data, human PK and, if applicable, pharmacodynamics (PD) data, clinical immunogenicity data, and other clinical safety and effectiveness data. At each step, Lim (2017) also pointed out that the applicant should evaluate the extent to which there is residual uncertainty about the biosimilarity of the proposed biosimilar product to the reference product and take actions if such uncertainty is identified. Lim (2017) also pointed out that the underlying presumption of an abbreviated development program is that a molecule that is shown to be structurally and functionally highly similar to a reference product is anticipated to behave like the reference product in clinical setting(s). Thus, the stepwise approach should start with extensive structural and functional characterization of both the proposed biosimilar product and the reference product. This analytical characterization will serve as the foundation of the proposed biosimilar development program. Based on these results analytical similarity assessment between the proposed biosimilar product and the reference product can be made to determine that the amount of residual uncertainty remaining with respect to both the structural and functional evaluation and the potential for clinically meaningful differences. It should be noted that in practice, additional data such as nonclinical and/or clinical data can also be used to address residual uncertainty.

In general, an applicant needs to provide information to demonstrate biosimilarity based on data directly comparing the proposed product with the US-licensed reference product. The BPCI Act defines the reference product as the single biological product licensed under section 351(a) of the PHS Act against which a proposed biosimilar product is evaluated in a 351(k) application. When an applicant's proposed biosimilar development program includes data generated using a non-US-licensed comparator to support a demonstration of biosimilarity to the US-licensed reference product, the applicant must provide adequate data or information to scientifically justify the relevance of these comparative data to an assessment of biosimilarity and establish an acceptable bridge to the US-licensed reference product. In practice, the type of bridging data needed often include data from analytical studies (e.g., structural and functional data) that directly compare all three products (i.e., the proposed biosimilar product, the reference product, and the non-US-licensed comparator product) and is likely to also include bridging clinical pharmacokinetics (PK) or pharmacodynamics (PD) study data for all three products.

In this chapter, two recent biosimilar regulatory submissions voted approval by the ODAC meeting held in Silver Spring on July 13, 2017 are reviewed as case studies of the abbreviated licensure pathway developed by the FDA. In the next section, commonly adopted strategies for biosimilar regulatory submissions are outlined. Sections 10.3 and 10.4 provide detailed discussions (by focusing on analytical similarity assessment) of recent biosimilar regulatory submissions of Avastin (sponsored by Amgen) and Herceptin (sponsored by Mylan), respectively. Section 10.5 provides some concluding remarks for the chapter.

10.2 Sponsor's Strategy for Regulatory Submission

In practice, commonly considered strategy for regulatory submission depends upon whether there is single reference or multiple references. As indicated in the previous section, a reference product is defined as the single biological product licensed under section 351(a) of the PHS Act, which is also referred to as US-licensed reference product. Multiple references include a US-licensed reference product and other non US-approved reference products (e.g., EU-approved reference product).

Single Reference Strategy – When there is single reference, the sponsors usually follow the FDA recommended stepwise approach for obtaining the totality-of-the-evidence to demonstrate that the proposed biosimilar product is highly similar to the reference product in terms of safety, purity, and potency of the product by the following the steps:

Step 1: Obtain extensive analytical data to support (i) a demonstration that the proposed biosimilar product and the reference product are highly similar, and (ii) a demonstration that the proposed biosimilar product can be manufactured in a well-controlled and consistent manner that meet appropriate quality standards.

Step 2: Conduct a single-dose pharmacokinetic (PK) study to support PK similarity between the proposed biosimilar product and the reference product.

Step 3: Conduct a clinical study or studies to support clinical similarity in terms of safety (immunogenicity) and efficacy between the proposed biosimilar product and the reference product.

Step 4: Provide scientific justification for extrapolation of data to support biosimilarity in each of the additional indications for which the sponsor is seeking licensure.

Multiple References Strategy – When there are multiple references, e.g., a US-licensed reference product and a non-US-licensed product such as EU-approved reference product, the sponsors often consider the following strategy for obtaining the totality-of-the-evidence to demonstrate that the proposed biosimilar product is highly similar to the reference product in terms of safety, purity, and potency:

Step 1: Obtain extensive analytical data intended to support (i) a demonstration that the proposed biosimilar product and the US-licensed reference product are highly similar, (ii) a demonstration that the proposed biosimilar product can be manufactured in a well-controlled and consistent manner that meets appropriate quality standards, and (iii) a justification of the relevance of the comparative data generated using an EU-approved reference to support a demonstration of biosimilarity of the proposed biosimilar product to the US-licensed reference product.

Step 2: Conduct a single-dose PK study providing a three-way comparison of the proposed biosimilar product, US-licensed reference product, and EU-approved reference product to (i) support PK similarity of the proposed biosimilar product and the US-licensed reference product, and (ii) provide the PK portion of the scientific bridge to support the relevance of the comparative data generated using EU-approved reference product to support a demonstration of the biosimilarity of the proposed product to the US-licensed reference product.

Step 3: Conduct a comparative clinical study between the proposed biosimilar product and EU-approved product in patients with a selected indication to support the demonstration of no clinically meaningful differences in terms of safety, purity, and potency between the proposed biosimilar product and the US-licensed reference product. This clinical study is usually a randomized, double-blind, parallel group study conducted in sufficient patients with the disease under study who are randomized (usually 1:1) to receive the proposed biosimilar product or EU-approved reference product. Biosimilarity is evaluated based on some valid study endpoints.

Step 4: Provide scientific justifications for extrapolation of data to support biosimilarity in each of the additional indications for which the sponsor is seeking licensure.

TABLE 10.1

List of Up-to-Date Regulatory Guidance

Year (Version)	Title of the Guidance
2014 (Draft)	Reference Product Exclusivity for Biological Products filed Section 351(a) of the PHS Act
2015 (Draft)	Biosimlars: Additional Questions and Answers Regarding Implementation of the Biologics Price Competition and Innovation Act of 2009
2015 (Final)	Scientific Considerations in Demonstrating Biosimilarity to a Reference Product
2015 (Final)	Quality Considerations in Demonstrating Biosimilarity to a Reference Product
2015 (Final)	Biosimilars: Questions and Answers Regarding Implementation of the Biologics Price Competition and Innovation Act of 2009
2015 (Final)	Formal Meetings between the FDA and Biosimilar Biological Product Sponsors or Applicants
2016 (Draft)	Labeling for Biosimilar Products
2016 (Final)	Clinical Pharmacology Data to Support a Demonstration of Biosimilarity to a Reference Product
2017 (Draft)	Considerations in Demonstrating Interchangeability With a Reference Product
2017 (Final)	Nonproprietary Naming of Biological Products
2017 (Draft)	Statistical Approaches to Evaluate Analytical Similarity

Remarks – It should be noted that appropriate study design and statistical methods for data analysis should be employed for a valid assessment of biosimilarity between the proposed biosimilar product and the reference product regardless of whether there is a single reference or multiple references for the submission. Lim (2017) suggested that regulatory guidance should be carefully reviewed during the preparation and assembling of the submission package. A list of up-to-date regulatory guidance is given in Table 10.1.

10.3 Avastin Biosimilar Regulatory Submission

Amgen submitted a Biologics License Application (BLA# 761028) under section 351(k) of the PHS Act for ABP215, a proposed biosimilar to US-licensed Avastin (bevacizumab) of Genentech. Genentech's Avastin (BLA# 125085) was initially licensed by FDA on February 26, 2004. Amgen's submission was discussed and voted approval by the Oncologic Drugs Advisory Committee (ODAC) meeting held within FDA in Silver Spring on July 13, 2017.

In this section, the case of Amgen's Avastin biosimilar regulatory submission is studied by focusing on similarity assessment of data collected from the analytical studies. In what follows, Amgen's strategy for biosimilar submission is outlined, followed by the introduction of the mechanism of action of the US-licensed reference product, analytical data generation, results of analytical similarity assessment, and FDA's assessment of analytical data.

Amgen's Strategy for Biosimilar Submission – Amgen adopted a strategy for biosimilar submission with multiple references (i.e., US-licensed Avastin and EU-approved bevacizumab). The application consisted of the following:

1. Extensive analytical data intended to support (i) a demonstration that ABP215 and US-licensed Avastin are highly similar; (ii) a demonstration that ABP215 can be manufactured in a well-controlled and consistent manner that is sufficient to meet appropriate quality standards; and (iii) a justification of the relevance of the comparative data generated using EU-approved bevacizumab to support a demonstration of biosimilarity of ABP215 to US-licensed Avastin;

2. A single-dose pharmacokinetic (PK) study providing a three-way comparison of ABP215, US-licensed Avastin, and EU-approved bevacizumab intended to (i) support PK similarity of ABP215 and US-licensed Avastin and (ii) provide the PK portion of the scientific bridge to support the relevance of the comparative data generated using EU-approved bevacizumab to support a demonstration of the biosimilarity of ABP215 to US-licensed Avastin;

3. A comparative clinical study (Study 20120265) between ABP215 and EU-approved bevacizumab in patients with advanced/metastatic non-small-cell lung cancer (NSCLC) to support the demonstration of no clinically meaningful differences in terms of response, safety, purity, and potency between ABP215 and US-licensed Avastin. This was a randomized, double-blind, parallel group study conducted in 642 patients with previously untreated NSCLC who were randomized (1:1) to receive carboplatin and paclitaxel with ABP215 or EU-approved bevacizumab (15 mg/kg dose every 3 weeks for up to 6 cycles). The primary endpoint of Study 20120265 was the risk ratio of the overall response rate (ORR). The study met its primary endpoint, as the risk ratio of ORR fell within the pre-specified margin. In addition to meeting the primary endpoint, the study showed that cardinal anti-VEGF effects (e.g., hypertension) were similar between arms;

4. A scientific justification for extrapolation of data to support biosimilarity in each of the additional indications for which Amgen is seeking licensure.

Introduction of US-licensed Reference Product – Vascular endothelial growth factor (VEGF) plays an important pathophysiologic role in mechanism of action of Avastin (Ellis, 2008). VEGF family members, A (VEGFA), B, C, D, and placental growth factor, belong to a superfamily of proteins that are classified as cysteine knot growth factors based on structure (Muller, 1997) are known to be responsible for regulating vasculogenesis, angiogenesis, and lymphangiogenesis under both normal and pathophysiological conditions. Specifically, VEGFA provides several functions that are important for angiogenesis and include induction of endothelial cell proliferation and survival, increase in vascular permeability, and chemotaxis and homing of bone marrow cells for hematopoiesis (Ferrara, 2004). The main receptors that bind VEGFA and mediate vasculogenesis/angiogenesis and chemotaxis/hematopoiesis are VEGF2 (receptor 2), which is also known as KDR or Flk-1 (mouse) and VEGF1 (receptor 1), which is also known as Flt-1, respectively (Ferrara, 2004). Also, see Figure 10.1.

Ferrara (2010) indicated that VEGFA can exist in several isoforms and exert local as well as distal signaling events. Arcondeguy (2013) showed that VEGFA isoforms are the result of alternative splicing of eight exons. The most commonly expressed VEGFA isoforms are 165, 121, 189, and 206, where numbers represent the amino acid lengths following cleavage of 26 N-terminal residues of the signal peptide (Figure 10.2).

ABP215 is a recombinant humanized IgG1 monoclonal antibody that targets human VEGFA and prevents the interaction of VEGFA to its receptors. Goel (2013) indicated that the targeting of VEGFA by ABP215 results in the

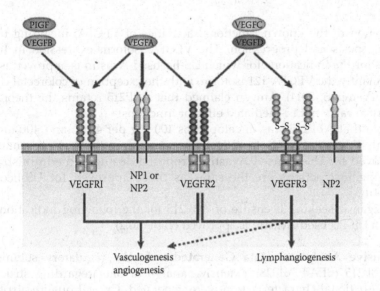

FIGURE 10.1
Vascular endothelial growth factor family members and receptors. (From Ellis, L.A. *Nat Rev Cancer*, 579–591, 2008.)

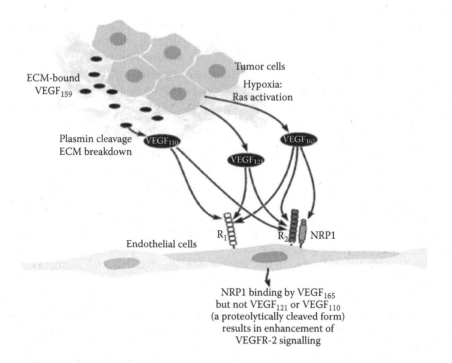

FIGURE 10.2
VEGF isoforms and their interaction with VEGFRs. (From Ferrara, N. *Endocr Rev*, 581–611, 2004.)

inhibition of the known functional activities of VEGFA, including tumor angiogenesis and progression. The VEGFA isoform expression in tumor types for the indications for which US-licensed Avastin is approved is predominantly the VEGFA 121 isoform, with the exception of colorectal cancer type (Vempati, 2014). Amgen claimed that ABP215 retains the theoretical ability to carry out Fc-mediated effector functions.

Note that ABP215 was developed as 100 mg per 4 mL and 400 mg per 16 mL single-use vials to reflect the same strength and presentations approved for US-licensed Avastin. Proposed dosing and administration labeling instructions are the same as those approved for US-licensed Avastin.

Amgen is seeking licensure of ABP215 for the following indications for which US-licensed Avastin is approved (Table 10.2).

Extensive Analytical Data Generated – In the regulatory submission of ABP215 (BLA# 761028), extensive analytical data regarding structural and functional characterization were generated. Critical quality attributes (CQAs) that are related to bevacizumab structure mechanism of action were identified which are summarized in Table 10.3.

TABLE 10.2

Approved Indications for Avastin

No.	Indication
1.	Metastatic colorectal cancer, with intravenous 5-fluorouracil–based chemotherapy for first- or second-line treatment.
2.	Metastatic colorectal cancer, with fluoropyrimidine-irinotecan- or fluoropyrimidine oxaliplatin-based chemotherapy for second-line treatment in patients who have progressed on a first-line Avastin-containing regimen.
3.	Non-squamous non-small-cell lung cancer, with carboplatin and paclitaxel for first line treatment of unresectable, locally advanced, recurrent, or metastatic disease.
4.	Glioblastoma, as a single agent for adult patients with progressive disease following prior therapy.
5.	Metastatic renal cell carcinoma with interferon alfa.
6.	Cervical cancer, in combination with paclitaxel and cisplatin or paclitaxel and topotecan in persistent, recurrent, or metastatic disease.

TABLE 10.3

List of Critical Quality Attributes Evaluated in ABP215 Regulatory Submission

Structure/Function/Activity	Critical Quality Attributes
Primary structure	Intact molecular weight Amino acid sequence Disulfide bonds
Higher order structure	Secondary structure Tertiary structure Thermal stability
Glycosylation	Afucosylation Galactosylation High mannose Sialylation
Biological Activities (Fab-mediated)	Inhibition of HUVEC VEGFA binding Binding kinetics for VEGFA isoforms Binding specificity
Biological activities (Fc-mediated)	FcRn Fcg receptors C1q Antibody-dependent cellular cytotoxicity Complement-dependent cytotoxicity
Product related species	Charge variants Size variants
Drug product attributes	Protein content Sub-visible particles Deliverable volume Appearance, pH, osmolality
Stability	Degradation profiles under accelerated and stress conditions

Abbreviations: HUVEC = human umbilical vein endothelial cell proliferation; VEGFA = vascular endothelial growth factor family member A.

TABLE 10.4

Product Lots Used Data Analysis

Product	Number of Lots	CQA Assessment	Statistical Analysis
ABP 215 DP	19	Tier 1	Equivalence test
ABP 215 DS	13*		
US-licensed Avastin	27	Tier 2	Quality range
EU-approved bevacizumab	29	Tier 3	Graphical comparison

Abbreviations: DP = drug product; DS = drug substance.
*13 independent DS lots were used to derive 19 DP lots.

These CQAs were classified into three tiers according to their criticality or risk ranking relevant to clinical outcomes and tested on some selected reference and test (ABP 215) lots using a tiered approach as recommended by the FDA (see Table 10.3).

These CQAs were tested on 19 lots of ABP215, 27 lots of US-licensed Avastin, and 29 lots of EU-approved bevacizumab. Tier 1 CQAs, Tier 2 CQAs, and Tier 3 CQAs were evaluated using equivalence test, quality range approach, and graphical comparison, respectively (Table 10.4).

Note that Tier 1 CQAs such as assays that assessed the primary mechanism of action were tested based on the endpoints of (i) percent relative potency as assessed by proliferation inhibition bioassay, and (ii) VEGF-A binding by enzyme-linked immunosorbent assay (ELISA).

Analytical Similarity Assessment – As it can be seen from Table 10.4, a total of 19 ABP215, 27 US-licensed Avastin, and 29 EU-approved bevacizumab lots were used in the analytical similarity assessment. However, not all lots were used. Amgen indicated that the number of lots used to evaluate each quality attribute was determined based on their assessment of the variability of the analytical method and availability of US-licensed Avastin and EU-approved bevacizumab. Both the 100 mg/vial and 400 mg/vial strengths were used in the analytical similarity assessment.

For illustration purpose, consider equivalence testing for the most critical quality attributes related to the mechanism of action, such as binding to VEGFA. Figure 10.3 provides scatter plots of VEGF-A binding by ELISA for US-licensed Avastin, ABP215, and·EU-approved bevacizumab.

Pairwise comparisons of ABP215 to EU-approved bevacizumab and EU-approved bevacizumab to US-licensed Avastin were performed for the purpose of establishing the analytical portion of the scientific bridge necessary to support the use of the data derived from the clinical studies that used the EU-approved bevacizumab as the comparator. The results are summarized in Table 10.5.

Table 10.5 shows that the 90% confidence interval for the mean difference in VEGFA binding by ELISA between ABP215 and US-licensed Avastin

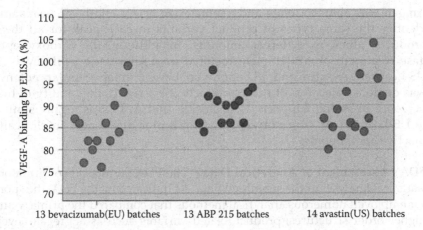

FIGURE 10.3
Scatter plots of VEGF-A binding by ELISA. (Courtesy of BLA# 761068.)

TABLE 10.5

Equivalence Testing Results for the VEGF-A Binding by ELISA

Comparison	# of Lots	Mean Difference (%)	90% Confidence Interval (%)	Equivalence Margin (%)	Equivalent
ABP215 vs. US	(13, 14)	0.63	(−2.93, 4.18)	(−9.79, +9.79)	Yes
ABP215 vs. EU	(13, 13)	4.85	(1.23, 8.45)	(−9.57, +9.57)	Yes
EU vs. US	(13, 14)	−4.22	(−8.47, 0.03)	(−9.57, +9.79)	Yes

is (−2.93%, 4.18%) which falls entirely within the equivalence margin of (−9.79%, 9.79%). This result supports a demonstration that ABP215 is highly similar to US-licensed Avastin. On the other hand, the 90% confidence interval for the mean difference in VEGFA binding by ELISA between ABP215 and EU-approved bevacizumab is (1.24%, 8.45%) which is also within the equivalence limit of (−9.79%, 9.79%). The result supports analytical portion of the scientific bridge to justify the relevance of EU-approved bevacizumab data from the comparative clinical study.

However, several quality attribute differences were noted. These notable differences include differences in glycosylation content (galactosylation and high mannose), FcgRIIIa (158) binding, and product related species (aggregates, fragments, and charge variants). Actions were taken to address these notable differences. For example, for glycosylation and FcγRIIIa (158V) binding differences, *in vitro* cell based ADCC and CDC activities were assessed and were not detected for all three products. In addition, clinical pharmacokinetic data further addressed the residual uncertainty and showed that differences between the three products were unlikely to have clinical impact. For differences in charge variants,

Amgen was able to isolate and characterize acidic and basic peaks and identify the same types of product variants in each peak for all three products, albeit in different amounts. In addition, the carboxypeptidase treatment of ABP215 also resulted in similar basic peak levels as US-licensed Avastin and EU-approved bevacizumab. Similar potency was demonstrated for all three products. As a result, the totality-of-the-evidence provided supports a conclusion that ABP215 is highly similar to US-licensed Avastin notwithstanding minor differences in clinically inactive components.

FDA's Assessment of Analytical Data – The FDA performed confirmatory statistical analyses of the submitted data. As indicated by the FDA, the sponsor employed numerous analytical methods that compared the primary and higher order structures, product-related variants such as aggregate levels and charge variants, process-related components such as host cell DNA, and biological functions to support a demonstration that ABP215 is highly similar to US-licensed Avastin. In addition, the sponsor supported the analytical portion of the scientific bridge to justify the relevance of data obtained from the use of EU-approved bevacizumab as the comparator product in clinical studies.

The analytical data submitted supports a demonstration that ABP215 is highly similar to US-licensed Avastin. All three products demonstrated similar binding affinities to VEGFA and similar potency, which are product-quality attributes associated with the mechanism of action for ABP215 and US-licensed Avastin. Lastly, the higher order structure determinations showed the presence of similar secondary and tertiary structures and further support the binding and potency results. The impurity profiles also demonstrated that ABP215 has acceptably low levels of impurities that are similar to US-licensed Avastin.

Some quality attributes were found to be slightly different between products but unlikely to have clinical impact, and they do not preclude a demonstration that ABP215 is highly similar to US-licensed Avastin. For example, the differences in charge variants for ABP215 were due to lower levels of the product variants in US-licensed Avastin and are likely due to the age difference between ABP215 and US-licensed Avastin at the time of the analytical similarity assessment. Furthermore, the differences in the glycan species were shown to not affect PK in clinical studies and no effector functions were observed *in vitro* that could be impacted by differences in the level of glycoforms.

The analytical similarity data comparing ABP215, EU-approved bevacizumab and US-licensed Avastin supported the relevance of clinical data derived from using EU-approved bevacizumab as the comparator to support a demonstration of biosimilarity of ABP215 to US-licensed Avastin.

Remarks – The Applicant used a non-US-licensed comparator (EU-approved bevacizumab) in the comparative clinical study intended to support a demonstration of no clinically meaningful differences from US-licensed Avastin. Accordingly, the Applicant provided scientific justification for the relevance of that data by establishing an adequate scientific bridge between EU-approved bevacizumab, US-licensed Avastin, and ABP215. Review of an extensive battery of test results provided by the Applicant confirmed the adequacy of the scientific bridge and hence the relevance of comparative clinical data obtained with EU-approved bevacizumab to support a demonstration of biosimilarity to US-licensed Avastin. This battery of tests included both analytical studies and a comparative PK study in humans.

In considering the totality of the evidence, the data submitted by the Applicant support a demonstration that ABP215 is highly similar to US-licensed Avastin, notwithstanding minor differences in clinically inactive components, and support a demonstration that there are no clinically meaningful differences between ABP215 and US-licensed Avastin in terms of the safety, purity, and potency of the product.

The Applicant has also provided an extensive data package to address the scientific considerations for extrapolation of data to support biosimilarity to other conditions of use and potential licensure of ABP215 for each of the indications for which US-licensed Avastin is currently licensed and for which the Applicant is eligible for licensure.

10.4 Herceptin Biosimilar

Mylan submitted a Biologics License Application (BLA# 761074) under section 351(k) of the PHS Act for MYL-1401O, a proposed biosimilar to US-licensed Herceptin of Genentech. Genentech's Herceptin (BLA# 103792) was initially licensed by FDA on September 25, 1998. Mylan's submission was discussed and voted approval by the Oncologic Drugs Advisory Committee (ODAC) meeting held within FDA in Silver Spring on July 13, 2017.

In this section, the case of Mylan's Herceptin biosimilar regulatory submission is studied by focusing on similarity assessment of data collected from the analytical studies. In what follows, Mylan's strategy for biosimilar submission is outlined followed by introduction to pathophysiology of HER2 and mechanism of action of trastuzumab, analytical data generation, results of analytical similarity assessment, and FDA's assessment of analytical data.

Mylan's Strategy for Biosimilar Submission – Mylan adopted a strategy for biosimilar submission with multiple references (i.e., US-Herceptin and EU-Herceptin). The application consisted of the following:

1. Extensive analytical data intended to support (i) a demonstration that MYL-1401O and US-Herceptin are highly similar; (ii) a demonstration that MYL-1401O can be manufactured in a well-controlled and consistent manner that is sufficient to meet appropriate quality standards; and (iii) a justification of the relevance of the comparative data generated using EU-Herceptin to support a demonstration of biosimilarity of MYL-1401O to US-Herceptin.

2. A single-dose pharmacokinetic (PK) study providing a three-way comparison of MYL-1401O, US-Herceptin, and EU-Herceptin intended to (i) support PK similarity of MYL-1401O and US-Herceptin, and (ii) provide the PK portion of the scientific bridge to support the relevance of the comparative data generated using EU-Herceptin to support a demonstration of the biosimilarity of MYL-1401O to US-Herceptin.

3. A comparative clinical study between MYL-1401O and EU-Herceptin in patients with untreated metastatic HER2 positive breast cancer to support the demonstration of no clinically meaningful differences in terms of safety, purity, and potency between MYL-1401O and US-Herceptin. This was a randomized, double-blind, parallel group study conducted in 493 patients with previously untreated breast cancer who were randomized (1:1) to receive MYL-1401O or EU-Herceptin (loading dose 8 mg/kg in cycle 1 followed by maintenance dose 6 mg/kg 3 week cycles up to 8 cycles; if stable disease after 8 cycles, can continue combination treatment at investigator's discretion). The primary endpoint of MYL-Her 3001 study was the risk ratio of the overall response rate (ORR). The study met its primary endpoint, as the risk ratio of ORR fell within the pre-specified margin of (0.81, 1.24).

4. A scientific justification for extrapolation of data to support biosimilarity in each of the additional indications for which Mylan is seeking licensure.

Pathophysiology of HER2 and Mechanism of Action of Trastuzumab – The HERs are a family of transmembrane tyrosine kinases that consists of the following known isoforms: HER1, HER2, HER3, and HER4, which are also known as epidermal growth factor receptors ErbB1, ErbB-2, ErbB-3, and ErbB-4, respectively. The receptors are composed of an extracellular binding domain, a lipophilic transmembrane domain, and an intracellular tyrosine kinase domain (with the exception of HER3). The HER family is known to regulate cell survival responses including, but not limited to, cell proliferation, adhesion, and differentiation. The activity of the HER family receptors is typically initiated by ligand binding, followed by homodimerization or heterodimerization, phosphorylation of tyrosine kinase residues, and downstream signal transduction; however, activation can also result from receptor mutation or overexpression. Of the four family members, the HER2 isoform is the preferred dimerization partner. However, it has no known ligand and can undergo ligand-independent dimerization and activation. Cleavage of the HER2 extracellular domain (shedding) results in a constitutively active phosphorylated signaling remnant. Receptor dimerization leads to tyrosine kinase phosphorylation followed by activation of the PI3K/AKT survival pathway and the RAS/RAF/MEK/MAPK pathway. HER2 messenger RNA or protein overexpression and/or gene amplification occurs in over 20% of breast cancer and in gastric cancer, leading to the use of anti-HER2 therapy for these indications. To provide a better understanding, a graphical presentation of HER activation is given in Figure 10.4.

Trastuzumab is a humanized IgG1κ monoclonal antibody directed against an epitope on the extracellular juxta membrane domain of HER2. Multiple mechanisms of action have been proposed for trastuzumab, including inhibition of HER2 receptor dimerization, increased destruction of the endocytic portion of the HER2 receptor, inhibition of extracellular domain shedding, and activation of cell-mediated immune defenses such as ADCC activity. Trastuzumab has not been shown to inhibit the dimerization of HER2 with

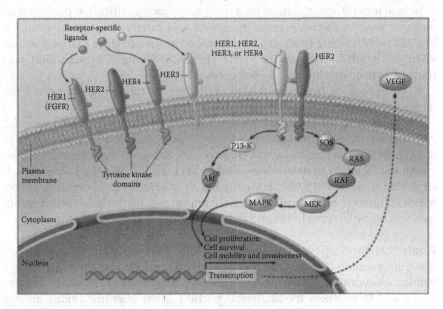

FIGURE 10.4
HER activation. (From Hudis, C.A. *New England Journal of Medicine*, 357, 39–51, 2007.

the other isoforms; therefore, signaling through the other three receptor isoforms is maintained in the presence of the antibody. Studies have supported a mechanism by which trastuzumab is bound to the HER2 receptor and taken up by the target cell through endocytosis and subsequently degrades the receptor leading to a downregulation of downstream survival signaling, cell cycle arrest, and apoptosis. Trastuzumab has also been shown to block the cleavage/shedding of the HER2 receptor extracellular domain, thereby preventing the formation of the activated truncated which has been correlated with a poor prognosis based on the detection of the released extracellular domain of HER2 in the serum of metastatic breast cancer patients. In addition, the initiation of ADCC activity plays a role in the mechanism of action of trastuzumab; it appears that natural killer (NK) cells are important mediators of ADCC activity in the context of trastuzumab-treated breast cancer. Receptor-ligand binding between trastuzumab and HER2 lead to the recruitment of immune cells (e.g., NK cells, macrophages, neutrophils) that express FcγRIIIa or certain other Fc receptors. The effector cell Fc receptor binds to the Fc region of the antibody, triggering release of cytokines and recruitment of more immune cells. These immune cells release proteases that effectively lyse the HER2 expressing target cell, resulting in cell death. Trastuzumab has also been demonstrated to inhibit angiogenesis and proliferation by reducing the expression of pro-angiogenic proteins, such as vascular endothelial growth factor (VEGF) and the transforming growth factor beta (TGF-β) in a mouse model.

Note that MYL-1401O is produced using a mammalian cell line expanded in bioreactor cultures followed by a drug substance purification process that includes various steps designed to isolate and purify the protein product. The MYL-1401O drug product was developed as a multi-dose vial containing 420 mg of lyophilized powder, to reflect the same strength, presentation, and route of administration as US-Herceptin (420 mg).

Mylan is seeking licensure of MYL-1401O for the following indications for which US-Herceptin is approved (Table 10.6).

The purpose of the Oncologic Drugs Advisory Committee (ODAC) meeting is to discuss whether the totality-of-the-evidence presented support licensure of MYL-1401O as a biosimilar to US-Herceptin. Based on the information provided, the ODAC is to determine whether MYL-1401O meet the following criteria: that (i) MYL-1401O is highly similar to US-Herceptin, notwithstanding minor differences in clinically inactive components, (ii) there are no clinically meaningful differences between MYL-1401O and US-Herceptin.

Analytical Similarity Assessment – The analytical similarity assessment was performed to demonstrate that MYL-1401O and US Herceptin are highly similar, notwithstanding minor differences in clinically inactive components, and to establish the analytical portion of the scientific bridge among MYL-1401O, US-Herceptin, and EU-Herceptin to justify the relevance of the comparative clinical and nonclinical data generated using EU-Herceptin.

TABLE 10.6

Approved Indications for US-Herceptin

No.	Indication
1.	Adjuvant breast cancer
	a. As part of a treatment regimen consisting of doxorubicin, cyclophosphamide, and either paclitaxel or docetaxel
	b. With docetaxel and carboplatin
	c. As a single agent following multi-modality anthracycline based therapy
2.	Metastatic breast cancer (MBC)
	a. In combination with paclitaxel for first-line treatment of HER2-overexpressing metastatic breast cancer
	b. As a single agent for treatment of HER2-overexpressing breast cancer in patients who have received one or more chemotherapy regimens for metastatic disease
3.	Metastatic gastric cancer
	a. In combination with cisplatin and capecitabine or 5-fluorouracil, for the treatment of patients with HER2 overexpressing metastatic gastric or gastroesophageal junction adenocarcinoma who have not received prior treatment for metastatic disease

The similarity assessments were based on pairwise comparisons of the analytical data generated by the Applicant or their contract laboratory using several lots of each product. The FDA performed confirmatory statistical analyses of the data submitted, which included results from an assessment of up to 16 lots of MYL-1401O, 28 lots of US-Herceptin, and 38 lots of EU-Herceptin. All lots of each product were not included in every assessment; the number of lots analyzed in each assay was determined by the Applicant based on the availability of test material at the time of analysis, orthogonal analytical techniques, variability of the analytical method, method qualification, and use of a common internal reference material.

The expiration dates of the US-Herceptin and EU-Herceptin lots included in the similarity assessment spanned approximately 6 years (2013 to 2019), and the MYL-1401O lots used for analysis were manufactured between 2011 and 2015.

The analytical similarity exercise included a comprehensive range of methods (listed in Table 10.1), which included orthogonal methods for the assessment of critical quality attributes. A number of assays were designed to specifically assess the potential mechanisms of action of trastuzumab, including Fc-mediated functions. All methods were validated or qualified prior to the time of testing and were demonstrated to be suitable for the intended use.

FDA's Assessment of Analytical Data – The MYL-1401O drug product was evaluated and compared to US-Herceptin and EU-Herceptin using a battery of biochemical, biophysical, and functional assays, including assays that addressed each major potential mechanism of action. The analytical data submitted support the conclusion that MYL-1401O is highly similar to US-Herceptin. The amino acid sequences of MYL-1401O and US-Herceptin are identical. A comparison of the secondary and tertiary structures and the impurity profiles of MYL-1401O and US-Herceptin support the conclusion that the two products are highly similar. HER2 binding, inhibition of proliferation, and ADCC activity, which reflect the presumed primary mechanisms of action of US-Herceptin, were determined to be equivalent.

Some tests indicate that subtle shifts in glycosylation (sialic acid, high mannose, and NG-HC) exist and are likely an intrinsic property of the MYL-1401O product due to the manufacturing process. High mannose and sialic-acid containing glycans can impact PK, while NG-HC is associated with loss of effector function through reduced FcγRIIIa binding and reduced ADCC activity. However, FcγRIIIa binding was similar among products and ADCC activity was equivalent among products. The residual uncertainties related to the increases in total mannose forms and sialic acid and decreases in NG-HC were addressed by the ADCC similarity and by the PK similarity between MYL-1401O and US-Herceptin as discussed in the section on Clinical Pharmacology below. Additional subtle differences in size and charge related variants were detected; however, these variants generally remain within the quality range criteria. Further, the data submitted by the Applicant support the conclusion that MYL-1401O and US Herceptin can function through the same mechanisms of action for the indications for which Herceptin is currently approved, to the extent that the mechanisms of action are known or can reasonably be determined. Thus, based on the extensive comparison of the functional, physicochemical, protein, and higher order structure attributes, MYL-1401O is highly similar to US-Herceptin, notwithstanding minor differences in clinically inactive components.

In addition, the three pairwise comparisons of MYL-1401O, US-Herceptin and EU-Herceptin establish the analytical component of the scientific bridge among the three products to justify the relevance of comparative data generated from clinical and nonclinical studies that used EU-Herceptin to support a demonstration of biosimilarity of MYL-1401O to US-Herceptin.

10.5 Concluding Remarks

In both case studies, there are notable differences in some CQAs in Tier 1. The sponsors provided scientific rationales and/or justifications indicating that the notable differences have little or no impact on clinical outcomes based on known mechanism of action (MOA). The scientific rationales/justifications were accepted by the ODAC panel with some reservation because no data were collected to support the relationship between the analytical similarity and clinical similarity. In other words, it is not clear whether a notable change in CQAs is translated to a clinically meaningful difference in clinical outcome. Thus, in practice, the relationship between Tier 1 CQAs (which are considered most relevant to clinical outcomes) and clinical outcomes should be studied whenever possible.

In both biosimilar regulatory submissions, pairwise comparisons among the proposed biosimilar product, US-licensed product, and EU-approved product were performed to support (i) a demonstration that the proposed biosimilar product and the US-licensed reference product are highly similar, and (ii) a justification of the bridging of the relevance of the comparative data generated using EU-approved reference to support a demonstration of bio-similarity of the proposed biosimilar product to the US-licensed reference product under the following primary assumptions:

1. Analytical similarity assessment is predictive of PK/PD similarity in terms of drug absorption.
2. PK/PD similarity is predictive of clinical similarity (in terms of safety and efficacy).
3. Analytical similarity assessment is predictive of clinical similarity.

These primary assumptions, however, are difficult (if not impossible) to be verified. Thus, analytical similarity assessment using pairwise comparisons among the proposed biosimilar, US-licensed product, and EU-approved product is critical. However, at the ODAC meeting for evaluation of Avastin and Herceptin held on July 13th, the ODAC panel indicated that pairwise comparisons suffer from the following disadvantages: (i) each comparison was made based on the data collected from the two products to be compared and hence did not fully utilize all of the data collected, (ii) the three comparisons used different products as reference product. For example, when comparing the proposed product with the US-licensed product and when comparing the EU-approved product with the US-licensed product, the US-licensed product was used as the reference product. On the other hand, when comparing the proposed biosimilar product with the EU-approved product, the EU-approved product is used as the reference product. Thus, the ODAC panel suggested a simultaneous confidence interval approach, which fully utilizes all data collected from the three products and the same reference product (i.e., the US-licensed product), be considered for providing a more accurate and reliable comparison among the proposed biosimilar product, US-licensed product, and EU-approved product.

11

Practical and Challenging Issues

11.1 Introduction

For biosimilar product development, FDA recommends a stepwise approach be used for obtaining totality-of-the-evidence for demonstration that a proposed biosimilar product is highly similar to its corresponding innovative biological product. The stepwise approach starts with similarity assessment for critical quality attributes for functional/structural characterization at various stages of the manufacturing process. Recently, FDA circulated a draft guidance on analytical similarity assessment (FDA, 2017b) to assist sponsors in implementing analytical similarity testing using the FDA's recommended tiered approach. That is, equivalence tests for CQAs in Tier 1, a quality range approach for CQAs in Tier 2, and a raw data and graphical presentation for CQAs in Tier 3. Following analytical similarity assessment, the stepwise approach suggested PK/PD similarity (pharmacological activities such as drug absorption profile) and clinical similarity (safety, tolerability, and efficacy) be performed for obtaining the totality-of-the-evidence for demonstration of highly similarity between the proposed biosimilar product and the innovative product.

The recommended stepwise approach for obtaining the totality-of-the-evidence has been widely accepted by the sponsors since it was introduced by the 2012 FDA draft guidance, which was subsequently finalized in 2015 (FDA, 2015a). However, several practical issues have been raised. These practical issues include, but are not limited to, (i) the confusion between the use of a 90% confidence interval for assessment of bioequivalence for generic drugs and biosimilarity for biosimilar products and the use of a 95% confidence interval for clinical investigation of new drugs, (ii) the interpretation of the totality-of-the-evidence in the regulatory approval pathway, and (iii) the inconsistencies of test results between tiers. The purpose of this chapter is not only to provide some insights on these issues, but also to clarify confusion whenever possible.

In addition to these practical issues, there are many specific questions that are commonly asked by the sponsors during the implementation of analytical tests for similarity assessment. Some of these questions related to the selection of test lots and reference lots for analytical similarity assessment. For example, the selection of drug product (DP) and/or drug substance (DS) lots as some DP lots may be made from the same DS lots. Thus, it is a concern whether the selected lots for analytical similarity assessment are independent for a fair and unbiased/reliable assessment of the true difference between the test product and the reference product. Also, a commonly asked question is that how many test lots and reference lots are considered sufficient for providing totality-of-the-evidence for demonstrating high similarity. This chapter also attempts to address these specific questions.

In the next section, some insights and clarification between the use of 90% confidence interval for assessment generic/biosimilar drugs and the use of 95% confidence interval approach for assessment of new drugs are discussed. Section 11.3 provides interpretation of the totality-of-the-evidence from academic and/or pharmaceutical perspectives. The inconsistencies of test results for CQAs from Tier 1 and other Tiers are studied in Section 11.4. Section 11.5 discusses the potential use of individual bioequivalence for biosimilarity assessment in biosimilar studies. Section 11.6 deals with some commonly asked questions during the implementation of analytical similarity assessment. Some concluding remarks are given in the last section of this chapter.

11.2 Hypotheses Testing versus Confidence Interval Approach

For regulatory review and approval of drug products, a commonly asked question is why the FDA considers $(1 - \alpha) \times 100\%$ confidence interval (CI) (or 95% CI if $\alpha = 0.05$) for new drug products but $(1 - 2\alpha) \times 100\%$ CI (or 90% CI if $\alpha = 0.05$) for generic/biosimilar drug products. FDA has been challenged by sponsors for adopting different levels of significance, i.e., 5% versus 10% for new drugs and generic/biosimilar drug products, respectively. This challenge, however, is conceptually incorrect. The confusion between the use of 95% CI for new drugs and 90% CI for generic/biosimilar drug products is probably due to the mixed-up use of the concepts of hypotheses testing and confidence interval approach for evaluation of drug products. In this section, we attempt to clarify the confusion.

11.2.1 Interval Hypotheses Testing

First, for approval of generic and biosimilar drug products, the FDA requires bioequivalence and biosimilarity assessment should be performed, respectively. Bioequivalence assessment for generic drugs and biosimilarity assessment for biosimilar drug products are usually performed by testing the following *interval* hypotheses for equivalence or similarity:

$$H_0 : \mu_T/\mu_R \leq -0.8 \text{ or } \mu_T/\mu_R \geq 1.25 \text{ vs. } H_a : 0,8 < \mu_T/\mu_R < 1.25, \quad (11.1)$$

where 0.8 (80%) and 1.25 (125%) are lower and upper equivalence or similarity limits. For the above interval hypotheses, the intention is to reject the null hypothesis of bio*in*equivalence (or dis-similarity) and conclude the alternative hypothesis of bioequivalence or similarity. Schuirmann (1987) proposed a two one-sided tests (TOST) procedure for testing the above interval hypotheses. The idea is to test one-side to determine whether the test product is inferior to the reference product at the α level of significance and then test the other side to determine whether the test product is not superior to the reference product at the α level of significance after the non-inferiority of the test product has been established. Note that FDA recommended that Schuirmann's two one-sided tests procedure should be used for testing the above interval hypotheses for bioequivalence or biosimilarity (FDA, 1992, 2003). Chow and Shao (2002) pointed out that Schuirmann's two one-sided tests procedure is a size-α test. In other words, the overall type 1 error is well controlled at the α level of significance. Chow and Liu (1992) indicated that Schuirmann's two one-sided tests procedure is *operationally* (algebraically) equivalent to confidence interval approach in many cases under certain assumptions. In other words, we claim bioequivalence or biosimilarity if the constructed $(1 - 2\alpha) \times 100\%$ confidence interval falls completely within the bioequivalence or biosimilarity limits. As a result, interval hypotheses testing (i.e., two one-sided tests procedure) for bioequivalence or biosimilarity has been mixed up with the use of $(1 - 2\alpha) \times 100\%$ confidence interval approach since then.

11.2.2 Confidence Interval Approach

For clinical investigation of new drugs, a typical approach is to testing the following point hypotheses for equality:

$$H_0 : \mu_T/\mu_R = 1 \text{ vs. } H_a : \mu_T/\mu_R \neq 1. \quad (11.2)$$

We then reject the null hypothesis of no treatment difference and conclude that there is statistically significant treatment difference. The statistical difference is then evaluated to determine whether such a difference is of clinically meaningful difference. In practice, a power calculation is usually performed for determination of sample size for achieving a desired power (e.g., 80%) for detecting a clinically meaningful difference (or treatment effect) under the alternative hypothesis that such a difference truly exists.

For point hypotheses testing for equality, a two-sided test (TST) at the α level of significance is usually performed. The TST at the α level of significance is equivalent to the $(1 - \alpha) \times 100\%$ confidence interval. In practice, the $(1 - \alpha) \times 100\%$ confidence interval approach is often used to assess treatment effect instead of hypotheses testing for equality.

11.2.3 Remarks

Hypotheses testing procedure for evaluation of drug products is the official test procedure recommended by the FDA. However, the confidence interval approach is often mis-used for evaluation of the drug products, regardless of the framework of hypotheses testing. The concept of hypotheses testing procedure (e.g., point hypotheses testing for *equality* and interval hypotheses testing for *equivalence*) and confidence interval approach are very different. For example, hypotheses testing procedure focuses on the control of type II error (i.e., power), while the confidence interval approach is based on type I error. Sample size requirements based on the control of type I error or type II error are very different. To provide a better understanding, Table 11.1 summarizes the fundamental differences between hypotheses testing for equality (new drugs) and hypotheses testing for equivalence (generic drugs and biosimilar products).

As it can be seen from Table 11.1 that TST and TOST are official test procedures recommended by the FDA for evaluation of new drugs and generic/biosimilar drugs, respectively. Both TST and TOST are size-α test procedures. In other words, the overall type I error rates are well controlled. In practice, it is suggested that the concepts of hypotheses testing (based on type II error) and confidence interval approach (based on type I error) should not be mixed up to avoid possible confusion in evaluation of new drugs and generic/biosimilar products.

TABLE 11.1

Comparison of Statistical Methods for Assessment of Generic/Biosimilar Drugs and New Drugs

Characteristics	Generic/Biosimilar Drugs	New Drugs
Hypotheses testing	Interval hypotheses	Point hypotheses
FDA recommended approach	TOST	TST
Control of α	Yes	Yes
Confidence interval approach	Operationally equivalent $(1 - 2\alpha) \times 100\%$ CI 90% CI if $\alpha = 5\%$	Equivalent $(1 - \alpha) \times 100\%$ CI 95% CI if $\alpha = 5\%$
Sample size requirement	Based on TOST	Based on TST

Abbreviations: TOST = two one-sided tests. TST = two-sided test.

11.3 Totality-of-the-Evidence

11.3.1 Primary Assumptions of Stepwise Approach

For approval of a proposed biosimilar product, the United States (US) Food and Drug Administration (FDA) requires that totality-of-the-evidence be provided to support a demonstration that the proposed biosimilar product is highly similar to the US-licensed product, notwithstanding minor differences in clinically inactive components, and that there are no clinically meaningful differences between the proposed biosimilar product and the US-licensed product in terms of the safety, purity, and potency of the product.

To assist the sponsor in biosimilar product development, FDA recommends a stepwise approach for obtaining the totality-of-the-evidence for demonstrating biosimilarity between the proposed biosimilar product and its innovative drug product in terms of safety, purity, and efficacy (Chow, 2013; FDA, 2015, 2017; Endrenyi et al., 2017). The stepwise approach starts with similarity assessment in critical quality attributes (CQAs) in analytical studies, followed by the similarity assessment in pharmacological activities in pharmacokinetic and pharmacodynamic (PK/PD) studies and similarity assessment in safety and efficacy in clinical studies. For analytical similarity assessment in CQAs, FDA further recommends a tiered approach which classifies CQAs into three tiers depending upon their criticality or risk ranking relevant to clinical outcomes. For determination of criticality or risk ranking, FDA suggests establishing a predictive (statistical) model based on either mechanism of action (MOA) or PK relevant to clinical outcome. Thus, the following assumptions are made for the stepwise approach for obtaining the totality-of-the-evidence.

  i. Analytical similarity is predictive of PK/PD similarity;
 ii. Analytical similarity is predictive of clinical outcomes;
iii. PK/PD similarity is predictive of clinical outcomes.

These assumptions, however, are difficult (if not impossible) to verify in practice. For assumptions (i) and (ii), although many *in vitro* and *in vivo* correlations (IVIVC) have been studied in the literature, the correlations between specific CQAs and PK/PD parameters or clinical endpoints are not fully studied and understood. In other words, most predictive models are not well established or are established but not validated. Thus, it is not clear how a (notable) change in a specific CQA can be translated to a change in drug absorption or clinical outcome. For (iii), unlike bioequivalence assessment for generic drug products, there does not exist *Fundamental Biosimilarity Assumption* indicating that PK/PD similarity implies clinical similarity in terms of safety and efficacy. In other words, PK/PD similarity or dis-similarity may or may not lead to clinical similarity. Note that the assumptions (i) and (iii) combined does not lead to the validity of assumption (ii) automatically.

The validity of assumptions (i)-(iii) is critical for the success of obtaining totality-of-the-evidence for assessing biosimilarity between the proposed biosimilar and the innovative biological product. This is because the validity of these assumptions ensures the relationships among analytical, PK/PD, and clinical similarity assessment and consequently the validity of the overall biosimilarity assessment.

11.3.2 Relationships among Analytical, PK/PD, and Clinical Similarity

Figure 11.1 illustrates relationships among analytical, PK/PD, and clinical assessments in the stepwise approach for obtaining the totality-of-the-evidence in biosimilar product development.

In practice, for simplicity, CQAs, PK/PD responses, and clinical outcomes are usually assumed to be linearly correlated. For example, let x, y, and z be the test result of a CQA, PK/PD response, and clinical outcome, respectively. Under assumptions (i)-(iii), we have

Model 1. $y = a_1 + b_1 x + e_1$;

Model 2. $z = a_2 + b_2 y + e_2$;

Model 3. $z = a_3 + b_3 y + e_3$;

where e_1, e_2, and e_3 follow a normal distribution with mean 0 and variances σ_1^2, σ_2^2, and σ_3^2, respectively. In practice, each of the above models is often difficult, if it is not impossible, to be validated due to lack of insufficient data collected during the biosimilar product development. Under each of the above models, we may consider the criterion for examination of the closeness between an observed response and its predictive value to determine whether the respective model is a good predictive model. As an example, under model (1), we may consider the following two measures of closeness,

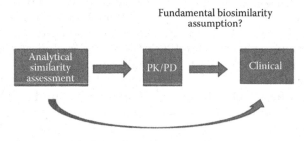

FIGURE 11.1
Relationships among analytical, PK/PD, and clinical assessment.

which are based on either the *absolute* difference or the *relative* difference between an observed value y and its predictive value $\hat{y}$

Criterion I. $\quad p_1 = P\left\{|y - \hat{y}| < \delta\right\}$,

Criterion II. $\quad p_2 = P\left\{\left|\dfrac{y - \hat{y}}{y}\right| < \delta\right\}$.

It is desirable to have a high probability that the difference or the relative difference between y and $\hat{y}$, given by p_1 and p_2, respectively, is less than a clinically meaningful difference δ.

Suppose there is a well-established relationship between x (e.g., test result of a given CQA) and y (e.g., PK/PD response). Model 1 indicates that a change in CQA, say Δ_x corresponds to a change of $a_1 + b_1\Delta_x$ in PK/PD response. Similarly, model (2) indicates that a change in PK/PD response, say Δ_y, corresponds to a change of $a_2 + b_2\Delta_y$ in clinical outcomes. Models 2 and 3 allows us to evaluate the impact of the change in CQA (i.e., x) on PK/PD (i.e., y) and consequently clinical outcome (i.e., z). Under models 2 and 3, we have

$$a_2 + b_2 y + e_2 = a_3 + b_3 x + e_3.$$

This leads to

$$a_1 = \frac{a_3 - a_2}{b_2}, \; b_1 = \frac{b_3}{b_2}, \text{ and } e_1 = \frac{e_3 - e_2}{b_2}.$$

with

$$b_2^2 \sigma_1^2 = \sigma_3^2 + \sigma_2^2$$

or

$$\sigma_1 = \frac{1}{b_2}\sqrt{\sigma_2^2 + \sigma_3^2}.$$

In practice, the above relationships can be used to verify primary assumptions as described in the previous section provided that models 1-3 have been validated. Suppose models 1-3 are well-established, validated, and fully understood. A commonly asked question is whether PK/PD studies and/or clinical studies can be waived if analytical similarity and/or PK/PD similarity have been demonstrated.

Note that the above relationships hold only under linearity assumption. When there is a departure from linearity in each one of models 1-3, the above relationships are necessarily modified.

11.3.3 Practical Issues

For biosimilar product development and regulatory review and approval, FDA recommends a stepwise approach by performing analytical similarity assessment, PK/PD similarity test, and clinical similarity assessment in terms of safety, tolerability, and efficacy for obtaining the totality-of-the-evidence. FDA's recommended stepwise approach focuses on three major domains, namely, analytical, PK/PD, and clinical similarity, which are highly correlated under models (1-3). Some pharmaceutical scientists interpret the stepwise approach as a scoring system (perhaps, with appropriate weights) that includes the domains of analytical, PK/PD, and clinical similarity assessment. In this case, the totality-of-the-evidence can be assessed based on information regarding biosimilarity obtained from each domain. In practice, for each domain, we may consider either FDA's recommended binary response (i.e., similar or dis-similar) or the use of the concept of biosimilarity index (Chow et al., 2011) to assess similarity information and consequently the totality-of-the-evidence across domains.

For the FDA's recommended approach, Table 11.2 provides possible scenarios when performing analytical similarity assessment, PK/PD similarity test, and clinical similarity assessment.

As it can be seen from Table 11.2, if the proposed biosimilar product passes similarity test in all domains, FDA considers the sponsor has provided totality-of-the-evidence for demonstration of highly similarity between the proposed biosimilar and the innovative biological product. On the other hand, if the proposed biosimilar product fails to pass any of the suggested similarity assessments (i.e., analytical similarity, PK/PD similarity, and clinical similarity), then the regulatory agency will reject the proposed biosimilar product.

TABLE 11.2

Assessment of Totality-of-the-Evidence

No. of Dis-Similarities	Analytical Similarity Assessment	PK/PD Similarity Assessment	Clinical Similarity	Overall Assessment
0	Yes	Yes	Yes	Yes
1	Yes	Yes	No	No
1	Yes	No	Yes	*
1	No	Yes	Yes	*
2	Yes	No	No	No
2	No	Yes	No	No
2	No	No	Yes	No
3	No	No	No	No

*Scientific rationale are necessary provided.

In practice, it is uncommon to see that the proposed biosimilar may fail in one of the three suggested similarity assessments, namely analytical similarity, PK/PD similarity, and clinical similarity assessments. In this case, the regulatory agency may have a hard time granting approval of the proposed biosimilar product. A typical example is that notable differences in some CQAs between the proposed biosimilar product and the innovative biological product may be observed in analytical similarity assessment. In this case, the sponsors often provide scientific rationales/justifications to indicate that the notable differences have little or no impact on clinical outcomes. This is probably the most debatable issue between FDA and the Advisory Committee during the review/approval process of the proposed biosimilar product, because it is not clearly stated in the FDA guidance whether a proposed biosimilar product is required to passes all similarity tests regardless if they are Tier 1 CQAs or Tier 2/Tier 3 CQAs before the regulatory agency can grant approval of the proposed biosimilar product. In this case, if FDA and the ODAC panel accept sponsors' scientific rationales and justifications that the notable differences have little or no impact on the clinical outcomes, the proposed biosimilar is likely to be granted approval.

This, however, has raised an interesting question whether the proposed biosimilar product is required to pass all similarity tests (i.e., analytical similarity, PK/PD similarity, and clinical similarity) for regulatory approval.

11.3.4 Examples

For illustration purpose, consider two FDA recent biosimilar regulatory submissions, i.e., Avastin biosimilar (ABP215 sponsored by Amgen) and Herceptin biosimilar (MYL-1401O sponsored by Mylan). These two regulatory submissions were reviewed and discussed at an Oncologic Drug Advisory Committee (ODAC) meeting held on July 13th, 2017 in Silver Spring, Maryland. Table 11.3 briefly summarizes the results of the review based on the concept of totality-of-the-evidence.

TABLE 11.3

Examples of Assessment of Totality-of-the-Evidence

Regulatory Submission	Innovative Product	Proposed Biosimilar	Analytical Similarity	PK/PD Similarity	Clinical Similarity
BLA 761028 (Amgen)	Avastin	ABP215	Notable differences observed in glycosylation content and Fc γ RIIIa binding	Pass	Pass
BLA 761074 (Mylan)	Herceptin	MYL-1401O	Subtle shifts in glycosylation (sialic acid, high mannose, and NG-HC	Pass	Pass

With header spanning "Totality of the Evidence" over Analytical Similarity, PK/PD Similarity, and Clinical Similarity.

For ABP215, a proposed biosimilar to Genentech's Avastin, although ABP215 passed both PK/PD similarity and clinical similarity tests, several quality attribute differences were noted. These notable differences include glycosylation content, FcγRllla binding, and product related species (aggregates, fragments, and charge variants). The glycosylation and FcγRllla binding differences were addressed by means of *in vitro* cell based on antibody dependent cell-mediated cytotoxicity (ADCC) and complement dependent cytotoxicity (CDC) activity, which were not detected for all products (ABP215, US-licensed Avastin, and EU-approved Avastin). The ODAC panel considered the submission to have provided totality-of-the-evidence for demonstration of highly similarity between ABP215 and the US-licensed Avastin, notwithstanding minor differences in clinically inactive components, and supported that there were no clinically meaningful differences between ABP215 and the US-licensed Avastin in terms of the safety, purity, and potency of the product.

For MYL-1401O, a proposed biosimilar to Genentech's Herceptin, although MYL-1401O passed both PK/PD similarity and clinical similarity tests, there were subtle shifts in glycosylation (sialic acid, high mannose, and NG-HC). However, the residual uncertainties related to increase in total mannose forms and sialic acid and decrease in NG-HC were addressed by ADCC similarity and by the PK similarity. Thus, the ODAC panel determined that the submission had provided totality-of-the-evidence to support a demonstration of highly similarity between MYL-1401O and US-licensed Herceptin, notwithstanding minor differences in clinically inactive components, and supported that there were no clinically meaningful differences between MYL-1401O and US-licensed Herceptin in terms of the safety, purity, and potency of the product.

11.3.5 Remarks

For regulatory approval of new drugs, Section 314.126 of 21 CFR states that substantial evidence needs to be provided to support the claims of new drugs. For regulatory approval of a proposed biosimilar product, the FDA requires totality-of-the-evidence be provided to support a demonstration of biosimilarity between the proposed biosimilar product and the US-licensed drug product. In practice, it should be noted that there is no clear distinction between the substantial evidence in new drug development and the totality-of-the-evidence in biosimilar drug product development.

As discussed in the previous section regarding the two recent regulatory submissions, it is not clear whether totality-of-the-evidence of highly similarity can only be achieved if the proposed biosimilar product has passed all similarity tests across different domains of analytical, PK/PD, and clinical assessment. When notable differences in CQAs (in Tier 1) are observed, the notable differences may be ignored if the sponsors are able to provide scientific rationales/justification to rule out that the observed differences have an impact on clinical outcomes. This, however, is somewhat controversial because Tier 1 CQAs are considered most relevant to clinical outcomes depending upon their criticalities or risk rankings that impact the clinical outcomes. The criticalities and/or risk rankings may be determined using model 3. If a notable difference is considered having little or no impact on the clinical outcome, then the CQA should not be classified into Tier 1 in the first place. This controversy could be due to classification of CQAs based on subjective judgement rather than objectively statistical modeling.

In the two examples concerning biosimilar regulatory submissions of ABP215 (Avastin biosimilar) and MYL-1401O (Herceptin biosimilar), the sponsors also seek for approval across different indications. There have been tremendous discussions regarding whether totality-of-the-evidence observed from one indication or a couple of indications can be used to extrapolate to other indications, even different indications that have similar mechanisms of action. The ODAC panel expressed their concern about extrapolation without collecting any clinical data from other indications and encouraged further research on scientific validity of extrapolation and/or generalizability of the proposed biosimilar product.

11.4 Inconsistencies between Tiered Approaches

11.4.1 *In Vitro* Bioequivalence Testing versus Analytical Testing

In practice, a frequently asked question is "What is the difference between *in vitro* bioequivalence testing for generic drug products and equivalence testing for analytical similarity in CQAs in biosimilar studies?" In 2003, FDA published guidance for *in vitro* bioequivalence testing for locally acting drug products such as nasal spray drug products (FDA, 2003b). On the other hand, FDA circulated a draft guidance for analytical similarity assessment for biosimilar drug products in September 2017 (FDA, 2017b). These guidances outline required *in vitro* or analytical tests for assessment of bioequivalence for generic drug products or analytical similarity assessment for biosimilar products, respectively. The differences in some characteristics such as statistical methods and evaluation criteria between *in vitro* bioequivalence testing and analytical similarity assessment are summarized in Table 11.4.

As it can be seen from Table 11.4, FDA 2013 guidance suggested profile/ non-profile analysis should be used for assessment of *in vitro* bioequivalence for locally acting generic drug products such as nasal spray drug products. FDA further suggested bioequivalence limits of (90%, 111%) be used. On the other hand, for analytical similarity assessment of biosimilar products, FDA 2017 draft guidance recommends a tiered approach depending upon the criticality or risking ranking of the identified CQAs relevant to clinical

TABLE 11.4

Comparison of Various Types of Equivalence Testing

Characteristics	*In vitro* BE Testing	Analytical Similarity Assessment
Primary assumptions	$\mu_{T_i} = \mu_{R_i}$ for all i $\sigma_{T_i} = \sigma_{R_i}$ for all i	$\mu_{T_i} \neq \mu_{R_i}$ for some i $\sigma_{T_i} \neq \sigma_{R_i}$ for some i $\mu_T - \mu_R \propto \sigma_R$
Focus on	Population mean	Population mean (Tier 1) Population (Tier 2)
Variability	< 10%	Vary
Criterion	(90%, 111%)	$EAC = \pm 1.5^* \sigma_R$
Analysis	Profile/non-profile	Hypothesis/CI

outcomes. FDA suggested a similarity margin of 1.5 σ_R, which depends upon the variability associated with the reference product.

11.4.2 Primary Assumptions for Tiered Approach

As indicated in Tsong et al. (2015), the Tier 1 equivalence test is considered more rigorous than the Tier 2 quality range approach, which is in turn more rigorous than the Tier 3 raw data and graphical comparison. Thus, for a given CQA, if the CQA passes the Tier 1 equivalence test, one would expect that the CQA will also pass the Tier 1 quality range approach. In practice, this is not necessary true due to different primary assumptions made for these tiered approaches.

In practice, statistical methods (e.g., profile/non-profile analysis) and bioequivalence margin for assessment of *in vitro* bioequivalence cannot be directly applied to analytical similarity assessment for biosimilar products due to fundamental differences in primary assumptions made. For *in vitro* bioequivalence testing, since generic drug products contain identical active ingredient(s), it is reasonable to assume that they have the same means and variances (i.e., $\mu_T = \mu_R$ and $\sigma_T = \sigma_R$), Unlike generic drug products, biosimilars are expect to have different means and variances (i.e., $\mu_T \neq \mu_R$ and $\sigma_T \neq \sigma_R$), In addition, for analytical similarity assessment, the following assumptions are also made: (i) the maximum mean shift allowed for a Tier 1 equivalence test is $\mu_T - \mu_R = \dfrac{1}{8}\sigma_R$ and (ii) the difference between means is proportional to σ_R. It should be noted that Tier 1 equivalence test focuses on population means of test and reference products, while the Tier 2 quality range approach focuses on population of test and reference products.

11.4.3 Inconsistencies between Different Tiered Tests

Under different assumptions, there is no guarantee that passing the Tier 1 equivalence test will also pass the Tier 2 quality range approach and vice

TABLE 11.5

Probabilities of Inconsistencies

Tier 1 Equivalence Test	Tier 2 Quality Range Approach	
	Pass	Fail
Pass	p_{11}	p_{12}
Fail	p_{21}	p_{22}

versa, although these tests are conducted based on test results obtained from the same test and reference lots under study. This is mainly due to the fact that Tier 1 equivalence tests focus on comparing *population means* of the test and the reference products, while the Tier 2 quality range approach focuses on comparing *populations* of the test and the reference product. In practice, as CQAs in Tier 1 or Tier 2 are considered more or less relevant to clinical outcomes, respectively, it is of interest to evaluate inconsistencies regarding the passages between Tier 1 equivalence tests and the Tier 2 equality range approach (Chow et al., 2016).

Chow et al. (2016) indicated that for a given CQA, the inconsistencies between a Tier 1 equivalence test and a Tier 2 quality range approach can be assessed by means of clinical trial simulation as follows. Let p_{ij} be the probability of passing the ith tier test given that the CQA has passed the jth tier test. Thus, we have the following 2×2 contingency table (Table 11.5) for comparison between the Tier 1 equivalence test and the Tier 2 quality range approach.

As can be seen, the probability of inconsistencies is $p = p_{12} + p_{21}$, which depends upon ε and C, where $\varepsilon = \mu_T - \mu_R$ and $C = \sigma_T/\sigma_R$. Chow et al. (2016) suggested that the probability of inconsistencies between Tier 1 equivalence tests and the Tier 2 quality range approach be evaluated at various combinations of $\varepsilon = 0$ (no mean shift), $\varepsilon = \frac{1}{8}\sigma_R$ (the maximum mean shift allowed by the FDA) and $\varepsilon = \frac{1}{4}\sigma_R$ (the worst possible scenario of mean shift), and $C = 0.8$ (deflation of variability associated with test product as compared to that of the reference product), $C = 1.0$ ($\sigma_T = \sigma_R$), and $C = 1.2$ (inflation of variability associated with test product as compared to that of the reference product) to provide a complete picture of the relative performance of the Tier 1 equivalence test and Tier 2 quality range approach.

11.5 Individual Bioequivalence

As discussed in Section 11.4.2, statistical methods and evaluation criterion for assessment of *in vitro* bioequivalence cannot be directly applied to analytical similarity assessment for biosimilar products due to the primary assumptions made. In other words, generic drugs are assumed to have same means and variances (i.e., $\mu_T = \mu_R$ and $\sigma_T = \sigma_R$ and biosimilar drug products are expected to have different means and variances (i.e., $\mu_T \neq \mu_R$ and $\sigma_T \neq \sigma_R$. For analytical similarity assessment of biosimilar products, however, we only focus on testing similarity in means and ignore possible heterogeneity in variance between drug products. Thus, in practice, it is suggested that similarity in variability should also be established in addition to the establishment of similarity in mean between products. To account for both mean and variability, in the early 1990s, FDA proposed the following individual bioequivalence criterion (IBC) for assessment of individual bioequivalence:

$$\theta = \frac{(\mu_T - \mu_R)^2 + \sigma_D^2 + \left(\sigma_{WT}^2 - \sigma_{WR}^2\right)}{\max\left(\sigma_0^2, \sigma_{WR}^2\right)}, \quad (11.3)$$

where μ_T and μ_R are means for the test product and the reference product, respectively, σ_{WT}^2 and σ_{WR}^2 are within-subject variances for the test product and the reference product, respectively, σ_D^2 is the variability due to the interaction between subject and product, and σ_0^2 is a regulatory constant. If we only consider σ_{WR}^2 in the denominator, the above criterion (11.3) becomes

$$\theta = \frac{(\mu_T - \mu_R)^2 + \sigma_D^2 + \left(\sigma_{WT}^2 - \sigma_{WR}^2\right)}{\sigma_{WR}^2}. \quad (11.4)$$

The above criterion can be linearized as follows:

$$\lambda = (\mu_T - \mu_R)^2 + \sigma_D^2 + \left(\sigma_{WT}^2 - \sigma_{WR}^2\right) - \theta\sigma_{WR}^2 < 0. \quad (11.5)$$

We then reject the null hypothesis of bioinequivalence and conclude individual bioequivalence if the 95% upper limit of the constructed confidence interval is less than 0.

As it can be seen from (11.4), the IBC not only take variabilities (including within-subject variabilities of the test and reference product and the variability due to subject-by-product interaction) into consideration, but also adjusts for the within-subject variability associated with the reference product. One of the major criticisms is that IBC is an aggregated moment-based criterion which may suffer from the potential masking effects between means and variance components. In other words, the difference in mean could be offset by the inflation/deflation of the corresponding variance components.

Despite the potential drawbacks, Haidar et al. (2008) proposed a scaled average bioequivalence (SABE) criterion based on the first term of (11.3) for highly variable drug products. Following a similar idea, Chow et al. (2015) considered the first two terms of (11.3) and proposed scaled criterion for drug interchangeability (SCDI) for assessment of drug interchangeability. Based on simulation studies, the SCDI criterion was found to out-perform the usual BE criterion and SABE criterion. Thus, for assessment of analytical similarity, PK/PD similarity, and clinical similarity, it is suggested either IBC, SABE, or SCDI be considered in order to account for possible heterogeneity in variability between test and reference product in addition to the assessment of similarity in means.

However, it should be noted that IBC can only be applied under a two-sequence, three-period crossover dual design, e.g., (RTR, TRT) or under a replicated 2 × 2 crossover design, e.g., (RTRT, TRTR), which provide estimates of within-subject variabilities of the test and reference products. In the 2017 FDA guidance on interchangeability, FDA recommends the switching designs of (RTR, RRR) (two switches) and (RTRT, RRRR) (three switches) for evaluation of relative risk with/without switching (two switches) and alternation (three switches) for interchangeable products. Under study design of either (RTR, RRR) or (RTRT, RRRR), the IBC, SABE, and/or SCDI can be applied.

11.6 Commonly Asked Questions from the Sponsors

Despite the fact that FDA has published several guidance to assist the sponsors in biosimilar product development, there are still many questions which either require clarification or remain unsolved. In this section, commonly asked questions regarding analytical similarity assessment in biosimilar product development are briefly described below.

Difference between 351(a) BLA and 351(k) BLA of PHS Act – Basically, 351(a) BLA of PHS Act deals with stand alone biologics licenses. No person shall introduce or deliver for introduction into interstate commerce any biological product unless (A) a biologics license under this subsection or subsection (k) is in effect for the biological product; and (B) each package of the biological product is plainly marked with (i) the proper name of the biological product contained in the package; (ii) the name, address, and applicable license number of the manufacturer of the biological product; and (iii) the expiration date of the biological product.

On the other hand, 351(k) BLA of PHS Act deals with licensure of biological products as biosimilar or interchangeable products. For 351(k), the sponsor is required to provide the following information demonstrating that

1. The biological product is biosimilar to a reference product based upon data derived from

   a. analytical studies that demonstrate that the biological product is highly similar to the reference product notwithstanding minor differences in clinically inactive components;

b. animal studies (including the assessment of toxicity); and

c. a clinical study or studies (including the assessment of immuno-genicity and pharmacokinetics or pharmacodynamics) that are sufficient to demonstrate safety, purity, and potency in 1 or more appropriate conditions of use for which the reference product is licensed and intended to be used and for which licensure is sought for the biological product.

2. The biological product and reference product utilize the same mechanism or mechanisms of action for the condition or conditions of use prescribed, recommended, or suggested in the proposed labeling, but only to the extent the mechanism or mechanisms of action are known for the reference product.

3. The condition or conditions of use prescribed, recommended, or suggested in the labeling proposed for the biological product have been previously approved for the reference product.

4. The route of administration, the dosage form, and the strength of the biological product are the same as those of the reference product.

5. The facility in which the biological product is manufactured, processed, packed, or held meets standards designed to assure that the biological product continues to be safe, pure, and potent.

Note that for an interchangeable product, an application (or a supplement to an application) submitted under this subsection may include information demonstrating that the biological product meets the standards described in paragraph 351(k)(4) of PHS Act.

Types of meetings with FDA – How many types of formal meetings can occur between the sponsors and FDA for discussion of biosimilar product development?

As indicated in the FDA guidance entitled *Formal Meetings Between the FDA and Biosimilar Biological Product Sponsors or Applicants*, there are five types of formal meetings that can occur during biosimilar product development. These types of formal meetings between the sponsors or applicants and FDA include (i) biosimilar initial advisory meeting, (ii) biosimilar product development (BPD) type 1 meeting, (iii) BPD type 2 meeting, (iv) BPD type 3 meeting, and (v) BPD type 4 meeting.

Biosimilar initial advisory meeting is an initial assessment limited to (i) a general discussion regarding whether licensure under section 351(k) of the PHS Act may be feasible for a particular product and (ii) general advice on the expected content of the development program. BPD type 1 meeting is a meeting that is necessary for an otherwise stalled BPD program to proceed. Thus. BPD type 1 meetings could include discussion for clinical holds, special protocol assessment meetings, discussion on an important safety issue, and dispute resolution meetings as described in 21 CFR 10.75 and 312.48 and in section IV of the Biosimilar User Fee Act (BsUFA). A BPD type 2 meeting, on the other hand, is a meeting to discuss a specific issue or questions where the FDA will provide targeted advice regarding an ongoing BPD program. A BPD type 3 meeting is an in-depth data review and advice meeting regarding an on-going BPD program. A BPD type 4 meeting is a meeting to discuss the format and content of a biosimilar biological product application or supplement to be submitted under section 351(k) of the PHS Act. More details regarding these types of meetings with FDA can be found in FDA (2015e).

Selection of reference and biosimilar lots – Since available test lots from US-licensed reference products are usually fewer (e.g., 5 lots) than reference lots (perhaps, from different geographic locations, such as EU-approved products), it is an interesting question whether reference lots from all geographic locations can be combined to derive a statistical test for analytical similarity assessment.

For analytical similarity assessment, the 2017 FDA draft guidance suggested a minimum of 10 lots from both biosimilar (test) lots and reference lots should be tested for analytical similarity. In practice, it is a concern that there is a difference between reference lots from different geographical locations (e.g., EU-approved product and US-licensed product) even though they were manufactured by the same sponsor. Thus, it should be confirmed that there is no clinically meaningful difference between reference lots from different geographical locations before they can be pooled to derive equivalence acceptance criterion (EAC) for similarity assessment.

Reference lots with different expiration dating periods – At the time of analytical similarity assessment, there could be differences in expiration dating period (shelf-life) of reference lots and in-house lots. A question that is commonly asked is whether the data generated from such different aged lots can be used for analytical similarity assessment.

In practice, all reference and test lots under study need to be within their respective shelf-lives during the conduct of the study. If some lots expire during the conduct of the study, their data may not be accepted by the regulatory agency for analytical similarity assessment. In some cases, if a significant mean shift occurs with respect to their corresponding shelf-lives, similarity assessment before and after the shift should be carefully evaluated and comparison should be made with proper adjustment if necessary.

DP lots versus DS lots – For biosimilar product development, if several lots of drug product (DP) are made from one batch of drug substance (DS), or several lots of DS are used for a single lot of DP, it is an interesting question whether all of these lots (both DS and DP lots) can be used for analytical similarity assessment.

In this case, regulatory agencies usually suggest that *independent* lots be used for analytical similarity assessment. For example, if several DP lots are made from the same DS batch, select one lot for testing analytical similarity.

For another question regarding DP lots versus DS lots, if the DP composition and concentration are similar to that of DS, can DS be used for similarity assessment in place of DP? This is a controversial issue. Different reviewers may have different opinions on this. For example, "A is similar to B, B is similar to C, and C is similar to D" does not necessarily imply that "A is similar to D."

Biosimilar product with multiple strengths – If there are several presentations (e.g., different doses or strengths), then similarity needs to be demonstrated individually for each equivalent presentation of reference product versus biosimilar product, or they will all be grouped together for statistical representation.

Like bioequivalence assessment, if there is a well-established dose proportionality or linearity relationship between dose and response, the demonstration of similarity individually for each equivalence presentation may not be necessary.

Imbalanced reference lots versus test lots – In the 2017 FDA draft guidance, FDA recommends at least 10 lots from each product should be used for analytical similarity assessment. In practice, usually more reference lots than test lots are available. In this case, how should a sponsor handle the imbalanced reference lots and test lots to meet regulatory requirement for a valid analytical similarity assessment?

In the past year, several methods for selection of reference and test lots, when available lots are imbalanced, have been proposed. These methods include randomly selecting the proportion of lots and reference matching (either 1:1 matching or 2:1 matching). However, none of these methods are statistically or scientifically justifiable, because these methods are purely based on simulation studies under certain assumptions, which may be difficult to verify. Most recently, Dong et al. (2017) proposed the following rule of thumb for selection of reference and test lots for analytical similarity assessment: Let n_T and n_R be the number of test lots and the number of reference lots available, respectively, where $n_R > n_T$. The rule of thumb for determining the number of reference lots and test lots required for similarity assessment is given below:

$$n_R^* = \min(1.5 * n_T, n_R),$$

where n_R^* is the suggested number of reference lots for analytical similarity assessment.

Tiered approach – Does the FDA expect categorization of quality attributes into Tiers 1, 2, and 3 at an early stage when not enough US-licensed reference product lots are studied?

As indicated in the FDA 2017 draft guidance, FDA's recommended tiered approach includes (i) equivalence test for CQAs in Tier 1 (which are considered most clinical relevant), (ii) quality range approach for CQAs in Tier 2 (which are less clinical relevant), and (iii) raw data and graphical comparison for CQAs in Tier 3 (which are least relevant to clinical outcome). The categorization of CQAs to different tiers depends upon their criticality or risk ranking relevant to clinical outcomes based on mechanism of action (MOA) and/or pharmacokinetics (PK) and pharmacodynamics (PD) relevant to clinical outcomes under appropriate statistical models.

Switch of Tiers – Is it acceptable to FDA to switch the tier categories (i.e., switch from Tier 1 to Tier 2 or Tier 2 to Tier 3 for some quality attributes) in the regulatory submission during the biosimilar product development?

In the tiered approach, it is recognized that a given attribute which passes the Tier 1 equivalence test is not guaranteed to pass the Tier 2 quality range test and vice versa. Thus, regulatory agencies requires CQAs be identified and categorized prior to performing analytical similarity assessment. Regulatory agency will not accept switching tier categories, especially if there is a failure to pass tests on certain tier categories.

If there are process changes in reference lots which lead to a notable change in product performance, the following questions need to be addressed: (i) for a given attribute, is the change considered a minor change or a significant change relevant to clinical outcomes? (ii) how is this change relative to criticality or risking ranking? and (iii) if this change is considered a major (or significant) change in process, the process will need to be re-validated. Whether the CQAs can switch from one category to another depend upon the answers to the above questions.

Quality range approach for Tier 2 CQAs – How would one differentiate the k-factor in Tier 2? That is, when is it appropriate to make it $\pm$2SD or $\pm$3SD?

If $k = 2$SD is chosen, it is expected that about 95% of test values from test lots will fall within 2SD below and above the mean of the reference lots. On the other hand, if $k = 3$SD is chosen, it is expected that about 99% of test values from test lots will fall within 3SD below and above the mean of the reference lots. FDA suggests that k be chosen between 2 to 3. It should be noted that the quality range approach for attributes in Tier 2 is useful especially if the test product is highly similar to the reference product in terms of their means and standard deviations.

Note that for the Tier 2 quality range approach, the primary assumption is that $\mu_T \approx \mu_R$ and $\sigma_T \approx \sigma_R$ In the interest of balance, we assume $n_T = n_R$. Thus, we would expect that about 95% (99%) of test values will fall within $\pm$2SD ($\pm$3SD) mean of test values of the reference lots (i.e., quality range).

In vitro **and** *in vivo* **correlation (IVIVC)** – There are some well-established, structure-function related attributes like ADCC and specific glycan species such as G0. For such attributes, does one ideally categorize them in the same Tier due to similar risk/criticality assessment? Or can they still be classified in different tiers due to inherent test variation differences between the two methods (e.g., Glycan G0 in Tier 2 and ADCC in Tier 3)?

As indicated in the FDA draft guidance, FDA suggests that classification of attributes be done according to criticality or risk ranking relative to clinical outcomes. In this case, it is reasonable to classify ADCC and Glycan G0 in the same tier if they have similar risk/criticality assessment. FDA, however, also suggests criticality or risk ranking be assessed via an appropriate statistical model such as in vitro-in vivo correlation (IVIVC) study. In practice, the classification of attributes into Tiers 1, 2, and 3 is subjective and may be biased and hence misleading. Thus, it is a good idea to account for variation differences when classifying attributes into appropriate tiers to avoid possible selection bias for obtaining totality-of-the-evidence in demonstrating biosimilarity between products.

Product changes or shifts – When there are notable product changes or shifts, the following three questions are commonly asked: (i) Can the removal of an out-of-trend (OOT) data point be acceptable for a particular attribute while retaining other attributes where it is within the statistical corridor? (ii) If such a removal is not permitted, can it be retained as data but not included in the statistical calculation with a supporting discussion on OOT? (iii) If such OOT data points favor a wider statistical corridor, is it justified for the biosimilar sponsor to capitalize on such data points to fit the biosimilar product?

Regarding the first question, removal of an out-of-trend (OOT) or *outlier* is very controversial in clinical research. An identified outlier could mean (i) the model is *incorrect* (distribution assumption is incorrect) or (ii) the identified outlier is indeed an outlier. In practice, an identified outlier could be due to (i) relatively small sample size, or (ii) large variability associated with the test value (i.e., the observed test value could be purely by chance alone). As indicated by FDA, an identified outlier (which is usually identified under a certain model) should not be removed from the analysis unless the identified outlier is due to transcription error or any other reasons unrelated to the evaluation of the test treatment under investigation.

For the second question, when there are outliers, appropriate transformation such as log-transformation is usually employed in order to remove heterogeneity before data analysis. Alternatively, one may consider a nonparametric method for a more robust assessment of the test treatment under investigation.

As to the final question, the sponsor may consider having a few replicates and take the median of the test values. On the other hand, OOT data points could also favor a narrower statistical corridor. In this case, it may not to the best interest of the biosimilar sponsors. In practice, both situations may happen. This deficiency may be due to the fact that the FDA's recommended equivalence acceptance criterion (EAC) is *not* a fixed approach rather than a *random approach* (i.e., random EAC depending upon the observed test values of the selected reference lots).

Single test versus multiple tests – How many independent analyses need to be performed to obtain a single data point? Is triplicate analysis recommended for all tests or only for highly variable assays such as potency?

FDA draft guidance recommends a single test value from each reference lot should be obtained for establishing an equivalence acceptance criterion (EAC) for performing an equivalence test with CQAs in Tier 1. The CQAs in Tier 1 are considered most relevant to clinical outcomes according to their criticality or risking ranking. Wang and Chow (2017) showed that FDA's recommended method is an unbiased method for characterizing the variability associated with the reference lot. Wang and Chow (2017) also evaluated the method with multiple test values for each reference lot. It should be noted that multiple test values from each reference lot provide the opportunity for characterizing between-lot and within-lot variabilities associated with the reference product. As to the question regarding how many independent test values should be done, it depends upon the desired power the sponsor would like to achieve with a pre-selected (or available) number of reference lots.

Late stage method change – Several questions are raised when a late stage (analytical) method change occurs: (i) How does a sponsor handle a method change during a late stage after having analyzed multiple reference lots and in-house lots with the old method? (ii) Is it essential for the sponsor to retrospectively re-analyze all previous lots and show results of the old corridor and new corridor? (iii) What if the old lots are no longer available or expired? (iv) What if the two corridors (using the old and new method) show differences in the upper and lower limits? and (v) If the test values are different before and after a method change, then they cannot be bridged, but only justified by demonstrating accuracy.

When there is a method change during late stage product development, it is necessary to conduct a comparability study to validate the new method against the old method in terms of sensitivity, specificity, and validation performance characteristics such as accuracy, precision, and ruggedness as described in the USPNF. If there are significant differences between the old method and the new method, it is necessary to conduct a retrospective re-analysis of previous lots.

Suppose that retrospective re-analysis is essential but the old lots are no longer available or expired. In this case, it is suggested that a sensitivity (or robustness) analysis be conduct to provide a complete picture of the new method as compared to the old method. The upper and lower limits can then be modified accordingly, using a relatively conservative approach.

When there is a major (or significant) change in analytical method, the new analytical method needs to be validated against the old analytical method in terms of sensitivity and specificity and analytical method performance characteristics such as accuracy, precision, and ruggedness.

Can a sponsor increase the number of test lots until they pass the equivalence test? – In practice, if a given quality attribute fails to pass the equivalence test marginally, the sponsor often tests more lots (i.e., increases the sample size) until the given quality attribute passes the equivalence test based on the combined analysis of the originally tested lots and the additional lots. This strategy, however, may not be accepted if there is evidence that the mean shift has exceed the maximum mean shift allowed (i.e., $\frac{1}{8}\sigma_R$) for demonstration of similarity between the proposed biosimilar products.

11.7 Concluding Remarks

To assist sponsors for biosimilar product development, FDA has published several guidances since the passage of the BPCI Act in 2009. In these guidances, FDA recommends a stepwise approach for obtaining totality-of-the-evidence for demonstration of similarity (including analytical similarity, PK/PD similarity, and clinical similarity) between a proposed biosimilar product and its innovative biological product in terms of safety, purity, and potency. This chapter outlines several practical issues (especially regarding analytical similarity assessment) that are commonly encountered during the implementation of the stepwise approach.

This chapter first clarifies the confusion between hypotheses testing procedure and confidence interval approach for testing equality for new drugs and establishment of bioequivalence or similarity for generic/biosimilar drug products. It should be noted that concepts of hypotheses testing (based on type II error or power) and confidence interval approach (based on type I error) are very different. The official method for assessment of new drugs and generic/biosimilar drugs recommended by the FDA is hypotheses testing. For assessment of new drugs, a two-sided test (TST) is typically performed to test for equality of means. Rejection of the null hypothesis of no difference indicates that a significant difference between treatment groups is observed. Sample size is selected for achieving a desired power for detecting a clinically meaningful difference if such a difference truly exists. On the other hand, for assessment of generic/biosimilar drug products, a two-one-sided test procedure (TOST) is usually performed for establishment of bioequivalence or similarity at a pre-specified level of significance. In practice, however, the confidence interval approach is often used to assess bioequivalence or similarity under the framework of hypotheses testing. It should be noted that although TOST is operationally equivalent to the confidence interval approach in many cases, they are not equivalent. Thus, mis-use of confidence interval for assessment of bioequivalence or similarity under the framework of hypotheses testing could be biased and hence misleading.

This chapter also provides some insight regarding totality-of-the-evidence under the primary assumptions described in Section 11.3 from an industrial or academic perspective. Suppose totality-of-the-evidence consists of evidence from the domains of analytical similarity, PK/PD similarity, and clinical similarity. If the proposed biosimilar product passes similarity tests in all domains, FDA considers there is sufficient totality-of-the-evidence for demonstration of high similarity between the proposed biosimilar and the innovative biological product. On the other hand, if the proposed biosimilar product fails to pass any of the suggested similarity assessments (i.e., analytical similarity, PK/PD similarity, and clinical similarity), then the regulatory agency will reject the proposed biosimilar product. If the proposed biosimilar product only passes one or two similarity tests, it is debatable whether the sponsor has provided sufficient totality-of-the-evidence for demonstration of high similarity between the proposed biosimilar product and the innovative biological product. It is then suggested a scoring system with different weights be applied for obtaining a composite score or index for totality-of-the-evidence. This, however, requires further research.

For analytical similarity assessment, FDA recommends a tiered approach for identified quality attributes at various stages of the manufacturing process depending upon their criticality or risk ranking relevant to clinical outcomes. However, different tier tests often result in inconsistent conclusions regarding the passage of similarity. This is due to different assumptions made at different tiers. Basically, Tier 1 equivalence tests focus on equivalence in population means between the test product and the reference product, while the Tier 2 quality range approach focuses on the quality range of the population of the reference product. In practice, sponsors often attempt to re-assign specific quality attributes to different tiers if they fail to pass the tests in the current tiers. This exercise, however, is not considered acceptable to the regulatory agency. For equivalence tests in means, one of the most common criticisms is probably the ignorance of the heterogeneity in variances of the test and the reference products. As an alternative, it is suggested that IBC, SABE, and/or SCDI be considered to account for the possible heterogeneity in variances in equivalence tests. In addition, one may consider disaggregated criteria for establishment of bioequivalence or similarity. In other words, it is suggested to test for equivalence in population mean first for establishment of general similarity. Then, test for equivalence in variability for demonstration of high similarity once generally similarity has been established.

This chapter also presents several commonly asked questions during biosimilar product development. These questions include the key question regarding the selection of test and reference lots (either DP or DS lots) for analytical similarity assessment, in addition to practical issues regarding sample size requirement and testing procedure for the selected lots. In practice, it is suggested the FDA statistical, pharmacological, and clinical reviewers be consulted if questions or issues regarding the assessment of analytical similarity, PK/PD similarity, and clinical similarity shall be raised. It should be note that by Generic Drugs User Fees Act (GDUFA) and Biosimilar User Fees Act (BsUFA), the FDA offers initial advisory meeting and biosimilar product development (BPD) types 1-4 meetings to assist the sponsors in their biosimilar product development. The sponsors should take advantage and request appropriate meetings with the FDA for their biosimilar product development.

12

Recent Development

12.1 Introduction

As mentioned in Chapter 1, the FDA recommends a stepwise approach for obtained totality-of-the-evidence for demonstrating similarity between a proposed biosimilar product and its innovative biological product. The stepwise approach starts with analytical similarity assessment, followed by PK/PD and immunogenicity and clinical studies. To assist the sponsor in demonstrating similarity between the proposed biosimilar product and the innovative biological product in critical quality attributes identified at various stages of the manufacturing process, the FDA circulated a draft guidance entitled *Statistical Approaches to Evaluate Analytical Similarity* in September 2017 for public comments. As discussed in previous chapters, FDA recommends an equivalence test for Tier 1 CQAs which are considered most relevant to clinical outcomes, assuming that the difference in mean responses is proportional to the standard deviation of the reference product, while recommending a quality range approach for Tier 2 CQAs, which are considered less relevant to clinical outcomes under certain assumptions that both means and variances between the test reference and the reference product are highly similar. These methods, however, have been challenged and hence are considered inappropriate due to insensitivity with respect to the assumptions. Alternatively, Chow et al. (2016) proposed a so-called unified approach by comparing the coefficient of variation (or effect size adjusted for standard deviation), assuming that both test and reference product have similar effect size adjusted for standard deviation.

In addition, FDA also circulated a draft guidance entitled *Considerations in Demonstrating Interchangeability with a Reference Product* for public comments in early 2017. In the draft guidance, FDA indicates that the 2 × 4 crossover design of (RTRT, RRRR) is considered a useful switching design for assessment of relative risk with/without switching and/or alternations between the reference product and the test product. Chow et al. (2017), however, indicated that FDA's recommended switching design is a special case of a complete N-of-1 trial design with four dosing periods. In this chapter, in addition to the FDA's recommended switching design, several useful switching studies such as Balaam's design, 2 × 3 dual design, replicated 2 × 2 design, and complete N-of-1 trial designs with two, three, and four periods are briefly described. Relative performances between the FDA recommended (RTRT, RRRR) trial design and complete N-of-1 trial design with 4 dosing periods are also discussed in this chapter.

In recent years, as more biosimilar drug products become available, it is a common practice that the provider (e.g., pharmacist or insurance company) may switch from the reference product (more expensive) to an approved biosimilar product (less expensive) based on factors unrelated to clinical/medical considerations. We will refer to this switch as a non-medical switch (NMS). In practice, NMS is the switching of a patient's medicine, often at the behest of a third party, for reasons other than the patient's health and safety. A non-medical switch raises the concern of possible reduction of efficacy and/or increase of adverse events (safety). In this chapter, valid statistical design and analysis of non-medical studies are discussed.

Recently, FDA circulated a guidance entitled *Statistical Approaches to Evaluate Analytical Similarity* for public comments in September, 2017. The majority of these comments are on (i) lack of scientific justification for the recommended similarity margin, (ii) FDA's current thinking regarding maximum allowed difference between the proposed biosimilar product and the reference product (iii) the validity of the Tier 2 quality range approach, (v) the potential use of a tolerance interval approach and/or min-max approach, and (vi) the recommendation of at least 10 lots per product when performing analytical similarity assessment.

The remaining sections of this chapter are organized as follows. In the next section, a unified approach proposed by Chow et al. (2016) is briefly outlined. Following FDA's guidance on interchangeability, several useful switching designs, including complete N-of-1 trial design are discussed in Section 12.3. Section 12.4 reviews practical issues of post-approval non-medical switches. Review and comments on FDA's draft guidance on analytical similarity assessment are given in Section 12.5. Some concluding remarks are provided in the last section of this chapter.

12.2 Comparing Means versus Comparing Variances

As indicated in Chapter 1, current regulatory requirements for assessment of biosimilarity between a proposed biosimilar (test) product and its innovative biological (reference) product focus on comparing means, which ignores the variabilities associated with the test and reference products (FDA, 2003a, 2015a; Chow and Liu, 2008; Chow, 2013). Unlike bioequivalence assessment for generic drug products, the method of comparing means (i.e., bioequivalence assessment based on average bioavailability) may not be suitable for biologic products due to some fundamental differences between generic (small-molecule) drug products and biological (large-molecule) products. It is a concern that comparing means may not be sufficient for demonstration of high similarity between the test product and the reference product, because biosimilar drug products are known to be highly variable. In practice, large variability is often more likely to occur during the manufacturing process of biosimilar products due to variability in the biologic mechanisms, inputs, and relatively large number of complex steps in the process.

12.2.1 Generally Similar versus Highly Similar

Variability in the products could lead to lower efficacy or potential safety risks for some patients and have a significant impact on clinical outcomes of biosimilar products. Therefore, in addition to comparing means, comparing variances (i.e., equivalence assessment in variances) should be an important part of the demonstration of highly similarity between the test product and the reference product. For example, we may consider the two products are *generally* similar if they pass a biosimilarity test in means between the two products. Once the generally biosimilarity has been established, we can further test for biosimilarity in variability between the two products. We can then claim that the two products are *highly* similar if they pass the biosimilarity test in variability between the two products. The strategy for demonstration of general and high similarity based on comparing means and comparing variances is highlighted in Table 12.1.

12.2.2 Similarity Test in Variability

For similarity tests in variability, the method described in Chow (2013) is useful. Consider a parallel design for evaluating the biosimilarity in variability of the test product with the reference product. Let's denote independent samples of T_i and R_j be the observations of T and R with $i = 1,$

TABLE 12.1

Strategy for Demonstration of Similarity

Similarity Test	Similarity Margin	Determination
Mean	(80%, 125%)	Generally similar
Variance	(θ_L, θ_U)[a]	Highly similar[b]

[a] θ_L and θ_U are often chosen as $\theta_L = 2/3$ and $\theta_U = 3/2$, respectively.
[b] Highly similarity can only be established after the generally similarity has been established.

..., n_T and $j = 1, ..., n_R$. Similarly, consider the following two one-sided hypotheses of

$$H_{01} : \theta_L \geq \frac{V_T}{V_R} \quad \text{vs.} \quad H_{\alpha 1} : \theta_L < \frac{V_T}{V_R},$$

$$H_{02} : \theta_U \leq \frac{V_T}{V_R} \quad \text{vs.} \quad H_{\alpha 2} : \theta_U > \frac{V_T}{V_R}.$$

To test the above two one-sided hypotheses, the extended F-test procedure described in Chow (2013) can be applied. Define

$$\begin{cases} L_j = \sqrt{\theta_L} R_j, & j = 1, ..., n_R \\ U_j = \sqrt{\theta_U} R_j, & j = 1, ..., n_R \end{cases}$$

so that $Var(L_j) \equiv V_L = \theta_L V_R$ and $Var(U_j) \equiv V_U = \theta_U V_R$. Then, it can be verified that the above two one-sided hypotheses are equivalent to

$$H_{01} : V_L \geq V_T \quad \text{vs.} \quad H_{\alpha 1} : V_L < V_T$$
$$H_{02} : V_U \leq V_T \quad \text{vs.} \quad H_{\alpha 2} : V_U > V_T$$

Apply the one-sided F-test for equal variances, H_{01} is rejected at the α level of significance if

$$F_L = \frac{\displaystyle\sum_{i=1}^{n_T} (T_i - \bar{T})^2 / (n_T - 1)}{\displaystyle\sum_{j=1}^{n_R} (L_j - \bar{L})^2 / (n_R - 1)} = \frac{s_T^2}{s_L^2} > F(\alpha; n_T - 1, n_R - 1)$$

and H_{02} is rejected if

$$F_U = \frac{\displaystyle\sum_{i=1}^{n_T} (T_i - \bar{T})^2 / (n_T - 1)}{\displaystyle\sum_{j=1}^{n_R} (U_j - \bar{U})^2 / (n_R - 1)} = \frac{s_T^2}{s_U^2} < F(1 - \alpha; n_T - 1, n_R - 1)$$

We then conclude that V_T and V_R are equivalent with significance level of α if both H_{01} and H_{02} are rejected.

12.2.3 Remarks

The proposed strategy for demonstration of highly similarity between a proposed biosimilar product and its innovative biological product is a disaggregated approach. In other words, we first test for similarity in means to establish general similarity. Once the generally similarity has been established, we can then test for similarity in variances for highly similarity. It should be noted that in practice, θ_L and θ_U are often chosen as $\theta_L = 2/3$ and $\theta_U = 3/2$, respectively. The relative performances between the method based on comparing means and the method based on the combined comparing means and comparing variances require further research.

12.3 Switching Design

12.3.1 Introduction

Biological products as therapeutic agents are made from a variety of living cells or living organisms. The application of biologics ranges from treating disease and medical conditions to preventing and diagnosing diseases. In practice, due to the expensive cost of access to the original biologics, the alternative products (follow-on biologics or biosimilars) are often produced by biopharmaceutical and biotech companies. The popularity of biosimilars brings not only advantages but also several critical issues. For benefits, considering that patents of the early biological products will expire in the next few years, biosimilars with the reduction of cost are more affordable to the general patient population. However, biosimilar products need to be demonstrated to be equivalent to the originator product in terms of their therapeutic effect in terms of safety and efficacy before they can be released in the market place. According to FDA, a biological product may be demonstrated to be biosimilar if data shows that, among other things, the product is *highly similar* to an already-approved biological product. To this end, the relevant properties of the biosimilar products with those of the original drugs are compared. If a proposed biosimilar product has been demonstrated to be highly similar to the originator product, the proposed biosimilar product will be approved by the regulatory authority. An approved biosimilar product can be used as a substitute for the originator product for treating patients with the diseases under study. In practice, as more and more biosimilar products become available in the market place, an interesting question has been raised. The question is whether approved biosimilar products can be used interchangeably: either between the approved biosimilar products and the originator product or among the approved biosimilar products.

As indicated in the Biologic Price Competition and Innovation (BPCI) Act passed by the US Congress in 2009, a proposed biosimilar product is considered *interchangeable* if (i) it is highly similar to the originator (reference) product and can be expected to produce the same clinical result as the reference product in any given patient and, (ii) if the biological product is administered more than once to an individual, the risk in terms of safety of diminished efficacy of alternating or switching between the use of the biological product and the reference product is not greater than the risk of using the reference product without such alternation or switch. Thus, the assessment of drug interchangeability of biosimilar products includes the evaluation of (i) the risk in terms of safety and diminished efficacy of alternating or switching between use of the biological product and the reference product, (ii) the risk of using the reference product without such alternation or switch, and (iii) the relative risk between switching/alternation and without switching/alternation. For this purpose, a crossover design is necessarily employed. In recent years, three types of hybrid parallel-crossover designs are commonly used to address drug interchangeability, which are the parallel plus 2 × 2 crossover design, the parallel plus 2 × 3 crossover design, and the parallel plus 2 × 4 crossover design. These study designs are motivated from a complete N-of-1 design with two, three, and four dosing periods, respectively. For example, the 2 × 4 hybrid parallel-crossover design, denoted by (RRRR, RTRT), where R is the reference product and T is the proposed biosimilar (test) product, has been proposed by many sponsors for addressing drug interchangeability in terms of switching and alternation. This study is also referred to as physician intuition design, which is a partial design of the complete N-of-1 design with four dosing periods.

12.3.2 Concept and Criteria for Drug Interchangeability

As indicated by Chow (2013), drug interchangeability for biosimilar products includes the concepts of potential risk of safety (e.g., increase incidence rate of adverse reaction or event) and efficacy (e.g., reduced efficacy) due to switching and alternation of the use of biosimilar products. Thus, the evaluation of relative risk with and without switching and alternation of biosimilar products is necessary.

The term interchangeable or interchangeability, as described in section 351(k)(4) of the PHS Act, refers that "the biological product may be substituted for the reference product without the intervention of the health care provider who prescribed the reference product." To evaluate the drug interchangeability, the conduction of the switching study or studies will be expected, whose aim is to assess the changes in treatment that result in two or more alternating exposures (switch intervals) to the proposed interchangeable product and to the reference product. According to FDA new draft guideline, a switching study is referred to a randomized two-arm-period study, with one arm incorporating the switch between the proposed interchangeable product and the reference product (i.e., switch from T to R) and the other arm remaining unswitched (i.e., receiving only R), where T is the proposed interchangeable product and R is the reference product. The number of switches should be taken into consideration regarding to the clinical condition to be treated. A single switch is referred to as the switch from one product to another, including (T to R), (R to T), (R to R) and (T to T). On the other hand, alternation is referred as multiple switches, such as (T to R to T), (T to T to T), (T to R to T to R) and (T to T to T to T). The switching arm is generally expected to include at least two separate exposure periods to each of the two products (i.e., switch from T to R to T to R).

For evaluation of the potential risk of switching and/or alternations based on the assessment of biosimilarity between products, several criteria for assessment of biosimilarity are useful. First, one may consider the well-established criteria for assessment of bioequivalence. Alternatively, the scaled average bioequivalence (SABE) criterion proposed by Haidar et al.[6] for highly variable drug products may be considered as most biosimilar products are considered highly variable drug products according to FDA's definition. FDA considers a drug product is a highly variable drug product if its intra-subject coefficient of variation is greater than 30%. Most recently, Chow et al. (2015) proposed a new criterion for drug interchangeability (SCDI) based on the criterion for assessment of individual bioequivalence (IBE) by taking the variability due to subject-by-drug interaction into consideration. The SCDI criterion is found to be superior to the classical bioequivalence criterion and SABE in many cases especially when there are notable large variabilities due to the reference product and/or the subject-by-drug interaction.

12.3.3 Hybrid Parallel-Crossover Design

As indicated earlier, an interchangeable biosimilar product is expected to produce the same clinical results as those of the reference product in any given patient with the disease under study (BPCI, 2009). To determine whether the proposed biosimilar product can produce the same clinical results in any given patient, a standard two-sequence, two-period crossover design, i.e., (TR, RT) is necessarily employed. In the 2×2 crossover design (TR, RT), if we replace the T in the first sequence at the first dosing period with R, the design becomes (RR, RT), which is referred to as a hybrid parallel-crossover design (i.e., the first sequence is considered parallel and the second sequence is crossover). This hybrid parallel-crossover design with two dosing periods allows the evaluation of potential risk with and without switching, i.e., the assessment of similarity between the first sequence (switch from R to R) and the second sequence (switch from R to T) after the switch.

In practice, there are three types of hybrid parallel-crossover designs that are commonly used for addressing drug interchangeability in terms of potential risk of switching and alternation. These three types of designs include the parallel plus 2×2 crossover design, the parallel plus 2×3 crossover design and the parallel plus 2×4 crossover design. As indicated by Chow, Song and Cui (2017), these study designs are motivated from a complete N-of-1 design with two, three, and four dosing periods, respectively. In what follows, these types of hybrid parallel-crossover designs and the corresponding complete N-of-1 trial design are described. In addition, the physician intuition design (RRRR, RTRT) will also be discussed.

Balaam Design – As mentioned before, Balaam design is a popular cross-over design, with four sequences and two periods (See Table 12.2).

Noticeable, the Balaam 4 × 2 design is a complete N-of-1 design with two periods. A complete N-of-1 design is a study design which contains all the possible arrangements for the switching and alternation among testing drugs and reference drugs under a fixed number of period. The Balaam 4 × 2 design is suitable for assessing the switching, whereas it cannot assess the alternation due to the limited number of periods in this study design.

Two-Sequence, Three-Period Crossover Design – The 2 × 3 dual crossover design can evaluate both switching and alternation (See Table 12.3), but this design has a problem with assessment of alternations within the same group of subjects.

To be noticed, the 2 × 3 dual crossover design is not a complete N-of-1 design, which is only a partial design derived from the N-of-1 design. For three periods, a complete N-of-1 design should be an 8 × 3 crossover design, with eight sequences included.

The 2 × 4 Replicated Crossover Design – The 2 × 4 replicated crossover design contains two sequences and four periods (See in Table 12.4), which is suitable for evaluating both switching and alternation, especially the alternations within the same group of subjects.

The 2 × 4 replicated crossover design is not a complete N-of-1 design either. With four periods, a complete N-of-1 design should contain 16 sequences of

TABLE 12.2

Balaam 4 × 2 Design

Group	Period I	Period II
1	T	T
2	R	R
3	T	R
4	R	T

TABLE 12.3

2 × 3 Dual Crossover Design

Group	Period I	Period II	Period III
1	T	R	T
2	R	T	R

TABLE 12.4

2 × 4 Replicated Crossover Design

Group	Period I	Period II	Period III	Period IV
1	T	R	T	R
2	R	T	R	T

treatments and 4 periods. This type of N-of-1 design will be discussed in detail later.

Complete N-of-1 Design – The potential limitations for the above three types of hybrid parallel crossover design lead to the application of the complete N-of-1 design. (See Table 12.5: with two sequences, three sequences, and four sequences.) Qualified subjects will be randomly assigned to the 16 treatment sequences to evaluate on the drug interchangeability.

Under the complete N-of-1 crossover design, all possible switching and alternations can be assessed and the results can be compared within the same group of patients and between different groups of patients.

Recently, a partial design from the complete N-of-1 design has been proposed and supported by many physicians for evaluating the switching between R and T and relative risk of with/without alternation, which is (RRRR, RTRT), equivalent to the first sequence and the sixth sequence in the above study design (See in Table 12.6).

The objectives of this article are to compare relative advantages and limitations between the N-of-1 design and (RRRR, RTRT) both theoretically and

TABLE 12.5

Complete N-of-1 Design

Group	Period I	Period II	Period III	Period IV
1	R	R	R	R
2	R	T	R	R
3	T	T	R	R
4	T	R	R	R
5	R	R	T	R
6	R	T	T	T
7	T	R	T	R
8	T	T	T	T
9	R	R	R	T
10	R	R	T	T
11	R	T	R	T
12	R	T	T	R
13	T	R	R	T
14	T	R	T	T
15	T	T	R	T
16	T	T	T	R

TABLE 12.6

FDA Recommended Design (RTRT, RRRR)

Group	Period I	Period II	Period III	Period IV
1	R	T	R	T
2	R	R	R	R

via simulation. The required sample size for the two crossover designs will be derived to achieve the expected power and significance level.

12.3.4 Statistical Model and Analysis

As can be seen, the hybrid parallel-crossover design, N-of-1 trial design, and the 2 × 4 physician intuition design (RTRT, RRRR) discussed in the previous section can be generally described as a J × K crossover design. Thus, in this section, drug effect and carryover effect will be assessed using the following statistical model under a general K-sequence and J-period crossover design comparing t formulations:

$$Y_{ijk} = \mu + G_k + S_{ik} + P_j + D_{(j,k)} + C_{(j-1,k)} + e_{ijk}, i = 1, 2, \cdots, J; k = 1, 2, \cdots, K$$

where μ is the overall mean, G_k is the fixed kth sequence effect, S_{ik} is random effect for the ith subject within the kth sequence with mean 0 and variance σ_S^2, P_j is the fixed effect for the jth period, $D(j, k)$ is the drug effect for the drug in the jth sequence at kth period, $C_{(j-1, k)}$ is the carry-over effect, and e_{ijk} is the random error with mean 0 and variance σ_e^2. Under the model, it is assumed that S_{ik} and e_{ijk} are mutually independent.

Analysis of Physician Intuition Design – To evaluate the interchangeability based on (RRRR, RTRT) study design, the Williams design method is applied to derive the statistics for test for biosimilarity/bioequivalence.

Under the William design method, the expected values of the sequence-by-period means is derived based on the above model, which adjusts for the first-order carryover effect (See in Table 12.7). Analysis of variance (ANOVA) table is given in Table 12.8.

Under the assumptions that $\sum_{k=1}^{K} G_k = 0, \sum_{j=1}^{J} P_j = 0, \sum_{t=R,T} D_t = 0,$ $\sum_{t=R,T} C_t = 0,$ all the unknown parameters are considered as the outcome

TABLE 12.7

Expected Values of the Sequence-by-Period Means for (RTRT, RRRR)

Group	Period I	Period II	Period III	Period IV
1	$\mu + G_1 + P_1 + D_R$	$\mu + G_1 + P_2 + D_R + C_R$	$\mu + G_1 + P_3 + D_R + C_R$	$\mu + G_1 + P_4 + D_R + C_R$
2	$\mu + G_2 + P_1 + D_R$	$\mu + G_2 + P_2 + D_T + C_R$	$\mu + G_2 + P_3 + D_R + C_T$	$\mu + G_2 + P_4 + D_T + C_R$

TABLE 12.8

Analysis of Variance Table for (RTRT, RRRR)

Source of Variation	Degree of Freedom
Intersubject	$n_1 + n_2 - 1$
Sequence	1
Residual	$n_1 + n_2 - 2$
Intrasubject	$n_1 + n_2$
Period	3
Formulation	1
Carryover	1
Residual	$n_1 + n_2 - 5$
Total	$2(n_1 + n_2) - 1$

need to estimate. Based on the above table, the design matrix X can be derived.

Since $Y = X\beta, \hat{\beta} = (X'X)^{-1}X'$ can be derived based on the ordinary least square method. Here Y refers to $(\mu, G_1, P_2, P_3, D_R, C_R)'$, and the design matrix and β based on Table 12.7 are shown below.

$$X = \begin{pmatrix} 1 & 1 & 1 & 1 & 1 & 1 & 1 & 1 \\ 1 & 1 & 1 & 1 & 0 & 0 & 0 & 0 \\ 1 & 0 & 0 & 0 & 0 & 1 & 0 & 0 \\ 0 & 1 & 0 & 0 & 0 & 1 & 0 & 0 \\ 0 & 0 & 1 & 0 & 0 & 0 & 1 & 0 \\ 1 & 1 & 1 & 1 & 1 & 0 & 1 & 0 \\ 0 & 1 & 1 & 1 & 1 & 1 & 0 & 1 \end{pmatrix}, \beta = \begin{pmatrix} \beta_{11} = \mu + G_1 + P_1 + D_R \\ \beta_{21} = \mu + G_1 + P_2 + D_R + C_R \\ \beta_{31} = \mu + G_1 + P_3 + D_R + C_R \\ \beta_{41} = \mu + G_1 + P_4 + D_R + C_R \\ \beta_{12} = \mu + G_2 + P_1 + D_R \\ \beta_{22} = \mu + G_2 + P_2 + D_T + C_R \\ \beta_{32} = \mu + G_2 + P_3 + D_R + C_T \\ \beta_{42} = \mu + G_2 + P_4 + D_R + C_R \end{pmatrix}$$

Then the estimated $\hat{D}$ can be obtained using constraint $\hat{D} = \hat{D}_T - \hat{D}_R$. The coefficients of $\hat{D}_T$ and $\hat{D}_R$ are the element of $(X'X)^{-1}X'$ which are given in Table 12.9.

Based on the estimated coefficients table,

$$\hat{D} = \frac{1}{2}\left[(2\bar{Y}_{11} - \bar{Y}_{21} - \bar{Y}_{41}) - (2\bar{Y}_{12} - \bar{Y}_{22} - \bar{Y}_{42}) \right]$$

$$E(\hat{D}) = \hat{D}_T - \hat{D}_R, Var(\hat{D}) = \frac{3}{2}\sigma_e^2\left(\frac{1}{n_1} + \frac{1}{n_2}\right) = \frac{3\sigma_e^2}{n}$$

TABLE 12.9

Coefficients for Estimates of Drugs in (RTRT, RRRR) - Adjusting for Carryover Effect

Sequence	D_R Period				D_T Period				D Period			
	I	II	III	IV	I	II	III	IV	I	II	III	IV
1	−1	0.5	0	0.5	1	−0.5	0	−0.5	2	−1	0	−1
2	1	−0.5	0	−0.5	−1	0.5	0	0.5	−2	1	0	1

Note: Coefficients are multiplied by 2.

To test the bioequivalence based on average bioequivalence method,

$$H_0 : |D_T - D_R| \le \theta \quad vs. \quad H_1 : |D_T - D_R| > \theta$$

The null hypothesis will be rejected and the bioequivalence will be demonstrated when the statistic

$$T_L = \frac{\hat{D} - \theta}{s\sqrt{\frac{1}{n}}} > t[\alpha, n_1 + n_2 - 5].$$

The corresponding confidence interval at α significance level is

$$\hat{D} \pm t[\alpha, n_1 + n_2 - 5] s \sqrt{\frac{3}{2} \left(\frac{1}{n_1} + \frac{1}{n_2} \right)}.$$

Similarly, the carryover effect coefficient estimates are derived and the corresponding statistic for testing bioequivalence is shown in Table 12.10. According to the estimated coefficients,

$$\hat{C} = \bar{Y}_{11} - \bar{Y}_{31} + \bar{Y}_{12} - \bar{Y}_{32}$$
$$E(\hat{C}) = C_T = C_R, Var(\hat{C}) = 2\sigma_e^2 \left(\frac{1}{n_1} + \frac{1}{n_2} \right) = \frac{4\sigma_e^2}{n}$$

TABLE 12.10

Coefficients for Estimates of Carryover Effect in (RTRT,RRRR)

Sequence	C_R Period				C_T Period				C Period			
	I	II	III	IV	I	II	III	IV	I	II	III	IV
1	−0.5	0	0.5	0	0.5	0	−0.5	0	1	0	−1	0
2	0.5	0	−0.5	0	−0.5	0	0.5	0	−1	0	1	0

The null hypothesis will be rejected and the bioequivalence will be demonstrated when the statistic

$$|T_C| = \frac{|\hat{C}|}{2s\sqrt{\dfrac{1}{n_1}+\dfrac{1}{n_2}}} > t\left[\frac{\alpha}{2}, n_1 + n_2 - 4\right].$$

When the washout is sufficient enougth during the trial, the first-order carry-over effect can be ignored. In this case,

$$E(\hat{D}) = D_T - D_R, Var(\hat{D}) = \sigma_e^2\left(\frac{1}{n_1}+\frac{1}{n_2}\right) = \frac{2\sigma_e^2}{n}.$$

Analysis of Complete N-of-1 Design – To evaluate the biosimilarity in the complete N-of-1 design, the analysis of variance table is given in Table 12.11, while the Williams design method is applied as well to derive the estimated coefficients for drug effect and carryover effect (See in Table 12.12).
Based on the estimated coefficients table,

$$E(\hat{D}) = D_T - D_R, Var(\hat{D}) = \frac{\sigma_e^2}{11n}$$

The null hypothesis will be rejected and the bioequivalence will be demonstrated when the statistic

$$T_L = \frac{\hat{D}-\theta}{s\sqrt{\dfrac{1}{11n}}} > t[\alpha, 16n - 5].$$

TABLE 12.11

Analysis of Variance Table for Complete N-of-1 Design

Source of Variation	Degree of Freedom
Intersubject	N-1
Sequence	15
Residual	N-16
Intrasubject	N
Period	3
Formulation	1
Carryover	1
Residual	N-5
Total	2N-1

TABLE 12.12

Coefficients for Estimates of Drug in Complete N-of-1
Design - Adjusting for Carryover Effect

Sequence	D Period			
	I	II	III	IV
1	3	−1	−1	−1
2	−3	−7	−7	17
3	−5	−9	15	−1
4	−11	−15	9	17
5	−5	15	−1	−9
6	−11	9	−7	9
7	−13	7	15	−9
8	−19	1	9	9
9	19	−1	−9	−9
10	13	−7	−15	9
11	11	−9	7	−9
12	5	−15	1	9
13	11	15	−9	−17
14	5	9	−15	1
15	3	7	7	−17
16	−3	1	1	1

Note: Coefficients are multiplied by 132.

The corresponding confidence interval at α significance level is

$$\hat{D} \pm t[\alpha, 16n - 5]s\sqrt{\frac{1}{11n}}.$$

Similarly, the carryover effect coefficient estimates are derived and the corresponding statistic for testing bioequivalence is shown in Table 12.13.
According to the estimated coefficients,

$$E(\hat{C}) = C_T - C_R, Var(\hat{C}) = \frac{4\sigma_e^2}{33n}.$$

When ignoring the first-order carryover effect,

$$E(\hat{D}) = D_T - D_R, Var(\hat{D}) = \frac{\sigma_e^2}{12n}.$$

Comparison – For a fixed sample size, e.g., N = 48, the number of patients required for each sequence in (RTRT, RRRR) is 24, whereas the number of patients

TABLE 12.13

Coefficients for Estimates of Carryover
Effect in Complete N-of-1 Design

Sequence	C Period			
	I	II	III	IV
1	12	-4	-4	-4
2	10	-6	-6	2
3	2	-14	-6	18
4	0	-16	-8	24
5	2	-6	18	-14
6	0	-8	16	-8
7	-8	-16	16	8
8	-10	-18	14	14
9	10	18	-14	-14
10	8	16	-16	-8
11	0	8	-16	8
12	-2	6	-18	14
13	0	16	8	-24
14	-2	14	6	-18
15	-10	6	6	-2
16	-12	4	4	4

Note: Coefficients are multiplied by 132.

required for each sequence in complete N-of-1 design is 3. Correspondingly, the variances of drug effects when adjusting for the carry-over effect are $\frac{\sigma_e^2}{8}$ and $\frac{\sigma_e^2}{33}$, respectively. The relative efficiency between two study designs is 24.24%, indicating that the efficiency of (RTRT, RRRR) design is 24.24% of complete N-of-1 design. When ignoring the carry-over effect, the variances of drug effects in (RTRT, RRRR) design is $\frac{\sigma_e^2}{12}$, and the variances of drug effects in complete N-of-1 design is $\frac{\sigma_e^2}{36}$. The relative efficiency between two study designs increases to 33.33%.

Therefore, the partial design (RTRT, RRRR) is less efficient than the complete design (N-of-1 design). When the washout is sufficient, the relative efficiency of partial design increases, but is still less than the complete design.

12.3.5 Sample Size Requirement

To calculate the expected sample size under the fixed power and significance level, Schuirmann's two one-sided t test procedure is applied. According to the ±20% rule, the bioequivalence is concluded if the average bioavailability of the test drug effect is within ±20% of that of the reference drug effect with a certain assurance. Therefore, the null hypothesis and alternative hypothesis are set as follows, with ∇ indicates the decision range, i.e., 20%.

$$H_0 : \mu_T - \mu_R < -\nabla\mu_R \ or \ \mu_T - \mu_R > \nabla\mu_R \quad vs. \quad H_a : -\nabla\mu_R \leq \mu_T - \mu_R \leq \mu_R$$

The power function

$$P(\theta)=F_v\left(\left[\frac{\nabla-\theta}{CV\sqrt{b/n}}\right]-t(\alpha,v)\right)-F_v\left(t(\alpha,v)-\left[\frac{\nabla+\theta}{CV\sqrt{b/n}}\right]\right)$$

where

$$\theta = \frac{\mu_T - \mu_R}{\mu_R}, CV = S/\mu_R,$$

μ_T and μ_R are the average bioavailability of the test and reference formulations, respectively; S is the squared root of the mean square error from the analysis of variance table for each crossover design, $[-\nabla\mu_R, \nabla\mu_R]$ is the bioequivalence limits, $t(\alpha, v)$ is the upper αth quantile of a t distribution with v degrees of freedom and F_v is the cumulative distribution function of the t distribution.

Accordingly, the exact sample size formula when

$$\theta = 0 \ \text{is} \ n \geq b\left[t(\alpha,v)+t\left(\frac{\beta}{2},v\right)\right]^2[CV/\nabla]^2;$$

the approximate sample size formula when $\theta > 0$ is

$$n \geq b\left[t(\alpha,v)+t(\beta,v)\right]^2[CV/(\nabla-\theta)]^2.$$

Setting $\nabla = 0.2$, to achieve the 80% or 90% power under 5% significance level, the required sample sizes for (RTRT, RRRR) design and complete N-of-1 design are shown in the Table 12.13 and Table 12.14.

Consider $\delta \in (0.8, 1.25)$ is the bioequivalence range of μ_T/μ_R, the hypothesis changes to

$$H_0 : \frac{\mu_T}{\mu_R} \leq 0.8 \ or \ \frac{\mu_T}{\mu_R} \geq x1.25 \ vs. \ H_a : 0.8 < \frac{\mu_T}{\mu_R} < 1.25$$

TABLE 12.14

Number of Subjects for Schuirmann's Two One-Sided Tests Procedure at
$\nabla = 0.2$ and the 5% Nominal Level in (RTRT, RRRR) Design

Power (%)	CV (%)	θ			
		0%	5%	10%	15%
80	10	32	40	80	300
	12	44	52	112	432
	14	56	68	152	588
	16	72	88	196	764
	18	88	112	244	968
	20	108	136	300	1192
	22	128	164	364	1440
	24	152	196	432	1716
	26	180	228	508	2012
	28	208	264	588	2332
	30	236	300	672	2676
	32	268	344	764	3044
	34	304	388	864	3436
	36	340	432	968	3852
	38	376	480	1076	4292
	40	416	532	1192	4752
90	10	40	52	108	416
	12	52	72	152	596
	14	68	96	208	812
	16	88	124	268	1056
	18	112	152	340	1336
	20	136	188	416	1648
	22	164	228	504	1996
	24	192	268	596	2372
	26	224	316	700	2784
	28	260	364	812	3228
	30	300	416	932	3704
	32	340	472	1056	4216
	34	380	532	1192	4756
	36	428	596	1336	5332
	38	476	664	1488	5940
	40	524	736	1648	6584

TABLE 12.15

Number of Subjects for Schuirmann's Two One-Sided Tests Procedure at $\nabla = 0.2$ and the 5% Nominal Level in Complete N-of-1 Design

Power (%)	CV (%)	θ			
		0%	5%	10%	15%
80	10	16	16	16	48
	12	16	16	16	64
	14	16	16	32	80
	16	16	16	32	96
	18	16	16	32	128
	20	16	32	48	160
	22	32	32	48	176
	24	32	32	64	224
	26	32	32	64	256
	28	32	48	80	288
	30	32	48	96	336
	32	48	48	96	384
	34	48	48	112	432
	36	48	64	128	480
	38	48	64	144	528
	40	64	80	160	592
90	10	16	16	16	64
	12	16	16	32	80
	14	16	16	32	112
	16	16	16	48	144
	18	16	32	48	176
	20	32	32	64	208
	22	32	32	64	256
	24	32	48	80	304
	26	32	48	96	352
	28	48	48	112	400
	30	48	64	128	464
	32	48	64	144	512
	34	48	80	160	592
	36	64	80	176	656
	38	64	96	192	736
	40	80	96	208	800

In case of the skewed distribution, the hypothesis are transformed to logarithmic scale,

$$H_0 \log\mu_T - \log\mu_R < \log(0.8) \ or \ \log\mu_T - \log\mu_R > \log(1.25)$$
$$vs. \ H_a : \log(0.8) \le \log\mu_T - \log\mu_R \le \log(1.25)$$

Then, the sample size formulas for different δ are given below,

$$n \ge b\left[t(\alpha,\upsilon)+t\left(\frac{\beta}{2},\upsilon\right)\right]^2 [CV/\ln 1.25]^2 \ if \ \delta = 1$$

$$n \ge b\left[t(\alpha,\upsilon)+t(\beta,\upsilon)\right]^2 [CV/\ln 1.25 - \ln\delta]^2 \ if \ 1 < \delta < 1.25$$

$$n \ge b\left[t(\alpha,\upsilon)+t(\beta,\upsilon)\right]^2 [CV/(\ln 0.8 - \ln\delta)]^2 \ if \ 0.8 < \delta < 1$$

To achieve 80% or 90% power under 5% significance level, the required sample size for (RTRT, RRRR) and complete N-of-1 designs are calculated in Table 12.15, Table 12.16 and Table 12.17.

An Example – To compare the sample sizes between two designs, for instance, we have a summary table (See in Table 12.18) of the number of subjects for the additive models between (RTRT, RRRR) design and complete N-of-1 Design with 80% power, CV = 20%, and $\theta = 5$ and 10%.

Table 12.19 summarizes the number of subjects for the multiplicative models between (RTRT, RRRR) design and complete N-of-1 Design with 80% power, CV = 20%, and $\delta = 0.90$ and 1.00 as follows:

Remarks – The drug interchangeability including switching and alternation can be assessed in both 2 × 4 parallel crossover design (RTRT,RRRR) and the complete N-of-1 design with four dosing periods. (RTRT,RRRR) design, as a partial design of the complete N-of-1 design, is suitable for evaluating the switch from R to R, switch from R to T, and switch from T to T. The analysis for assessing alternations of R to R to R, R to T to R, and T to R to T can also be conducted under this design. However, different from the partial design, the complete N-of-1 design brings a broader framework for switching and alternation. With 16 sequences, all possible switches and alternations can be analyzed to demonstrate the drug interchangeability. Comparing to (RTRT, RRRR) design, the complete N-of-1 design contains more information for testing and reference drugs, which provides the opportunity to comprehensively assess the drug biosmilarity and interchangeability.

The (RTRT, RRRR) design is less efficient than the complete N-of-1 design, in term of the variance of drug effect and relative efficiency. Adjusting for the first-order carry-over effect that may exists during the clinical trial, the variance of the difference between the testing drug effect and reference drug effect in (RTRT, RRRR) design is greater than that in the complete N-of-1 design.

TABLE 12.16

Number of Subjects for Schuirmann's Two One-Sided Tests Procedure at the 5% Nominal Level for the (0.8, 1.25) Bioequivalence Range in the Case of the Multiplicative Model in (RTRT, RRRR) Design

		δ							
Power (%)	**CV (%)**	**0.85**	**0.90**	**0.95**	**1.00**	**1.05**	**1.1**	**1.15**	**1.2**
80	10	104	30	16	14	16	26	56	226
	12	148	42	22	18	20	36	80	324
	14	200	56	28	24	26	48	108	438
	16	260	72	34	30	34	60	140	572
	18	330	90	44	36	42	76	176	724
	20	406	110	52	44	52	94	216	892
	22	492	132	64	52	62	112	260	1080
	24	584	156	74	62	72	134	310	1284
	26	684	184	88	72	86	156	364	1508
	28	794	212	102	84	98	180	420	1748
	30	910	244	116	96	112	206	482	2006
	32	1036	276	132	108	128	234	548	2282
	34	1170	312	148	122	144	264	620	2576
	36	1310	350	166	136	160	296	694	2888
	38	1460	388	184	152	178	330	772	3216
	40	1618	430	204	168	198	366	856	3564
90	10	142	40	20	16	20	34	76	310
	12	204	56	28	22	28	48	110	446
	14	276	76	36	28	36	64	148	606
	16	360	98	48	36	46	84	192	792
	18	456	122	60	46	58	104	242	1002
	20	562	150	72	56	70	128	298	1236
	22	680	182	86	66	84	154	360	1494
	24	808	216	102	78	100	184	428	1778
	26	948	254	120	92	116	216	502	2088
	28	1098	294	140	106	136	250	582	2420
	30	1260	336	160	120	154	286	668	2778
	32	1434	382	180	136	176	324	760	3160
	34	1618	430	204	154	198	366	856	3568
	36	1814	482	228	172	222	410	960	3998
	38	2022	538	254	192	246	456	1070	4456
	40	2240	596	282	212	274	506	1186	4936

TABLE 12.17

Number of Subjects for Schuirmann's Two One-Sided Tests Procedure at the 5% Nominal Level for the (0.8, 1.25) Bioequivalence Range in the Case of the Multiplicative Model in Complete N-of-1 Design

Power (%)	CV (%)	δ							
		0.85	0.90	0.95	1.00	1.05	1.1	1.15	1.2
80	10	32	16	16	16	16	16	16	64
	12	48	16	16	16	16	16	32	80
	14	64	16	16	16	16	16	32	112
	16	80	32	16	16	16	16	48	144
	18	96	32	16	16	16	32	48	192
	20	112	32	16	16	16	32	64	224
	22	128	48	32	16	16	32	64	272
	24	144	48	32	32	32	48	80	320
	26	176	48	32	32	32	48	96	368
	28	208	64	32	32	32	48	112	432
	30	224	64	32	32	32	64	128	496
	32	256	80	48	32	32	64	144	560
	34	288	80	48	32	48	80	160	640
	36	320	96	48	48	48	80	176	704
	38	368	96	48	48	48	96	192	784
	40	400	112	64	48	64	96	224	880
90	10	48	16	16	16	16	16	32	80
	12	64	16	16	16	16	16	32	112
	14	80	32	16	16	16	16	48	160
	16	96	32	16	16	16	32	48	208
	18	112	32	16	16	16	32	64	256
	20	144	48	32	16	32	32	80	304
	22	176	48	32	32	32	48	96	368
	24	208	64	32	32	32	48	112	448
	26	240	64	32	32	32	64	128	512
	28	272	80	48	32	48	64	144	592
	30	320	96	48	32	48	80	176	688
	32	352	96	48	48	48	80	192	768
	34	400	112	64	48	64	96	224	880
	36	448	128	64	48	64	112	240	976
	38	496	144	64	48	64	112	272	1088
	40	544	160	80	64	80	128	304	1200

TABLE 12.18

Summary for Additive Model

	θ	
	5%	10%
RRRR/RTRT	136	300
Complete N-of-1 Design	32	48

TABLE 12.19

Summary for Multiplicative Model

	δ	
	0.90	1.00
RRRR/RTRT	110	44
Complete N-of-1 Design	32	16

The relative efficiency in the partial design is 24.24% of that in the complete design (for a fixed sample size N = 48). When the washout period is long enough so that the carryover effect can be ignored, the complete N-of-1 design yields a smaller variance of drug effect and higher efficiency in the assessment of drug interchangeability. To achieve the expected power for evaluating the biosimilarity/interchangeability, the partial design requires a larger sample size than the complete design. Because of the smaller sample size required in complete N-of-1 design, conducting the drug trial under the complete design may save money, time, and be more efficient than the partial design. However, too many sequences in the complete design will cause too few subjects be allocated to each sequence. In case of the withdrawal of subjects, it will affect the evaluations in the complete design more than that in the partial design.

In this chapter, only one testing drug and one reference drug are involved in the evaluation for both designs. In practice, however, the testing drug and reference drug may be different. In the situation where more than one testing drug and reference drug are included, the above analysis can be generalized to evaluate the drug interchangeability but more complicated. The complete N-of-1 design is expected to perform better in terms of the relative efficiency and required sample size than the partial design since with more than two drugs in the crossover design, the partial design will lose much more information to assess the drug interchangeability.

12.4 Non-Medical Switching

For an approved biosimilar product, it is a common practice that the provider (pharmacist or insurance company) may switch from the reference product to the approved biosimilar product based on factors unrelated to clinical/medical consideration. In practice, it is a concern that this non-medical switch may present unreasonable risk (e.g., reduced efficacy or an increase of the incidence rate of adverse events) to the patient population with the diseases under study. In recent years, several observational studies and a national clinical study (NOR-SWITCH) were conducted to evaluate the risk of non-medical switches from a reference product to an approved biosimilar product. The conclusions from these studies, however, are biased and hence may be somewhat misleading due to some scientific and/or statistical deficiencies in design and analysis of the data collected. In this article, valid study designs and appropriate statistical methods are recommended for a more accurate and reliable assessment of potential risk of medical/non-medical switches between the proposed biosimilar product and the reference product. The results can be easily extended for evaluation of the potential risk of medical/non-medical switches among multiple biosimilar products and a reference product.

12.4.1 Introduction

When an innovative biologic drug product is going off patent protection, other pharmaceutical companies usually seek market authorization of similar biologic drug products. The similar biologic drug products are biosimilars (or follow-on biologics). In 2009, the United States (US) Congress passed the Biologic Price Competition and Innovation (BPCI) Act which gave the US Food and Drug Administration (FDA) the authority to approve biosimilar products. According to the BPCI Act, a biosimilar product is a biologic product that is highly similar to an innovative biologic (reference) product, notwithstanding minor differences in clinically inactive components, and which has no clinically meaningful differences in terms of safety, purity, and potency. A biosimilar drug can be generally used to substitute the innovative drug if it has been shown to be highly similar to the innovative drug. The FDA, however, does not indicate that (i) the approved biosimilar product and the reference product can be used interchangeably and (ii) two biosimilar products of the same innovative drug can be used interchangeably even though they are highly biosimilar to the same innovative drug.

Regarding drug interchangeability, the BPCI Act indicated that a biological product is to be interchangeable with the reference product if the information submitted in the application is sufficient to show (i) that the biological product is not only biosimilar to the reference product, but also that it can be *expected* to produce the *same clinical result* as the reference product in *any given patient*; and (2) that for a biological product that is administered more than once to an individual, the risk in terms of safety or diminished efficacy of alternating or switching between use of the biological product and the reference product is not greater than the risk of using the reference product without any such alternation or switch. In practice, it is not possible to show the same clinical result in any given patient (i.e., for every patient, we need to show that the proposed biosimilar will produce the same clinical result as that of the reference product). However, it is possible to demonstrate the same clinical result in any given patient *with certain assurance*. Along this line, the FDA circulated a draft guidance on drug interchangeability for public comments (FDA, 2017), although thus far FDA has not yet granted approval for drug interchangeability in recent regulatory submissions.

Non-medical reasons for switching a patient's medicine could include (i) to increase the profits of a private insurer; (ii) to reduce costs for a government agency, or employer; and (iii) an agreement between the payer and a particular manufacturer to favor that manufacturer's product. However, it is suggested that patients and their physicians should remain in control of their treatment decisions, rather than an insurer, government, pharmacy, or other third party. With this non-medical switch, it is a concern that the switch from the reference product to an approved biosimilar product may present unreasonable risk (e.g., reduced efficacy or increase of the incidence rate of adverse events) to the patient population, especially for those patients who have been receiving the reference product at a steady and efficacious level.

In recent years, several observational studies and a national clinical study (NOR-SWITCH) were conducted to evaluate the risk of non-medical switches from a reference product to an approved biosimilar product. The conclusions from these studies, however, are somewhat biased and hence may be misleading due to some scientific and/or statistical deficiencies in design and analysis of the data collected. In this article, valid study designs and appropriate statistical methods are recommended for a more accurate and reliable assessment of potential risk of medical/non-medical switches between the proposed biosimilar product and the reference product. The results can be easily extended for evaluation of the potential risk of medical/non-medical switches among multiple biosimilar products and a reference product.

12.4.2 Approaches for Evaluation of Non-Medical Switch

As indicated in the previous section, although none of recent regulatory submissions have been granted FDA approval for drug interchangeability, non-medical switches from a reference product to its approved biosimilar product inevitably occur due to certain considerations unrelated to clinical assessment/judgement. In practice, it is then of interest to evaluate whether such a switch will cause loss of efficacy and/or increase of adverse event rate. To address this issue, two approaches are commonly considered by conducting observational studies or clinical studies.

In this section, without loss of generality and for illustrational purpose, we will consider observational studies and clinical studies conducted for evaluation of the potential risk of non-medical switching of anti-TNF treatment, i.e., a switch from Remicade (reference) to Remsima (a proposed biosimilar product, also known as CT-P13).

Single arm observational studies – Several observational studies were conducted to evaluate the potential risk of non-medical switches from Remicade (reference) to Remsima (test). These observational studies are summarized in Table 12.20 and Table 12.21. The intention of conducting single arm observational studies for evaluation of potential risk in reduced efficacy and/or increase of adverse events due to a non-medical switch is good but the conclusions may be biased and hence misleading due to the limitations of single arm observational studies. These limitations and deficiencies are outlined below.

Descriptive statistics rather than statistics inference – The conclusions from all of these single arm observational studies were made based on descriptive statistics and/or graphical presentations on the data collected from a limited number of subjects. It is suggested that the confidence interval (CI) of the mean difference in primary study endpoint (e.g., disease activity)

TABLE 12.20

Reported Observational Studies

Experience Currently Limited to Celltrion's CT-P13 (Inflectra/Remsima)				
	Compare vs. Continued Ref. Product	**Switch x1 or Alternating**	**Sample Size of Switched Cohort (n)**	**Duration after Switch**
Buer L. et al. IDB	None	Single Switch	143	6 mos
Sieczkowska J. et al. IBD	None	Single Switch	32	8 mos
Smits L. et al. IBD	None	Single Switch	83	16 wks
Kolar M. et al. IBD	None	Single Switch	74	24 wks
Diaz Hernandez L. et al. IBD	None	Single Switch	72	6 mos
Fiorino G. et al. IBD	No: Compare vs. new/re-starts*	Single Switch	97	6 mos
Glintborg B. et al. Rheum conditions	None	Single Switch	647	3 mos
Nikiphorou E. et al. Rheum conditions	None	Single Switch	39	11 Months

Source: Buer L. et al. J Crohn's and Colitis Advance Access published September 22, 2016 10.1093/ecco-jcc/jjw166; Sieczkowska J, et al. Advanced access pub JCC 2015. DOI: 10.1093/ecco-jcc/jjv;Smits L. et al. ECCO 2016. Abstr DOPO30; Kolar M. et al. ECCO 2016. Abstr DOPO32; Diaz Hernandez L. et al. ECCO 2016. Abstr P449; Fiorino G. et al. DDW 2016 Oral Pres Abstr 439 (* "New/re-starts" include bio-naïve or previously TNFi exposed pts); Glintborg B. et al. EULAR 2016. Oral Pres Abstr THU0123; Nikiphorou E. Expert Opin, Biol. Ther. (2015)15(12):1677–1683

TABLE 12.21

Conclusions of Reported Observational Non-Medical Switching Studies of Anti-TNF Treatment

Study	Author's Main Conclusion
Buer L. et al.	Switching from Remicade® to Remsima™ was feasible and with few adverse events, including very limited antidrug antibody formation and loss of response
Sieczkowska J. et al.	Switching from IFX originator to its biosimilar seems to be a safe option in children with CD. Biosimilars after switch showed to be equally as effective as its originator
Smits L. et al.	No significant change in disease activity was observed 16 weeks after switching from Remicade® to CT-P13. Two patients developed new ADA with undetectable TL during follow-up. No SAEs were observed
Kolar M. et al.	Based on our results, switching of IBD patients from original to biosimilar IFX is effective and safe. Importantly no increase in immunogenicity was observed
Diaz Hernandez L. et al.	Switching to CT-P13 was effective in maintaining clinical remission at 6 months of treatment. No relevant AEs were observed. The use of the biosimilar supposed a cost savings in treatment
Fiorino G. et al.	No clear difference in safety was reported, however, a 5-fold increase in LOR after switch, and a trend towards more frequent primary failure in UC compared to CD patients was recorded. These findings should be evaluated with caution due to the short follow-up
Glintborg B. et al.	Disease activity was largely unaffected in the majority of patients 3 months after non-medical switch to biosimilar Remsima and comparable to the fluctuations observed in the 3 months prior to the switch. However, several patients (~6%) stopped treatment due to LOE or AE. This warrants further investigation before such a non-medical switch can be recommended
Nikiphorou E. et al.	Well tolerated in patients who maintained the treatment after 54 weeks and in patients who switched to CT-P13 after 54 weeks of IFX treatment

Source: Buer L. et.al. J Crohn's and Colitis Advance Access published September 22, 2016 10.1093/ecco-jcc/jjw166; Sieczkowska J et al. Advanced access pub JCC 2015. DOI: 10.1093/ecco-jcc/jjv; Smits L. et al. ECCO 2016. Abstr DOPO30; Kolar M. et al. ECCO 2016. Abstr DOPO32; Diaz Hernandez L. et al. ECCO 2016. Abstr P449; Fiorino G. et al. DDW 2016 Oral Pres Abstr 439 (* "New/re-starts" include bio-naïve or previously TNFi exposed pts); Glintborg B. et al. EULAR 2016. Oral Pres Abstr THU0123; Nikiphorou E. Expert Opin, Biol. Ther.(2015)15(12):1677–1683

should be obtained and sensitivity analysis should be performed before a valid statistical inference (conclusion) can be made.

Sample size justification – No sample size justifications were provided in these studies. As a result, whether the observed clinically meaningful difference truly exists or is purely by chance alone cannot be confirmed. In addition, little information regarding the variabilities associated with the reference product and the test products were provided.

Selection of non-inferiority margin – The primary objective is to show non-inferiority of CT-P13 as compared to Remicade, when switching from Remicade to CT-P13 across different indications. However, different indications may have different effect sizes. It is not clear what the non-inferiority margins for specific indications are.

Evaluation of potential risk of switching – In all of the studies, only single arm (R switch to T) is considered mainly because it is of interest to study the switch from R to T to determine (i) whether there is a loss of efficacy and (ii) the increase of adverse events. This single arm study, however, cannot fully address switchability between R and T. In other words, we need to address potential risk with and without such a switch. That is, we need to compare (R to T) as compared to (R to R).

It should be noted that the evaluation of potential risk of switching in terms of possible reduction of efficacy and/or increase of adverse events rate in these observational studies was performed by comparing the mean responses of the primary study endpoints between the proposed biosimilar product and the innovative biological drug product. In addition to the comparison of mean responses, it is also suggested that the comparison of variabilities associated with the observed responses be made because biosimilar products are known to be sensitive to environmental factors such as light and/or temperature. A small change or variation of critical quality attributes could translate to significant changes in clinical outcomes (i.e., safety and/or efficacy).

12.4.3 Clinical Studies

For clinical studies conducted for evaluation of the potential risk of non-medical from Remicade to Remsima, Table 12.22 lists the published results available in the literature. These clinical studies are briefly outlined below.

PLANETRA/PLANETAS studies – For the PLANETRA study (Yoo et al., 2016), a total of 302 patients with RA were studied under a 2 × 2 crossover, i.e., (TT, RT) design. Under the (TT, RT) design, 158 patients were in the TT (maintenance) group and 144 patients were in the RT (switch). Patients who had completed 54 weeks of treatment were analyzed in terms of ACR20, ACR50, ACR70, immunogenicity, and safety. Based on descriptive statistics, the investigators concluded that the approved biosimilar had comparable efficacy and tolerability as compared to the originator product. For PLANETAS (Park et al. 2016), a total of 174 patients with AS (ankylosing spondylitis) were studied under a similar 2 × 2 crossover (TT, RT) design. Under the (TT, RT) design, 88 patients were in the TT (maintenance) group and 86 patients were

TABLE 12.22

Summary of Clinical Studies for Non-Medical Switch

Study	Study Type	Indication	Treatment	Efficacy	Safety	ADA
PLANETAS extendion	OL 102-wk follow-up	AS	N = 174 (of original 250 randomized): • 88 continued (CT-P13 to CT-P13) • 86 switched (IFX to CT-P13)	ASAS20, ASAS40 and ASAS partial remission rates were similar between groups	Proportion of pts with ≥1 TEAE: • 48.9% continuers • 71.4% switchers • Mainly owing to fewer mild and moderates AEs	ADAs detected Wk 54: • 22.2% continuers • 26.2% switchers ADAs detected Wk 102: • 23.3% continuers • 27.4% switchers
PLANETRA extension	OL 102-wk follow-up	RA	N = 302 (of original 606 randomizes): • 158 continued (CT-P13 to Ct-P13) • 144 switched (IFX to CT-P13)	ACR20/50/70 response rates were maintained and similar in each group	Proportion pts with ≥1 AE or SAE: • Comparable between groups • 53.5% continuers • 53.8% switchers	ADA-positive pts comparable at Wk 54: • 49.1% continuers • 48.3% switchers Also at Wk 102: • 4.3% continuers • 44.8% switchers
NOR-SWITCH	1-sided transition	RA, SpA, PsA, UC, CD, Ps	Pts receiving IFX: • Switch to CT-P13 (same dose and frequency) • Or remain on IFX	TBD: • Disease worsening based on disease-specific assessment scores	TBD	TBD

in the RT (switch). Patients who had completed 54 weeks of treatment were analyzed in terms of ASAS20, ASAS40, and ASAS partial remission. Based on descriptive statistics, the investigators indicate that no negative effects on safety or efficacy in patients with AS were observed.

NOR-SWITCH study – A national, randomized, double-blind, parallel-group study was conducted to evaluate the efficacy and safety of switching from innovator infliximab (Remicade) to a biosimilar infliximab (Remsima) for N = 481 patients with one of the following diseases: ulcerative colitis (93 subjects), Crohn's disease (155 subjects), rheumatoid arthritis (78 subjects), spondyloarthritis (91 subjects), psoriatic arthritis (30 subjects), and psoriasis (35 subject). The primary study endpoint is disease worsening, which is measured based on the following criteria of individual diseases:

1. For rheumatoid arthritis and psoriatic arthritis, increase in DAS 28 of ≥1.2 from randomization, a minimum DAS 28 score of 3.2;

2. For spondyloarthritis, increase in ASDAS of ≥1.1 from randomization and a minimum ASDAS of 2.1;

3. For ulcerative colitis, increase in partial Mayo score of ≥3 points from randomization and a minimum partial Mayo score of ≥5 points;

4. For Crohn's disease, increase in HBI of ≥4 points from randomization and a minimum HBI score of 7 points;

5. For psoriasis, increase in PASI of ≥3 points from randomization and a minimum PASI score of 5;

6. Based on patient and investigator consensus on disease worsening: If a patient does not fulfil the formal definition, but experiences a clinically significant worsening according to both the investigator and the patient, which leads to a major change in treatment.

The study was designed for testing non-inferiority of Remsim (biosimilar or test product) as compared to Remicade (reference product) with a non-inferiority margin of 15% for achieving a 90% power for establishing non-inferiority assuming that 30% of subjects who receiving the reference product will occur disease worsening during 52 weeks. Based on the composite endpoint of pooled results, the investigators concluded that Remsima is highly similar to Remicade. The conclusion, however, is biased and may be misleading based on the following observation: the use of composite endpoint of pooled results is not statistically justifiable because the variabilities associated with patients' responses for different diseases are different.

12.4.4 Scientific Factors and Statistical Considerations

In order to have an accurate and reliable assessment of the potential risk of a non-medical switch in terms of possible reduced efficacy and/or increased incidence rate of adverse events, the following scientific factors and some statistical issues are necessarily considered during the stage of design and analysis for conducting non-medical switch studies.

12.4.4.1 Scientific Factors

Selection bias (multiple diseases) – For approval of a proposed biosimilar product, regulatory agencies such as US FDA do not require clinical studies be conducted on patients with specific diseases (indications) covered by the reference product. Instead, the sponsor may conduct clinical study (studies) on patients with one disease (separate diseases) and seek for approval for all diseases with scientific justification for extrapolations of other diseases. This has posed possible selection bias, especially when patients with different diseases respond to the proposed biosimilar product differently. In other words, we may show biosimilarity between the proposed biosimilar product and the reference product in some diseases but fail to show biosimilarity for other diseases. Besides, effect sizes for different diseases may be different from one disease to another. Selection bias certainly argues against the extrapolation approach with scientific justification without support of clinical data.

Confounding effects – When pooling several observational studies for a combined analysis, imbalance in demographics such as sex, age, race and patient characteristics are commonly seen. A serious imbalance in demographics and/or patient characteristics could cause a confounding effect between demographics and/or patient characteristics and the treatment effect. Consequently, the true treatment effect cannot be assessed accurately and reliably. In this case, the use of a propensity score is suggested.

Study endpoint selection – For evaluation of a non-medical switch, a composite endpoint by pooling the response rates across all diseases is often employed, regardless of (i) patients' distribution with respect to different diseases, (ii) whether the definitions of the responders under different diseases are different, (iii) that the variabilities associated with the responses under different diseases may be different, (iv) that the effect sizes for different diseases are different, and (v) that there is a possible treatment-by-disease interaction. As a result, the validity for the use of composite endpoints by pooling the response rates across different diseases is questionable and hence the conclusion may be misleading.

Non-inferiority margin – In practice, non-medical switch studies are often designed as non-inferiority trials in order to demonstrate that the proposed biosimilar product is not inferior to the reference product in terms of efficacy and safety. One of the major issues is then how to select the non-inferiority margin. The selection of non-inferiority margin not only has an impact on the sample size requirement, but also plays an important role for the success of the intended study. As mentioned earlier, effect sizes for different diseases may be different and hence non-inferiority margins for different diseases may be different. The US FDA recommends that its 2010 draft guidance on non-inferiority trials be consulted for selection of non-inferiority margin of the intended non-inferiority study. However, FDA's recommended approaches may result in different non-inferiority margins under different data sets available.

Sample size requirement – To accurately and reliably evaluate the potential risk of non-medical switches, it is suggested that statistical analysis for sample size calculation be performed to ensure that there is certain statistical assurance (e.g., sufficient power) for detecting a clinically meaningful difference (e.g., loss of efficacy, increase of incidence rate of adverse events, or risk/benefit assessment) at a pre-specified level of significance. With a limited number of subjects available, the observed clinically meaningful difference could be purely due to chance, especially when there is large variability associated with the observation. Sample size should be able to adjust for potential confounding and interaction effects when pooling several studies for a combined analysis.

12.4.4.2 Statistical Considerations

Bias and variability – In clinical research, bias and variability are related to accuracy and precision (reliability) of clinical data collected from the intended clinical study. Chow and Liu (2008) classified sources of bias and variation in clinical research into four categories: (i) expected and controllable (e.g., changes in laboratory testing procedures and/or diagnostic procedures), (ii) expected but not controllable (e.g., change in study dose and/or treatment duration), (iii) unexpected but controllable (e.g., patient noncompliance), and (iv) unexpected and not controllable (e.g., random error). In clinical research, it is not possible to avoid bias and variability in the real world. Thus, it is important to identify, eliminate (remove if possible), and control the bias and variability to an acceptable limit (in the sense that it will not have a significant negative impact on the statistical inference drawn).

Baseline comparability – Baseline comparability is referred to as comparison of baseline demographics such as gender, age, weight/height, or ethnic factors and patient characteristics such as patient severity and medical history for treatment balance. In clinical research, if significant differences in patient demographics and/or patient characteristics are observed, these differences may have contaminated the treatment effect and hence they should be included in the statistical model as baseline covariates for adjustment. In other words, analysis on endpoint (post-treatment) change from baseline is recommended in order to account for treatment imbalance.

The use of propensity score – In case there is evidence of confounding effects with demographics and/or patient characteristics, it is suggested that propensity score should be used to isolate the possible confounding effects for a more accurate and reliable biosimilarity assessment between the proposed biosimilar product and the reference product.

Control arm – One of major criticisms in single arm observational studies (i.e., R to T) is that there is no control arm (i.e., R to R). Without control arm, it is not possible to evaluate the potential risk of switch (i.e., R to T) because the risk should be assessed by comparing with and without switch, i.e., comparing (R to T) with (R to R). In the case of pooling several studies with control arm for a combined analysis, it is important to assess similarities and dissimilarities among the control arms before pooling, especially when a significant treatment-by-study interaction is observed. It is suggested that a test for poolability be performed before the data can be pooled for a combined analysis of statistical validity.

Carryover effect – Since the switch (e.g., from R to R or from R to T) occurs within individual subjects, the residual effect of R at previous dosing periods may carry over to the next dosing period (R or T) though there may be a sufficient length of washout between dosing periods. The carryover effects from R to R and from R to T may be different, which may have an impact on the assessment of the potential risk with/without the switch. Current 2×2 crossover designs such as (RR, RT) or (RT, TR) are unable to provide independent estimate of the possible carryover effect. To address the issue of carryover effect, a higher-order crossover design such as (TT, RR, RT, TR) or (RTR, TRT) may be useful.

Sensitivity analysis – Before a definite conclusion can be made, it is suggested that a clinical trial simulation in conjunction with sensitivity analysis be performed to provide a complete clinical picture of the non-medical switch. The sensitivity analysis should take the worst possible scenarios into consideration based on lower (upper) bound of a predictive confidence interval for the difference between the proposed biosimilar product and the reference product. In many cases, the benefit-risk ratio should also be take into consideration.

12.4.5 Design and Analysis of Switching Studies

Study designs – As indicated by BPCI Act, for a biological product that is administered more than once to an individual, the risk in terms of safety or diminished efficacy of alternating or switching between use of the biological product and the reference product should not be greater than the risk of using the reference product without such alternation or switch. Thus, an appropriate design for switching studies should be chosen in order to address (i) the risk in terms of safety or diminished efficacy of alternating or switching between use of the biological product and the reference product, (ii) the risk of using the reference product without such alternation or switch, and (iii) the relative risk between switching/alternating and not switching/ alternating. Note that in the recent FDA draft guidance, the switch is referred to as a single switch while alternation is referred to as multiple switches.

To determine whether the proposed biosimilar product can produce the same clinical results in any given patient, a standard two-sequence, two-period crossover design, i.e., (TR, RT) is necessarily employed. In the 2 × 2 crossover design (TR, RT), if we replace the T in the first sequence at the first dosing period with R, the design becomes (RR, RT), which is referred to as a hybrid parallel-crossover design (i.e., the first sequence is considered parallel and the second sequence is crossover). This hybrid parallel-crossover design with two dosing periods allows the evaluation of potential risk with and without switching, i.e., the assessment of similarity between the first sequence (switch from R to R) and the second sequence (switch from R to T) after the switch.

Statistical analysis – At the planning stage of non-medical clinical trials, under the study design, appropriate statistical methods should be developed under the null interval hypothesis of dis-similarity for achieving a desired power for establishment of biosimilarity between a proposed biosimilar and the reference product. At a pre-specified level of significance, power calculation for sample size should be performed under the alternative interval hypothesis of similarity at a pre-specified level of significance. It should be noted that we intend to reject the null hypothesis of dis-similarity and conclude the similarity between the proposed biosimilar product and the reference product. Since switching study designs with single switch or multiple switches (i.e., two switches or three switches) are special cases of a complete N-of-1 crossover trial design, statistical analysis can be performed using statistical methods described in the previous section.

Remarks – The potential risk of non-medical switches in terms of loss of efficacy and/or increase of the incidence rate of adverse reactions or adverse events needs to be carefully evaluated based on relevant clinical endpoints.

Single arm non-medical switch observational studies do not provide substantial evidence regarding the safety and efficacy of the proposed biosimilar product when switching from the reference product to the proposed biosimilar product.

NOR-SWITCH clinical studies attempted to evaluate the potential risk of non-medical switches from the reference product to an approved biosimilar product across various indications (diseases) of the reference product. The intention is good. However, there are several scientific and/or statistical deficiencies in design and analysis of the collected data. As a result, the conclusion made is biased and somewhat misleading.

As discussed in the previous section, in practice, there are three types of hybrid parallel-crossover designs that are commonly used for addressing drug interchangeability in terms of potential risk of switching and alternation. These three types of designs include the parallel plus 2 × 2 crossover design, the parallel plus 2 × 3 crossover design, and the parallel plus 2 × 4 crossover design. These study designs are special cases of a complete N-of-1 design with two, three, and four dosing periods, respectively. Valid study designs and appropriate statistical methods are strongly recommended for a more accurate and reliable assessment of the potential risk of medical/non-medical switches for consumer protection.

NOTE: that the recent FDA draft guidance does not address the question of non-medical switch post-approval. However, this issue has been raised and discussed at the ODAC meeting for review of two biosimilar regulatory submissions (i.e., Avastin biosimilar sponsored by Amgen and Herceptin biosimilar sponsored by Mylan) held on July 13th 2017 in Silver Spring, Maryland. Despite the lack of any FDA-approved interchangeable biosimilars, 26 states, including Puerto Rico, now have interchangeable biosimilar laws in place that restrict substitution.

12.5 FDA Draft Guidance on Analytical Similarity Assessment

In September 2017, FDA circulated a guidance entitled *Statistical Approaches to Evaluate Analytical Similarity* for public comments. This guidance is intended to provide advice on the evaluation of analytical similarity to sponsors or applicants who are interested in developing biosimilar products for licensure under section 351(k) of the Public Health Service (PHS). As indicated in the draft guidance, analytical similarity assessment is to support the demonstration that a proposed biosimilar product is highly similar to a reference product licensed under section 351(a) of the PHS Act. This guidance describes the type of information a sponsor should obtain regarding structural/functional and physicochemical attributes for evaluation of analytical similarity.

The draft guidance is well presented. However, there are a few obscure areas that need further clarification. These obscure shaded areas include (i) lack of scientific justification for the recommended similarity margin, (ii) FDA's current thinking regarding maximum allowed difference between the proposed biosimilar product and the reference product, (iii) validity of a Tier 2 quality range approach, (v) the potential use of a tolerance interval approach and/or min-max approach, and (vi) the recommendation of at least 10 lots per product when performing analytical similarity assessment. In what follows, comments and possible justifications/clarifications are briefly summarized.

Scientific justification for the recommended $1.5\sigma_R$ *similarity margin* – In the draft guidance, FDA recommended a similarity margin of $1.5\sigma_R$ for equivalence tests for CQAs in Tier 1 without providing scientific justification or rationales for the recommendation. This is one of the areas that requires clarification for scientific validity. The recommendation was made based on some extensive simulation studies under the assumptions that (i) the difference in mean response (i.e., $\mu_T - \mu_R$) is proportional to σ_R and (ii) the maximum mean shift between the test product and the reference product is $\frac{1}{8}\sigma_R$. That is, the worst possible scenario for similarity assessment may occur at $\mu_T - \mu_R = \frac{1}{8}\sigma_R$. Under these assumptions, and following similar ideas of scaled average bioequivalence criterion (Haidar et al., 2008), Chow et al. (2016) arrive at the same margin of $1.5\sigma_R$, which provides scientific justification to the FDA's recommended similarity margin.

Question regarding clinically relevant difference of $\frac{1}{8}\sigma_R$ – In a related question to the recommended similarity margin of $1.5\sigma_R$, is the maximum allowed mean shift $\frac{1}{8}\sigma_R = 12.5\%$ of σ_R considered a clinically relevant difference? Based on extensive simulation studies, FDA indicated that the difference between μ_T and μ_R (i.e., $\frac{1}{8} = 12.5\%$ of σ_R) is the maximum difference allowed for claiming high similarity, which is considered a clinically important difference. Sample size requirement should be performed for detecting such a difference for achieving an 80% power for analytical similarity assessment.

Comparing means versus comparing variances – One of the commonly asked questions regarding analytical similarity assessment is comparing mean responses versus comparing variances between the proposed biosimilar product and the reference product. This is mainly because biosimilar products are very sensitive to environmental conditions such as light and temperature. A small change in variation could translate to a notable difference in clinical outcomes. Thus, it is suggested that equivalence tests for analytical similarity should focus on comparing variances in addition to comparing mean responses or an alternative equivalence test which can account for variability associated with the product. While the comment regarding comparing variances is well taken, it should be noted that FDA's recommended margin of $1.5\sigma_R$, actually has taken variability into consideration under the assumptions that (i) the difference in mean response (i.e., $\mu_T - \mu_R$) is proportional to σ_R and (ii) the maximum mean shift between the test product and the reference product is $\frac{1}{8}\sigma_R$.

Concern regarding the Tier 2 quality range approach – The appropriateness of the FDA's recommended quality range approach for CQAs in Tier 2 has been challenged by many researchers because the probability of meeting the acceptance criteria of Tier 2 actually decreases with increasing sample size when sample size falls within a certain range. The quality range approach is a useful method for assessing similarity of non-Tier 1 quality attributes especially when both reference and test products have similar means and variances. In cases where there are notable shifts in means and/or heterogeneity in variances, the probability of meeting the acceptance criteria of Tier 2 decreases with increasing sample size when sample size falls within a certain range. In this case, the quality range method provides non-statistician such as biologists a visual examination of the seriousness of the mean shift and/or the degree of heterogeneity in variability associated with the proposed biosimilar product for assessment of similarity. Alternatively, one may consider using the quality range approach for similarity assessment based on standardized scores of the test and reference products when there are notable mean shift and/or heterogeneity in variability between the test and reference products (see also Section 12.2).

Potential use of tolerance interval and min-max approaches – Tolerance interval and min-max approaches are methods attempting to take variability into consideration. However, these methods were not recommended by the FDA without providing any scientific rationales. Perhaps, FDA is concerned about the potential increase of the false negative rate (accepting too many biosimilar products when in fact they are not biosimilar to the innovative biological drug product). Thus, it is suggested that scientific rationales be provided for further clarification of the concern.

Sample size requirement – FDA guidance indicates that a minimum of 10 lots from each product should be used for analytical similarity test without providing any scientific justification. Under the assumptions that (i) the difference in mean response (i.e., $\mu_T - \mu_R$) is proportional to σ_R, (ii) the maximum mean shift between the test product and the reference product is $\frac{1}{8}\sigma_R$, and (iii) the similarity margin is $1.5\sigma_R$, scientific justification for the recommended "a minimum of 10 lots per product" can be justified with respect to the variability associated with the reference product. In other words, we will be able to provide a statement such as the following:

Sample size of 10 lots from each product will yield an 80% power for establishment of similarity between the proposed biosimilar product and the reference product at the significance level of 5% assuming that (i) the difference in mean response (i.e., $\mu_T - \mu_R$) is proportional to σ_R, (ii) the maximum mean shift between the test product and the reference product is $\frac{1}{8}\sigma_R$, and (iii) the similarity margin is $1.5\sigma_R$.

12.6 Concluding Remarks

In this chapter, several recent developments in the past few years are briefly described. These recent developments include (i) a unified approach considering effect size adjusted for standard deviation for analytical similarity assessment proposed by Chow et al. (2016), (ii) several useful switching designs including complete N-of-1 trial design following the general principles as described in the FDA draft guidance on interchangeability, and (iii) the concern of off-label non-medical switch post-approval. In addition, several comments regarding the FDA's draft guidance on analytical similarity assessment are also discussed.

In biosimilar drug product development, however, there are still many research topics worth further consideration. For example, it is of interest to study the impact on equivalence tests for CQAs in Tier 1 based on random similarity margin rather than the fixed margin of $1.5\sigma_R$. Intuitively, one may consider a Bayesian approach using an appropriate prior on σ_R. For another example, current equivalence tests for both bioequivalence and biosimilarity assessment are performed based on comparing means and ignore possible heterogeneity in variances. Thus, in addition to comparing means, it is suggested that the assessment of similarity in variance should also be considered for establishing high similarity. In other words, similarity in means is considered generally similar and similarity in both means and variance is considered highly similar. The relative merits and disadvantages between comparing means and comparing variances require further research.

References and Further Reading

Arcondeguy, T. (2013). VEGF-A mRNA processing, stability and translation: A paradigm for intricate regulation of gene expression at the post-transcriptional level. *Nucleic Acids Research*, 7997–8010.

BPCI (2009). *Biologic Price Competition and Innovation Act*, Passed by the United States Congress, 2009.

Buer, L.C.T., Moum, B.A., Cvancarova, M., Warren, D.J., Medhus, A.W., and Hoivik, M.L. (2016). Switching from Remicade to Remsima is safe and feasible: A prospective, open-label study. European Crohn's and Colitis Organization (ECCO). Doi: 10.1093/ecco-jcc/jjw166.

Burdick, R. and Graybill, F. (1992). *Confidence Intervals on Variance Components*; Marcel Dekker: New York, 28–39.

Cardone, M.J. (1983). Detection and determination of error in analytical methodology. Part I. The method verification program. *Journal Association of Official Analytical Chemists*, 66, 1257–1281.

Cardone, M.J. (1983). Detection and determination of error in analytical methodology. Part II. Correction for corrigible systematic error in the course of real sample analysis. *Journal Association of Official Analytical Chemists*, 66, 1283–1294.

Caulcutt, R. and Boddy, R. (1983). *Statistics for Analytical Chemists*, Chapman and Hall, New York.

CD (2003). Commission Directive 2003/63/EC of 25 June 2003 amending Directive 2001/83/EC of the European Parliament and of the Council on the Community code relating to medicinal products for human use. Official Journal of the European Union L 159/46.

CDE/CFDA (2014). Draft guideline on Development and Evaluation of Biosimilars (Chinese Version). Center for Drug Evaluation, China Food and Drug Administration, Beijing, China, October 29, 2014.

Chen, K.W., Chow, S.C., and Li, G. (1997). A note on sample size determination for bioequivalence studies with higher-order crossover designs. *Journal of Pharmacokinetics and Biopharmaceutics*, 25, 753–765. doi: 10.1023/A:1025738019069.

Chirino, A.J. and Mire-Sluis, A. (2004). Characterizing biological products and assessing comparability following manufacturing changes. *Nature Biotechnology*, 22, 1383–1391.

Chow, S.C. (2010). *Generalizability probability of clinical results*. In Encyclopedia of Biopharmaceutical Statistics, Ed. Chow SC, Informa Healthcare, Taylor & Francis, London, 534–536.

Chow, S.C. (2011). Quantitative evaluation of bioequivalence/biosimilarity. *Journal of Bioequivalence and Bioavailability*, Suppl 1-002, 1–8.

Chow, S.C. (2013). *Biosimilars: Design and Analysis of Follow-on Biologics*. Chapman and Hall/CRC Press, Taylor & Francis, New York.

Chow, S.C. (2014). On assessment of analytical similarity in biosimilar studies. *Drug Designing*, 3, 119. doi:10.4172/2169-0138.

Chow, S.C. (2015). Challenging issues in assessing analytical similarity in biosimilar studies. *Biosimilars*, 5, 33–39.

Chow, S.C. and Chang, M. (2006). *Adaptive Design Methods in Clinical Trials*. Taylor & Francis, New York.

Chow, S.C., Chang, M., and Pong, A. (2005). Statistical consideration of adaptive methods in clinical development. *Journal of Biopharmaceutical Statistics*, 15, 575–591.

Chow, S.C., Endrenyi, L., Lachenbruch, P.A., and Mentre, F. (2014). Scientific factors and current issues in biosimilar studies. *Journal of Biopharmaceutical Statistics*, 24, 1138–1153.

Chow, S.C., Endrenyi, L., Lachenbruch, P.A., Yang, L.Y., and Chi, E. (2011). Scientific factors for assessing biosimilarity and drug interchangeability of follow-on biologics. *Biosimilars*, 1, 13–26.

Chow, S.C., Hsieh, T.C., Chi, E., and Yang, J. (2010). A comparison of moment-based and probability-based criterial for assessment of follow-on biologics. *Journal of Biopharmaceutical Statistics*, 20, 31–45.

Chow, S.C. and Shao, J. (2005). Inference for clinical trials with some protocol amendments. *Journal of Biopharmaceutical Statistics*, 15, 659–666.

Chow, S.C. and Liu, J.P. (2008). *Design and Analysis of Bioavailability and Bioequivalence Studies*. 3rd edition, Chapman Hall/CRC Press, Taylor & Francis, New York.

Chow, S.C. and Liu, J.P. (2010). Statistical assessment of biosimilar products, *Journal of Biopharmaceutical Statistics*, 20, 10–30.

Chow, S.C., Shao, J., and Hu, O.Y.P. (2002). Assessing sensitivity and similarity in bridging studies. *Journal of Biopharmaceutical Statistics*, 12, 385–400.

Chow, S.C., Shao, J., and Li, L. (2004). Assessing bioequivalence using genomic data. *Journal of Biopharmaceutical Statistics*, 14, 869–880.

Chow, S.C., Shao, J., and Wang, H. (2002). Individual bioequivalence testing under 2x3 crossover designs. *Statistics in Medicine*, 21, 629–648.

Chow, S.C., Shao, J., and Wang, H. (2008). Sample Size Calculations in Clinical Research. 2nd edition, Chapman and Hall/CRC Press, Taylor & Francis, New York.

Chow, S.C., Song, F.Y., and Bai, H. (2016). Analytical similarity assessment in biosimilar studies. *AAPS Journal*. doi:10.1208/s12248-016-9882-5.

Chow, S.C., Song, F.Y., and Bai, H. (2017). Sample size requirement in analytical studies for similarity assessment. *Journal of Biopharmaceutical Statistics*, 27, 233–238.

Chow, S.C., Song, F.Y., and Chen, M. (2016). Some thoughts on drug interchangeability. *Journal of Biopharmaceutical Statistics*, 26, 178–186.

Chow, S.C., Song, F.Y., and Cui, C. (2017). On hybrid parallel-crossover designs for assessing drug interchangeability of biosimilar products. *Journal of Biopharmaceutical Statistics*, 27, 265–271.

Chow, S.C., Song, F.Y., and Endrenyi, L. (2015). A note on Chinese draft guidance on biosimilar products. *Chinese Journal of Pharmaceutical Analysis*, 35(5), 762–727.

Chow, S.C., Xu, H., Endrenyi, L., and Song, F.Y. (2015). A new scaled criterion for drug interchangeability. *Chinese Journal of Pharmaceutical Analysis*, 35(5), 844–848.

Christl, L. (2015). Overview of regulatory pathway and FDA's guidance for development and approval of biosimilar products in US. Presented at FDA ODAC meeting, January 7; 2015, Silver Spring, Maryland. http://www.fda.gov /downloads/AdvisoryCommittees/CommitteesMeetingMaterials/Drugs /OncologicDrugsAdvisoryCommittee/UCM436387.pdf

CPMP (2009). Committee for Proprietary Medicinal Products. Guideline on comparability of medicinal products containing biotechnology-derived proteins as active substance: Non-clinical and clinical issues. EMEA/CPMP/3097/02 /Final8, 2009.

Crommelin, D., Bermejo, T., Bissig, M., Damianns, J., Kramer, I., Rambourg, P., Scroccaro, G., Strukelj, B., Tredree, R., and Ronco, C. (2005). Biosimilars, generic versions of the first generation of therapeutic proteins: Do they exist? *Contributions to Nephrology*, 149, 287–294.

Currie, L.A. (1968). Limits for qualitative detection and quantitative determination: Application to radiochemistry. *Analytical Chemistry*, 40, 586–593.

Dempster, A.P., Laird, N.M., and Rubin, D.B. (1977). Maximum likelihood from incomplete data via the EM algorithm. *Journal of the Royal Statistical Society Series B*, 39, 1–38.

Dong, X., Wang, Y.T., and Tsong, Y. (2017). Adjustment for unbalanced sample size for analytical biosimilar equivalence assessment. *Journal of Biopharmaceutical Statistics*, 27, 220–232.

Ellis, L.a. (2008). VEGF-targeted therapy: Mechanisms of anti-tumor activity. *Nature Reviews Cancer*, 579–591.

EMEA (2001). Note for Guidance on the Investigation of Bioavailability and Bioequivalence. The European Medicines Agency Evaluation of Medicines for Human Use. EMEA/EWP/QWP/1401/98, London, United Kingdom.

EMEA (2003a). Note for Guidance on Comparability of Medicinal Products Containing Biotechnology-derived Proteins as Drug Substance – Non Clinical and Clinical Issues. The European Medicines Agency Evaluation of Medicines for Human Use. EMEA/CHMP/3097/02, London, United Kingdom.

EMEA (2003b). Rev. 1 Guideline on Comparability of Medicinal Products Containing Biotechnology-derived Proteins as Drug Substance – Quality Issues. The European Medicines Agency Evaluation of Medicines for Human Use. EMEA /CHMP/ BWP/3207/00/Rev 1, London, United Kingdom.

EMA (2005a). Guideline on Similar Biological Medicinal Products. The European Medicines Agency Evaluation of Medicines for Human Use. EMEA/ CHMP/437/04, London, United Kingdom.

EMA (2005b). Guideline on Similar Biological Medicinal Products Containing Biotechnology-derived Proteins as Active Substance: Quality Issues. EMEA/ CHMP/BWP/49348, London, United Kingdom.

EMA (2005c). Draft Annex Guideline on Similar Biological Medicinal Products Containing Biotechnology-derived Proteins as Drug Substance – Non Clinical and Clinical Issues – Guidance on Biosimilar Medicinal Products containing Recombinant Erythropoietins. EMEA/CHMP/94526/05, London, United Kingdom.

EMA (2005d). Draft Annex Guideline on Similar Biological Medicinal Products Containing Biotechnology-derived Proteins as Drug Substance – Non Clinical and Clinical Issues – Guidance on Biosimilar Medicinal Products containing Recombinant Granulocyte-Colony Stimulating Factor. EMEA/CHMP/31329 /05, London, United Kingdom.

EMA (2005e). Draft Annex Guideline on Similar Biological Medicinal Products Containing Biotechnology-derived Proteins as Drug Substance – Non-Clinical and Clinical Issues – Guidance on Biosimilar Medicinal Products containing Somatropin. EMEA/CHMP/94528/05, London, United Kingdom.

EMA (2005f). Draft Annex Guideline on Similar Biological Medicinal Products Containing Biotechnology-derived Proteins as Drug Substance – Non Clinical and Clinical Issues – Guidance on Biosimilar Medicinal Products contain-ing Recombinant Human Insulin. EMEA/CHMP/32775/05, London, United Kingdom.

EMA (2006). Guideline on similar biological medicinal products containing biotechnology-derived proteins as active substance: Non-clinical and clinical issues. EMEA/CHMP/BMWP/42832, London, United Kingdom.

EMA (2009a). Non-clinical and clinical development of similar medicinal products containing recombinant interferon alfa. EMEA/CHMP/BMWP/102046/06, London, United Kingdom.

EMA (2009b). Guideline on non-clinical and clinical development of similar biological medicinal products containing low-molecular-weight-heparins. EMEA /CHMP/BMWP/118264/07, London, United Kingdom.

EMA (2010a). Concept paper on similar biological medicinal products containing recombinant follicle stimulation hormone. EMA/CHMP/BMWP/94899/2010, London, United Kingdom, 2010.

EMA (2010b). Draft guideline on similar biological medicinal prodcuts containing monoclonal antibodies. EMA/CHMP/BMWP/403543/2010, London, United Kingdom.

EMA (2011a). Concept paper on the revision of the guideline on similar biological medicinal product. EMA/CHMP/BMWP/572643, London, United Kingdom.

EMA (2011b). Concept paper on the revision of the guideline on similar biological medicinal products contain biotechnology-derived proteins as active substance: Quality issues. EMEA/CHMP/BWP/617111, London, United Kingdom.

EMA (2011c). Concept paper on the revision of the guideline on similar biological medicinal products containing biotechnology-derived proteins as active substance: Non-clinical and clinical issues. EMEA/CHMP/BMWP/572828, London, United Kingdom.

EMA (2011d). Guideline on similar biological medicinal products containing interferon beta. EMA/CHMP/BMWP/652000/2010, London, United Kingdom.

FDA (2001). Guidance on *Statistical Approaches to Establishing Bioequivalence*. Center for Drug Evaluation and Research, the US Food and Drug Administration, Rockville, Maryland, USA.

FDA (2003a). Guidance on *Bioavailability and Bioequivalence Studies for Orally Administrated Drug Products – General Considerations*, Center for Drug Evaluation and Research, the US Food and Drug Administration, Rockville, Maryland, USA.

FDA (2003b). Guidance on *Bioavailability and Bioequivalence Studies for Nasal Aerosols and Nasal Sprays for Local Action*, Center for Drug Evaluation and Research, the US Food and Drug Administration, Rockville, Maryland, USA.

FDA (2011). Guidance for Industry – *Process Validation: General Principles and Practices, Current Good Manufacturing Practices (CGMP), Revision 1*, the United States Food and Drug Administration, Rockville, Maryland, USA, 2011.

FDA (2015a). Guidance on *Scientific Considerations in Demonstrating Biosimilarity to a Reference Product*. Food and Drug Administration, Silver Spring, Maryland.

FDA (2015b). Guidance on *Quality Considerations in Demonstrating Biosimilarity to a Reference Protein Product*. Food and Drug Administration, Silver Spring, Maryland.

FDA (2015c). Biosimilars: *Guidance on Questions and Answers Regarding Implementation of the Biologics Price Competition and Innovation Act of 2009*, Food and Drug Administration, Silver Spring, Maryland.

FDA (2015d). Guidance for Industry - *Analytical Procedures and Methods Validation for Drugs and Biologics*. Food and Drug Administration, Silver Spring, Maryland.

FDA (2015e). Guidance for Industry – *Formal Meetings Between the FDA and Biosimilar Biological Product Sponsors or Applicants*. US Food and Drug Administration, Silver Spring, Maryland.

FDA (2017a). Guidance for Industry – *Considerations in Demonstrating Interchangeability With a Reference Product*. Food and Drug Administration, Silver Spring, Maryland. January, 2017.

FDA (2017b). Guidance for Industry – *Statistical Approaches to Evaluate Analytical Similarity*. Food and Drug Administration, Silver Spring, Maryland. September, 2017.

Federal Register (1995). International Conference on Harmonization; *Guideline on Validation of Analytical Procedures: Definitions and Terminology*, March 1, 1995; 11259–11262.

Ferrara, N. (2004). Vascular endothelial growth factor: Basic science and clinical progress. *Endocrine Reviews*, 581–611.

Ferrara, N. (2010). Binding to the extracellular matrix and proteolytic processing: Two key mechanisms regulating vascular endothelial growth factor action. *Molecular Biology of the Cell*, 687–690.

Fiorino, G., Manetti, N., Variola, A. et al. (2016). The PROSIT-BIO cohort of the IG-IBD: A prospective observational study of patients with inflammatory bowel disease treated with infliximab biosimilars. *Gastroenterology*, 150(4), Suppl. 1, p. S92.

Glintborg, B. (2016). Three months' clinical outcomes from a nationwide non-medical switch from originator ro biosimilar infliximab in patients with inflammatory arthritis. Presented at EULAR 2016. *Annals of Rheumatic Diseases*, 75, Suppl 2, p. 142.

Goel, H.L. (2013). VEGF targets the tumor cell. *Nature Reviews Cancer*, 871–882.

GPhA (2004). Biopharmaceuticals (Follow-on protein products): Scientific considerations for an abbreviated approcal pathway. Generic Pharmaceutical Association, December 8, 2004.

Gutierrez-Lugo, M.T. (2015). Chemistry, Manufacturing, and Controls. Presented at *ODAC Meeting on BLA 125553 for EP2006*, January 7, 2015, Silver Spring, Maryland.

Haidar, S.H., Davit, B., Chen, M.L., Conner, D., Lee, L., Li, Q.H., Lionberger, R., Makhlouf, F., Patel, D., Schuirmann, D.J., and Yu, L.X. Bioequivalence approaches for highly variable drugs and drug products. *Pharmaceutical Research* 2008; 25, 237–41.

HC (2010a). Authority of the Minister of Health. Guidance for sponsors: Information and submission requirements for subsequent entry biologics (SEBs). Canada.

HC (2010b). Health Canada. Guidance for sponsors: Information and submission requirements for subsequent entry biologics (SEBs). Canada.

Hernandez, L.D., Gonzalez, G.E.R., Gonzalez, M.V., Tardillo Marin, C.A., Diaz, C.Y.R., Presentations: Clinical: Therapy & Observation.

Hsieh, T.C., Chow, S.C., Liu, J.P., Hsiao, C.F., and Chi, E. (2010). Statistical test for evaluation of biosimilarity of follow-on biologics. *Journal of Biopharmaceutical Statistics*, 20, 75–89.

Hsieh, T.C., Chow, S.C., Yang, L.Y., and Chi, E. (2013). The evaluation of biosimilarity index based on reproducibility probability for assessing follow-on biologics. *Statistics in Medicine*, 32, 406–414.

Hudis, C.A. (2007). Trastuzumab-mechanism of action and use in clinical practice. *New England Journal of Medicine*, 357, 39–51.

ICH Q5C (1996). Q5C Guideline on Quality of Biotechnological Products: Stability Testing of Biotechnological/Biological Products. Center for Drug Evaluation and Research, Center for Biologics Evaluation and Research, the US Food and Drug Administration, Rockville, Maryland, USA.

ICH (1996). Guideline on Validation of Analytical Procedures: Methodology; November 6, 1996.

ICH Q6B (1999). Q6B Guideline on Test Procedures and Acceptance Criteria for Biotechnological/Biological Products. Center for Drug Evaluation and Research, Center for Biologics Evaluation and Research, the US Food and Drug Administration, Rockville, Maryland, USA.

ICH Q5E (2005). Q5E Guideline on Comparability of Biotechnological/Biological Products Subject to Changes in Their Manufacturing Process. Center for Drug Evaluation and Research, Center for Biologics Evaluation and Research, the US Food and Drug Administration, Rockville, Maryland, USA.

ICH Q8(R2) (2006). Q8(R2) Pharmaceutical Development, 2006.

ICH Q2R1 (1996). ICH Harmonized Tripartite Guideline, Validation of Analytical Procedures: Text and Methodology. International Conference on Harmonization of Technical Requirements for Registration of Pharmaceuticals or Human Use. Geneva, Switzerland.

JMP (2012). Quality and Reliability Methods. JMP Version 10.1, A Business Unit of SAS, SAS Campus Drive, Cary, NC 27513.

Johnson, N.L. and Kotz, S. (1970) *Distributions in Statistics --- Continuous Univariate Distribution 1*. Wiley, New York, New York.

Kang, S.H. and Chow, S.C. (2013). Statistical assessment of biosimilarity based on relative distance between follow-on biologics. *Statistics in Medicine*, 32, 382–392.

KFDA (2009). Korean guidelines on the evaluation of similar biotherapeutic products (SBPs). South Korea.

Keith O. Webber (2007). Biosimilars: Are we there yet? Presented at Biosimilars 2007, George Washington University, Washington, DC.

Kolar, M., Duricova, D., Brotlk, M., Hruba, V., Machkova, N., Mitrova, K., and Lukas, M. (2016). Sqitching of patients with inflammatory bowel disease from original infliximab (Remicade) to biosimilar iunfliximab (Remsima) is effective and safe. Presented at IFX & IFX Biosimilars.

Kozlowski, S. (2007). FDA Policy on follow on biologics. Presented at Biosimilars 2007, George Washington University, Washington, DC.

Lange, K.L., Little, R.J.A., Taylor, J.M.G. (1989). Robust statistical modeling using the t distribution. *Journal of the American Statistical Association*, 84, 881–896.

LeBrun, P. (2012). Bayesian design space applied to pharmaceutical. Unpublished dissertation. University de Liege https://www.google.com/url?sa=t&rct=j&q=&e src=s&frm=1&source=web&cd=3&ved=0ahUKEwijkrXoq8HKAhXI1x4KHXjo DEYQFggoMAI&url=https%3A%2F%2Forbi.ulg.ac.be%2Fbitstream%2F2268%2 F126503%2F1%2Fthesis.pdf&usg=AFQjCNG-IxFztbhvUfL_3qMwMqKXH4LvZ w&bvm=bv.112454388,d.dmo.

Lebrun, P., Giacoletti, K., Scherder, T, Rozet, E., and Boulanger, B. (2015). A quality by design approach for longitudinal attributes. *Journal of Biopharmaceutical Statistics*, 25, 247–259.

Liang, B.A. (2007). Regulating follow-on biologics. *Harvard Journal on Legislation*, 44, 363–373.

Liao, J.J.Z. and Darken, P.F. Comparability of critical quality attributes for establishing biosimilarity. *Statistics in Medicine* 2013; 32, 462–9.

Lim, S. (2017). Overview of the regulatory framework and FDA's guidance for development and approval of biosimilar products in the US. Presented at the Oncologic Drugs Advisory Committee (ODAC) meeting held in Silver Spring, July 13, 2017.

Liu, C.H. and Rubin, D.B. (1995). ML estimation of the t distribution using EM and its extensions, ECM and ECME. *Statistica Sinica*, 5, 19–39.

Løvik Goll, G. (2016). NOR-SWITCH study: A randomized, double-blind, parallel-group study to evaluate the safety and efficacy of switching from innovator infliximab to biosimilar infliximab compared with maintained treatment with innovator infliximab in patients with rheumatoid arthritis, spondyloarthritis, psoriatic arthritis, ulcerative colitis, Crohn's disease and chronic plaque psoriasis. http://www.lisnorway.no/

Massart, D.L., Dijkstra, A. and Kaufman, L. (1978). *Evaluation and Optimization of Laboratory Methods and Analytical Procedures*; Elsevier: New York.

McCamish, M. and Woollett, G. (2011). Worldwide experience with biosimilar development. *mAbs*, 3, 209–217.

MHLW (2009). Guidelines for the quality, safety and efficacy Assurance of follow-on biologics. Japan, 2009. (Yakushoku shinsahatu 0304007)

Miller, J.C. and Miller, J.N. (1988). *Statistics for Analytical Chemistry*, 2nd Ed., Wiley: New York.

Montgomery, D.C. (2008). *Introduction to Statistical Quality Control*, 6th ed., John Wiley & Sons, New York.

Morgan, B. (1992). *Analysis of Quantal Response Data*; Chapman and Hall, New York, 370–371.

Morrison, D. (1967). *Multivariate Statistical Methods*. McGraw-Hill, New York, 221–258.

Muller, Y.A. (1997). Vascular endothelial growth factor: Crystal structure and functional mapping of the kinase domain receptor binding site. *Proceedings of the National Academy of Sciences USA*, 71927197.

Nikiphorou, E., Kautiainen, H., Hannonen, P., Asikainen, J., Kokko, A., Rannio, T., and Sokka, T. (2015). Clinical effectiveness of CT-P13 (infliximab biosimilar) used as a switch from Remicade (infliximab) in patients with established rheumatic disease. Report of clinical experience based on prospective observational data. *Expert Opinion on Biological Therapy*, 15:12, 1677–1683, DOI: 10.1517/14712598.2015.1103733.

Oppenheimer, L., Capizzi, T., Weppelman, R., and Mehta, H. (1983). Determining the lowest limit of reliable assay measurement. *Analytical Chemistry*, 55, 638–643.

Park, W., Yoo, D.H., Miranda, P. et al. (2013). Efficacy and safety of switching from reference infliximab to CT-P13 compared with maintenance of CT-P13 in ankylosing spondylitis: 102-week data from the PLANETAS extension study. *Annals of the Rheumatic Diseases*, 72, 1605–1612.

Peterson, J.J. (2004). A Posterior predictive approach to multiple response surface optimization. *Journal of Quality Technology*, 36, 139–153.

Peterson, J.J. (2008). A Bayesian approach to the ICH Q8 definition of design space. *Journal of Biopharmaceutical Statistics*, 18, 959–975.

Peterson, J.J. (2009). What your ICH Q8 design space needs: A multivariate redictive distribution. *Pharmaceutical Manufacturing*, 8(10), 23–28.

Peterson, J.J. and Lief, K. (2010). The ICH Q8 definition of design space: A comparison of the overlapping means and the Bayesian predictive approaches. *Statistics in Biopharmaceutical Research*, 2, 249–259.

Peterson, J.J., Miró-Quesada, G., and del Castillo, E. (2009a). A Bayesian reliability approach to multiple response optimization with seemingly unrelated regression models. *Quality Technology and Quality Management*, 6(4), 353–369.

Peterson, J.J., Snee, R.D., McAllister, P.R., Schoefield, T.L., and Carella, A.J. (2009b). Statistics in pharmaceutical development and manufacturing (with discussion). *Journal of Quality Technology*, 41, 111–147.

Peterson, J.J. and Yahyah, M. (2009). A Bayesian design space approach to robustness and system suitability for pharmaceutical assays and other processes. *Statistics in Biopharmaceutical Research*, 1(4), 441–449.

Roger, S.D. (2006). Biosimilars: How similar or dissimilar are they? *Nephrology*, 11, 341–346.

Roger, S.D. and Mikhail, A. (2007). Biosimilars: Opportunity or cause for concern? *Journal of Pharmaceutical Science*, 10, 405–410.

Schellekens, H. (2004). How similar do 'biosimilar' need to be? *Nature Biotechnology*, 22, 1357–1359.

Schenerman, M.A., Axley, M.J., Oliver, C.N., Ram, K., and Wasserman, G.F. (2009). Using a risk assessment process to determine criticality of product quality attributes. In *Quality by Design for Biopharmaceutical: Principles and Case Studies*, ed by Rathore, A.S. and Mhatre, R., Wiley.

Schuirmann, D.J. (1987). A comparison of the two one-sided tests procedure and the power approach for assessing the equivalence of average bioavailability. *Journal of Pharmacokinetics and Biopharmaceutics*, 15, 657–680.

Shao, J. and Chow, S.C. (2002). Reproducibility probability in clinical trials. *Statistics in Medicine*, 21, 1727–1742.

Sieczkowska, J., Jarzebicka, D., Banaszkiewicz, A., Plocek, A., Gawronska, A., Toporowska- Kowalska, E., Oracz, G., Meglicka, M. and Kierkus, J. (2015). Switching between infliximab originator and biosimilar in prediatric patients with inflammatory bowel disease. Preliminary observation. European Crohn's and Colitis Organization (ECCO). DOI: 10.1093/2cco-jcc /jjv233.

Smits, L., Deriks, J., de Jong, J.D.D., van Esch, A., and Hoentjen, F. (2016). Elective switching from Remicade to biosimilar CT-P13 in inflammatory bowel disease patients: A prospective observational cohort study. Presented at IFX & IFX Biosimilars 2016.

Snedecor, G. and Cochran, W. (1967). *Statistical Methods*, 6th Ed., Iowa State Univ. Press: Ames, IA, 91–100.

SSAB. A communication package submitted to the FDA. Scientific Statistical Advisory Board on Biosimilars (SSAB) sponsored by Amgen, Thousand Oaks, California; 2010.

Stockdale, G. and Cheng, A. (2009). Finding design space and a reliable operating region using a multivariate Bayesian approach with experimental design. *Quality Technology and Quantitative Management*, 6(4), 391–408.

Suh, S.K. and Park, Y. (2011). Regulatory guideline for biosimilar products in Korea. *Biologicals*, 39, 336–338.

Thomas, A.L. (1982). Finding the observed information matrix when using the EM algorithm. *Journal of the Royal Statistical Society*, 44, 226–233.

Thomas, N. and Cheng, A. (2016). A Further Look at the Current Equivalence Test for Analytical Similarity Assessment. Presented at 2016 Regulatory-Industry Statistics Workshop.

Tothfalusi, L., Endrenyi, L., and Garcia Areta, A. (2009). Evaluation of bioequivalence for highly-variable drugs with scaled average bioequivalence. *Clinical Pharmacokinetics*, 48, 725–743.

Tsong, Y. (2015). Development of statistical approaches for analytical biosimilarity evaluation. Presented at *DIA/FDA Statistics Forum 2015*, April 20, 2015, Bethesda, Maryland.

Tsong, Y. Analytical Similarity Assessment. Presented at Duke-Industry Statistics Symposium, Durham, North Carolina. October 22–23, 2015.

Tsong, Y., Dong, X., and Shen, M. Development of statistical methods for analytical similarity assessment. *Journal of Biopharmaceutical Statistics*. 2015; DOI:10.1080/10543406.2015.1092038. http://dx.doi.org/10.1080/10543406.2015.1092038

USP/NF (2000). United States Pharmacopoeia 24 and National Formulary 19, United States Pharmacopeial Convention, Inc., Rockville, Maryland.

USP XXII. General Chapter 1225, Validation of Compendial Methods; 1982–1984.

Vempati, P. (2014). Extracellular regulation of VEGF: Isoforms, proteolysis, and vascular patterning. *Cytokine & Growth Factor Reviews*, 1–19.

Wang, J. and Chow, S.C. (2012). On regulatory approval pathway of biosimilar products. *Pharmaceuticals*, 5, 353–368; doi:10.3390/ph5040353.

Wang, T.R. and Chow, S.C. On establishment of equivalence acceptance criterion in analytical similarity assessment. Presented at Poster Session of the 2015 Duke-Industry Statistics Symposium, Durham, North Carolina, October 22–23, 2015.

Wei, G.C.G. and Tanner, M.A. (1990). A Monte Carlo implementation of the EM algorithm and the poor man's data augmentation algorithm. *Journal of the American Statistical Association*, 85, 699–704.

Winer, B. (1971). *Statistical Principles in Experimental Design*, 2nd Ed., McGraw-Hill, New York, 244–251.

Winkle, H.N. (2007). Quality by design (QbD). Keynote presentation at the 2007 PDA/FDA Joint Regulatory Conference, September 24–28, 2007, Washington, DC.

WHO (2005). World Health Organization Draft Revision on Multisource (Generic) Pharmaceutical Products: Guidelines on Registration Requirements to Establish Interchangeability, Geneva, Switzerland.

WHO (2009). Guidelines on evaluation of similar biotherapeutic products (SBPs). Geneva, Switzerland

Wu, P.S., Lin, M., and Chow, S.C. (2016). On sample size estimation and re-estimation adjusting for variability in confirmatory trials. *Journal of Biopharmaceutical Statistics*, 26, 44–54.

Yang, H. (2013). Setting specifications of correlated quality attributes. *PDA Journal of Pharmaceutical Science and Technology*, 67, 533–543.

Yang, H. (2016). Emerging Non-Clinical Biostatistics in Biopharmaceutical Development and Manufacturing. Chapman & Hall/CRC.

Yang, L.Y., Chow, S.C., Hsieh, T.C., and Chi, E. (2011). Bayesian approach for assessment of biosimilarity based on reproducibility probability. Submitted.

Yoo, D.H., Prodanovic, N., Jaworski, J. et al. (2013). Efficacy and safety of CT-P13 (biosimilar infliximab) in patients with rheumatoid arthritis: Comparison between switching from reference infliximab to CT-P13 and continuing CT-P13 in the PLANETRA extension study. *Annals of the Rheumatic Diseases*, 72, 1612–1620.

Yu, L.X. (2004). Bioinequivalence: Concept and definition. Presented at Advisory Committee for Pharmaceutical Science of the Food and Drug Administration. April 13–14, 2004, Rockville, Maryland.

Zheng, J., Yin, D., Yuan, M., and Chow, S.C. (2018). Simultaneous confidence interval methods for analytical similarity assessment. *Journal of Biopharmaceutical Statistics*. Submitted.

Index

Page numbers followed by f and t indicate figures and tables, respectively.

C

Printed in the United States
by Baker & Taylor Publisher Services